ADVANCES IN ONCOBIOLOGY

Volume 2 • 1998

BREAST CANCER

DEDICATION

To Dr. Helene Smith

ADVANCES IN ONCOBIOLOGY

BREAST CANCER

Series Editors: E. EDWARD BITTAR
Department of Physiology
University of Wisconsin Medical School
Madison, Wisconsin

GLORIA HEPPNER
Breast Cancer Program
Karmanos Cancer Institute
Detroit, Michigan

Guest Editors: WILLIAM P. PETERS
Breast Cancer Program
Karmanos Cancer Institute
Detroit, Michigan

DANIEL W. VISSCHER
Breast Cancer Program
Karmanos Cancer Institute
Detroit, Michigan

VOLUME 2 • 1998

JAI PRESS INC.

Stamford, Connecticut *London, England*

100 Prospect St.
Stamford, Connecticut 06901

JAI PRESS LTD.
38 Tavistock Street
Covent Garden
London WC2E 7PB
England

ISBN: 0-7623-0388-3

Manufactured in the United States of America

Transferred to digital printing in 2008.

CONTENTS

* *Deceased*

LIST OF CONTRIBUTORS

Robert D. Cardiff	Department of Pathology University of California, Davis Davis, California
Shanaz H. Dairkee	Geraldine Brush Cancer Research Institute California Pacific Medical Center San Francisco, California
Peter J. Dawson	Department of Pathology and Laboratory Medicine University of South Florida Tampa, Florida
Suzanne A.W. Fuqua	Department of Medicine University of Texas Health Science Center San Antonio, Texas
Gloria H. Heppner	Breast Cancer Program Karmanos Cancer Institute Wayne State University School of Medicine Detroit, Michigan
Francis G. Kern	Southern Research Insitute Birmingham, Alabama
Adrian V. Lee	Department of Medicine University of Texas Health Science Center San Antonio, Texas
David L. Miller	Department of Microbiology New York University Medical Center New York City, New York

Fred R. Miller	Breast Cancer Program Karmanos Cancer Institute Wayne State University School of Medicine Detroit, Michigan
Robert J. Munn	Department of Pathology School of Medicine University of California, Davis Davis, California
Steffi Oesterreich	Department of Medicine University of Texas Health Science Center San Antonio, Texas
Irma H. Russo	Breast Cancer Research Laboratory Fox Chase Cancer Center Philadelphia, Pennsylvania
Jose Russo	Breast Cancer Research Laboratory Fox Chase Cancer Center Philadelphia, Pennsylvania
*Helene S. Smith**	Geraldine Brush Cancer Research Institute California Pacific Medical Center San Francisco, California
T.H.M. Stewart	Department of Medicine Ottawa General Hospital Ottawa, Ontario
Daniel W. Visscher	Breast Cancer Program Karmanos Cancer Institute Wayne State University School of Medicine Detroit, Michigan
Susan M. Wykes	Department of Pathology Wayne Statue University School of Medicine Detroit, Michigan

* *Deceased*

PREFACE

Breast cancer research has never been in such an exciting and hopeful phase as today. From a clinical perspective, the discovery of genetic markers of risk in a proportion of familial breast cancer cases has opened up new vistas for understanding and ultimately preventing this disease. On the other hand, aggressive—even daring—therapies are being proven to be effective against advanced breast cancer. For the breast cancer experimentalist, this is also a time of great advance. Although animal and cell culture breast cancer models have proven to be of great use, there are now increasing opportunities to test the concepts developed in these models in actual clinical samples and cases. It is gratifying to see how well these concepts "translate" into the clinical setting. A very active area of research that is linking the laboratory to the clinic is the dissection of the biology and elucidation of the significance of proliferative breast disease and the identification of true, "high risk" or "preneoplastic" lesions within the previously ill-defined spectrum of fibrocystic or benign breast disease. One anticipates that discoveries made here will also lead to earlier detection, intervention and prevention of life-threatening cancer.

Even, however, as we look with optimism to the eventual eradication of breast cancer, we are once again forced to face the reality that we have not yet achieved our goal. Thus, we are saddened by the much too premature death of Dr. Helene Smith from breast cancer. Helene's work was at the forefront of efforts to understand the biology of human breast cancer at the molecular level. Her insight, open-mindedness, and refusal to sacrifice relevance for convenience will continue to set the standard for all breast cancer researchers. This volume is dedicated to her memory.

The authors of the chapters in this book were challenged to focus on the interface between research and the clinic, to show how basic investigations may explain, and hopefully improve, clinical outcome. We wish to thank them all for their thoughtful and informative contributions.

Gloria Heppner
Guest Editor

DIFFERENTIATION AND BREAST CANCER DEVELOPMENT

Jose Russo and Irma H. Russo

INTRODUCTION

The breast is the source of the most frequent malignancy in the female population. The knowledge that breast cancer risk is heavily influenced by both the degree of gland development and the reproductive history of the individual (Russo, J. and Russo, I.H., 1987a; Russo, J. and Russo, I.H., 1987b; De Waard and Trichopoulos, 1988; Russo, J. et al., 1992), requires a thorough understanding of how growth, pu-

Advances in Oncobiology
Volume 2, pages 1-10.

ISBN: 0-7623-0388-3

berty, pregnancy, lactation, and postmenopausal regression influence the development of this organ (Dabelow, 1957; Russo, J. and Russo, I.H., 1987a). Although the development of the human breast starts during embryonic life, the main spurt of growth initiates at puberty with lobule formation, but the completion of breast development and differentiation occurs only at the end of a full-term pregnancy (Russo, J. and Russo, I.H., 1987a). It has long been known that the risk of breast cancer shows a direct relationship with early menarche and nulliparity and an inverse relationship with early parity (MacMahon et al., 1970; Trapido, 1983; Vessey et al., 1985). However, case control studies have demonstrated that breast cancer risk increases with the age at which a woman bears her first child, indicating that the lengthening in the interval of time between menarche and the first pregnancy plays an important role for pregnancy to be protective. The increase in risk has been reported to occur when this interval is lengthened over 14 years (DeStavola et al., 1993). Thus, to be protective, pregnancy has to occur before age 30—indeed women first becoming pregnant after that age appear to have a risk above that of nulliparous women (MacMahon et al., 1970; Trapido, 1983; Vessey et al., 1985). Although multiparity appears to confer additional protection, the protective effect remains largely limited to the first birth. The protection conveyed by an early reproductive event persists at all subsequent ages, even until women become older than 75 years of age (Vessey et al., 1985). Although the ultimate mechanisms through which an early first full-term pregnancy protect the breast from cancer development are not known, a likely explanation has been provided by studies performed in an experimental animal model. The induction of rat mammary carcinomas with chemical carcinogens is inhibited by full-term pregnancy. The inhibition of the carcinogenic process is mediated by the differentiation of the mammary gland caused by the reproductive process. It can be postulated that differentiation activates specific genes that imprint the breast epithelium to respond differently to subsequent hormonal changes or to genotoxic influences. It is likely that similar mechanisms mediate the protection that an early first full-term pregnancy confers to women. There is no explanation for the higher risk of developing malignancies to nulliparous and late parous women. The fact that experimentally induced rat mammary carcinomas develop only when the carcinogen interacts with the undifferentiated and highly proliferating mammary epithelium of young nulliparous rats (Russo, J. et al., 1977; 1979; Russo, I.H. and Russo, J. 1878; Russo, J. and Russo, I.H., 1978; Russo, J., 1993), suggests that the breast of late parous and of nulliparous women might exhibit some of the undifferentiated and/or cell proliferative characteristics that predispose the tissue to undergo neoplastic transformation. The correlation of our findings in the experimental animal model with those obtained through the study of the development of the human breast support this postulate.

ARCHITECTURE OF THE NORMAL BREAST

The study of the normal breast requires a precise characterization of the source of material used since the concept of normality of the breast can be tinted by age, reproductive history, and specific hormonal conditions. The breast tissues that most fulfill the criteria of normality are those obtained by reduction mammoplasty performed for cosmetic reasons. For these research purposes, human breast samples obtained from bilateral or unilateral reduction mammoplasties performed in 33 patients were analyzed by quantitating the type of parenchymal structures present in them. The morphological analysis of the human breast has allowed us to determine that the mammary parenchyma is composed of lobular structures that have been characterized by their basic level of branching or number of ductules per lobular unit into four categories (Russo, J. and Russo, I.H., 1987a). Lobules type 1 (Lob 1), also called terminal ductal lobular units (TDLU), or virginal lobules, because they are present in the immature female breast before menarche, are the most undifferentiated ones. They are composed of clusters of 6–11 ductules per lobule. Lobules type 2 (Lob 2) evolve from the previous ones and have a more complex morphology; they are composed of a higher number of ductular structures per lobule. They progress to lobules type 3 (Lob 3), which are characterized by having an average of 80 ductules or alveoli per lobule; they are frequently seen in the breast of women under hormonal stimulation or during pregnancy. A fourth type of lobule, lobule type 4 (Lob 4) is present during late pregnancy and during the lactational period of the mammary gland, but it is not found in the breast of nulliparous postpubertal women. The Lob 4 is considered to be the maximal expression of glandular development and differentiation (Figure 1; Russo, J. and Russo, I.H., 1987a).

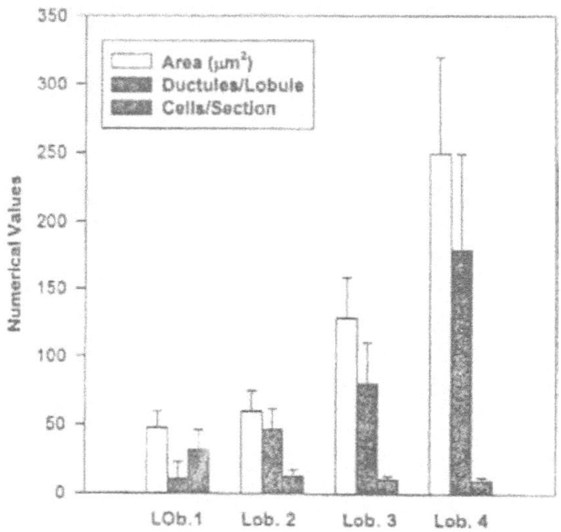

Figure 1. Histogram showing the characteristics of the lobular structures in the human breast based upon lobular area (in μm²), number of ductules per lobule, and number of cells per section in lobules type 1 (Lob 1), lobules type 2 (Lob 2), and lobules type 3 (Lob 3).

We studies breast samples from 24 parous women ranging in age from 29 to 33 years and nine nulliparous women, ranging in age from 16 to 28 years. An average of 100 grams of tissue were processed from every specimen. They were fixed in 10% neutral buffered formalin for a minimum of 24 h, defatted by submersion in acetone for two days, hydrated in decreasing concentrations of ethanol, and rehydrated and stained in 0.025% toluidine blue solution (Russo, J. and Russo, I.H., 1987a). A total of 652 slides consisting of an average of 12 slides per sample were examined and a total of 28,437 structures were classified and counted. In every specimen, the total numbers of Lob 1, Lob 2, and Lob 3 were counted and the relative percentage of every structure type was obtained (Figure 1).

In the overall population of breast tissues studied, Lob 1 constituted 22.45% of the total structures present, a value significantly lower ($p<0.04$) than the percentage of Lob 2 and Lob 3 ($p<0.07$), which represented 37% and 38%, respectively, of the structures counted. (Table I).

INFLUENCE OF AGE AND PARITY ON THE DEVELOPMENT OF THE HUMAN BREAST

The observations that in an experimental animal model the initiation of the neoplastic process is inversely related to the degree of differentiation of the mammary gland, which in turn is a function of age and reproductive history (Russo, J. and Russo, I.H., 1987c; Russo, J. et al., 1982; 1990), led us to postulate that the protective effect of pregnancy in women is due to the degree of differentiation of the breast (Russo, J. and Russo, I.H., 1987; Russo, J. et al., 1982; 1988;1990;1993). The study of the architecture of the breasts of women of different parity history revealed that the breasts of nulliparous women of all ages were composed predominantly of Lob 1 (Table 1). The breasts of premenopausal parous women, on the other hand, were composed of a high concentration of Lob 3., whereas Lob 1 were in a very low percentage (Table 1; Russo, J. et al., 1992). After the fourth decade of life in this group of women, the concentration of Lob 1 reached the same level observed in nulliparous women. The changes in the relative percentage of Lob 3 and Lob 1 observed with aging in the breast of parous women is due to the regression of Lob 3 to Lob 1, whereas in nulliparous women, Lob 1 never reached the degree of differentiation found in women who completed an early pregnancy. Thus, in the breasts of parous women, there was a negative correlation between the percentage of Lob 3 and aging (–0.90; Russo, J. et al., 1992). This correlation was not observed in nulliparous women, whose breasts' lobular composition remained unchanged with aging. The phase of the menstrual cycle did not seem to affect the distribution of lobular structures in the breast. These studies allowed us to determine that early parous women truly underwent lobular differentiation, which was evident at a younger age, whereas nulliparous women seldom reached the Lob 3 stage.

Table 1. Lobular Architecture of the Breast: Comparison of Percentages of Structures Found in Reduction Mammoplasties and Breast Biopsies[A]

Group	*No. Cases*	*Age*	*Lob 1 (%)*	*Lob 2 (%)*	*Lob 3 (%)*
RM (All)	33	29.4 ± 8.2	22.45 ± 23.7[1]	37.25 ± 28.61[3]	38.41 ± 34.22[5]
RM (Nulliparous)	9	22.9 ± 6.7	45.87 ± 27.40	47.17 ± 22.01	6.94 ± 7.01
RM (Parous)	24	31.9 ± 2.3	16.92 ± 8.26[7]	35.45 ± 3.14[9]	47.86 ± 33.4[11]
BB (All)	45	46.6 ± 1.5	65.66 ± 34.15[2]	24.64 ± 20.64[4]	9.68 ± 6.31[6]
BB (Nulliparous)	10	42.5 ± 10.3	70.99 ± 33.3	25.26 ± 24.74	3.75 ± 1.6
BB (Parous)	35	48.9 ± 11.8	65.25 ± 37.3[8]	21.10 ± 8.07[10]	13.62 ± 3.10[12]

Notes: [A] The percentages of lobular structures, lobules type 1 (Lob 1), lobules type 2 (Lob 2), and lobules type 3 (Lob 3) present in reduction mammoplasty (RM) specimens were compared with those present in breast biopsies (BB). [1-12] Indicate the groups in which differences are statistically significant: 1 vs. 2 $p < 0.0000005$; 3 vs. 4 $p < 0.04$; 5 vs. 6 $p < 0.00009$; 7 vs. 8 $p < 1 \times 10^{-7}$; 9 vs. 10 $p < 0.07$, and 11 vs. 12 $p < 0.0003$.

The proliferative activity of the breast epithelium was determined by immunocytochemistry with the monoclonal antibody against the Ki67 nuclear antigen. The number of cells exhibiting a positive immunocytochemical reaction was quantitated and expressed as the percentage of the total number of cells composing Lob 1, Lob 2, and Lob 3. The highest overall proliferative activity was observed in Lob 1 (2.15 ± 1.18) whereas it was intermediate in Lob 2 (0.22 ± 0.11) and lowest in Lob 3 (0.01 ± 0.01). In the menopausal breast, Lob 1 of nulliparous women exhibited a two-fold higher proliferative activity than Lob 1 of parous women (Russo, J. and Russo, I.H. 1997).

ARCHITECTURAL PATTERN OF THE ABNORMAL BREAST: PROLIFERATIVE BREAST DISEASE

The pattern of breast development was studied in 45 breast biopsies performed because of mammographic abnormalities or the presence of clinically suspicious breast masses. The patients ranged in age from 36 to 56 years, averaging 46.6 years of age (Table 1). Thirty five of the women were parous and 10 were nulliparous. All tissues were fixed in formalin, embedded in paraffin, sectioned at 5 μm thickness, and processed for light microscopy. An average of 12 histological sections per sample was examined. These histological sections were utilized for characterizing the type of lobular structures and the number and type of of pathological lesions present by applying criteria previously described (Figure 1). Twenty-one of the samples consisted of normal breast parenchyma with no signs of histopathological abnormalities; the remaining 24 samples exhibited mild to moderate ductal hyperplasia (DH), blunt duct adenosis (BDA), and/or sclerosing adenosis (SAD; Table 2). No significant differences in age at the time of biopsy, percentage of lobular structures present in the breast, and breast epithelial cell proliferation index were observed between nullipa-

Table 2. Lobular Structures Found in Breast Biopsies[A]

Group	*Age*	*n*	*Lob 1 (%)*	*Lob 2 (%)*	*Lob 3 (%)*
Normal Breast Control	42.28 ± 6.66	21	71.95 ± 7.09	23.02 ± 5.47	5.11 ± 3.17
Ductal Hyperplasia (DH)	44.06 ± 11.37	15	87.76 ± 3.16	11.85 ± 3.10	0.24 ± 0.94
BluntDuct Adenosis (BDA)	37.25 ± 11.64	4	58.39 ± 19.93	31.57 ± 11.75	9.82 ± 9.82
Sclerosing Adenosis (SAD)	37.00 ± 3.16	5	37.92 ± 15.52	48.95 ± 15.65	11.31 ± 9.97

Notes: [A] Percentages of lobular structures found in breast biopsies. The difference in the age of the patients in each group was not statistically significant. In the Normal Breast or Control Group, lobules type 1 (Lob 1) were significantly different from lobules type 2 (Lob 2) and 3 (Lob 3; p<0.0008). The percentage of Lob 2 was significantly different from that of Lob 3 (p<0.07). In the DH Group, the percentage of Lob 1 was significantly higher than the percentages of Lob 2 and Lob 3 (p<0.000001). The percentage of Lob 2 was also significantly higher than the percentage of Lob 3 (p<0.02). In both the BDA and SAD Groups, the percentage of Lob 2 was significantly higher than in the Normal Breast or Control Group (p<0.07) and the DH Group (p<0.001). The percentage of Lob 3 was significantly higher in the SAD Group (p<0.05) and in the BDA Group (p<0.05) than in those breast tissues containing DH or Normal Breast.

rous and parous women; therefore, the results of both groups were combined and parity history was not taken into consideration in the final analysis of results. Thus, the patient population was divided into four groups according to the histopathological characteristics of the breast tissues examined: No pathology present (Normal Breast or Control Group), ductal hyperplasia (DH Group), blunt duct adenosis (BDA Group), and sclerosing adenosis (SAD Group; Figures 2 and 3).

In the breast tissues of the Normal Breast and DH Groups, Lob 1 represented the highest percentage of all the structures counted (p<0.0008 and p <0.0001, respectively; Figure 2). The breast tissues containing BDA and SAD were also characterized by having a higher percentage of Lob 1, whereas the SAD Group had a higher percentage of Lob 2 (P<0.05; Table 2 and Figure 2). All the groups contained a significantly lower percentage of Lob 3 (P<0.05), although the relative percentage of Lob 3 was significantly lower in normal breast biopsies or in those diagnosed with DH than in the BDA and SAD groups (P<0.05; Figure 2).

The number of proliferating, or Ki67 positive epithelial cells, was maximal in Lob 1, decreasing progressively in Lob 2 and Lob 3 (p<0.001). The pattern was similar in the Normal Breast, DH, and SAD Groups (Figure 3). The differences in proliferative activity between Lob 1 and Lob 2 in the SAD group, however, were not significant, and in the BDA group, the rate of cell proliferation was higher in Lob 2 (P<0.01) than in Lob 1 (Figure 3). Although the proliferative activity was on average higher in Lob 2 than in Lob 3, the differences were not statistically significative (Figure 3). Lob 1 in the DH group had a significantly higher (p<0.01) proliferative activity than Lob 1 in the Normal Breast Group; however, the differences with the Lob 1 of the BDA and SAD groups were not significant. Lob 2 had a significantly higher proliferative index in the DH, BDA, and SAD groups than in the Normal Breast Group (P<0.01). Lob 3 had a significantly higher proliferative activity in the BDA and SAD groups than in the Normal Breast and DH groups, respectively (P<0.01; Figure 3).

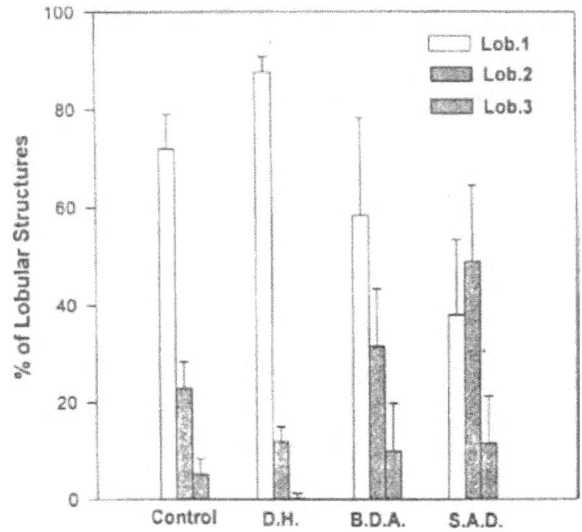

Figure 2. Architectural pattern of breast biopsies containing proliferative lesions. (Control, Normal Breast Group; D.H., ductal hyperplasia group; B.D.A., blunt duct adenosis group; S.A.D., sclerosing adenosis group.

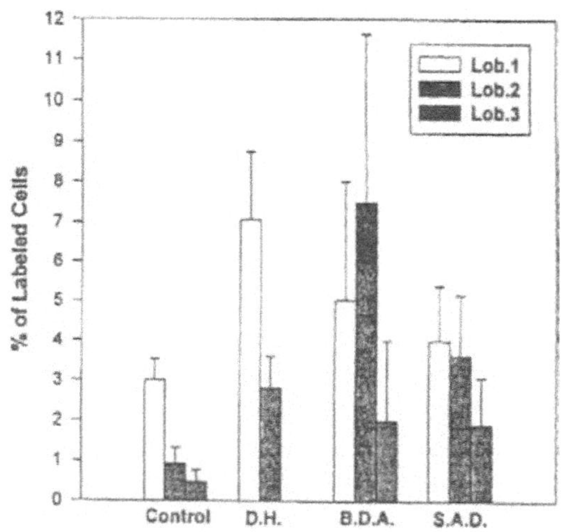

Figure 3. Proliferative activity in the lobular structures of breast biopsies. Percentage of cells positive for Ki67 antibody (ordinate). Patients groups as per figure 2.

Data presented here indicate that those breast tissues obtained from biopsies performed because of mammographic abnormalities or clinically suspicious masses have an architectural pattern different from that of normal breast tissues obtained from reduction mammoplasties. More important is the observation that even in those cases in which no pathology or only benign lesions were diagnosed, the pattern of breast development in tissues obtained by biopsy was more similar to those of the cancer-bearing breast, rather than to the population not requiring a biopsy. These observations support our previous observations (Russo, J. et al., 1990), which indicated that DH originates from Lob 1, the structure most frequently found in those breasts and the one with a significantly higher level of cell proliferation. The observations that Lob 2 and Lob 3 with high rates of cell proliferation are more frequently found in those breasts that also contain BDA and SAD might be an indication that these structures are the site of origin of these lesions. It is of importance to clarify that in breast tissues containing ductal hyperplasias parity seemed not to have influenced the pattern of development of the breasts because the lobular composition was more similar to that of the breasts of cancer-free nulliparous women or of the cancer-bearing parous and nulliparous women (Russo, J. et al., 1994). Breast tissues containing sclerosing adenosis and blunt duct adenosis, on the other hand, contained more differentiated structures (i.e., Lob 2 and Lob 3). However, it was difficult to evaluate the effect of parity because of the small number of cases studied.

PATHOGENESIS OF BREAST CANCER

One important concept that emerged from our study of breast development is that the TDLU, which had been originally identified by Wellings et al. (1975) as the site of origin of the most common breast malignancy, the ductal carcinoma, corresponds to a specific stage of development of the mammary parenchyma, the Lob 1. This observation is supported by comparative studies of normal and cancer-bearing breasts obtained at autopsy. It was found that the nontumoral parenchyma in cancer-associated breasts contained a significantly higher number of hyperplastic terminal ducts and atypical Lob 1 and ductal carcinomas in situ originating in Lob 1 than those breasts of women free of breast cancer, indicating that the Lob 1 is affected by preneoplastic as well as by neoplastic processes (Russo, J. et al, 1990).

More differentiated lobular structures have been found to be affected by neoplastic processes as well, although they give rise to tumors in which malignancy is inversely related to the degree of differentiation of the parent structure (Russo, J. et al., 1990; 1994; Russo, J. and Russo, I.H., 1987b; 1993; 1994). These observations led us to conclude that each specific compartment of the breast gives origin to a specific type of lesion. The finding that the most undifferentiated structures originate the most undifferentiated and aggressive neoplasms acquires relevance in the light that these structures are more numerous in the breast of nulliparous women who are, in turn, at a higher risk of developing breast cancer. We concluded that the Lob 1 found in the breast of nulliparous women never went through the process of differentiation, whereas the same structures, when found in the breast of postmenopausal parous women, did (Russo, J. and Russo, I.H., 1987a).

THE CANCER-BEARING BREAST IN NULLIPAROUS AND PAROUS WOMEN

The analysis of nontumoral breast tissues removed by lumpectomy or mastectomy after a diagnosis of cancer has been made revealed that, in nulliparous women, the architecture of the breast did not differ from that of nulliparous women free of mammary pathology, since in both populations the predominant lobular structure present was the Lob 1 (Russo, J. et al., 1994). In parous women, however, although the pathology-free population contains predominantly Lob.3 and has the lowest percentage of Lob 1, in those women who had developed breast cancer, Lob 1 is the predominant structure, appearing more similar to the breast of nulliparous women. These results support our hypothesis that the degree of breast development is of importance in the susceptibility to carcinogenesis, and that parous women who develop breast cancer might exhibit a defective response to the differentiating influence of the hormones of pregnancy (Russo, J. and Russo, I.H., 1987a; 1994).

ACKNOWLEDGMENTS

This work was supported by National Cancer Institute grants CA64896 and CA67238 and NIEHS grant ESO7280.

REFERENCES

Dabelow, A. (1957). In: Handbuch der Mikroskopischen Anatomic des Menschen (Bargmann, W., Ed.), pp. 277-485. Springer-Verlag, Berlin.

DeStavola, B.L., Wang, D.Y., and Allen, D.S. (1993) The association of height, weight, menstrual and reproductive events with breast cancer: Results from two projective studies on the island of Guersney. Cancer Causes and Control 4, 331-340.

De Waard, F., and Trichopoulos, D. (1988) A unifying concept of the aetiology of breast cancer. Int. J. Cancer 41, 666-669.

MacMahon, B., Cole, P., Liu, M., Lowe, C.R., Mirra, A.P., Ravinihar, B., Salber, E.J., Valaoras, V.G., and Yuasa, S. (1970). Age at first birth and breast cancer risk. Bull. World Hlth. Org. 43, 209-221.

Potten, C.S., Watson, R.J., Williams, C.T., Tickle, S., Roberts, S.A., Harris, H., and Howell, A. (1988). The effect of age and menstrual cycle upon proliferative activity of the normal human breast. Br. J. Cancer 58, 163-170.

Russo, J., and Russo, I.H. (1987a). In: The Mammary Gland. (Neville, M.C., and Daniel, C., Eds.). pp. 67-93. Plenum Publishing, Inc., New York.

Russo, J., and Russo, I.H. (1987b). Biological and molecular bases of mammary carcinogenesis. Lab. Invest. 57, 112-137.
Russo, J., and Russo I.H. (1987c). In: Cellular and Molecular Biology of Mammary Cancer. (Medina, D., Kidwell, W., Heppner, G., and Anderson, E., Eds.), pp. 399-417, Plenum Press, New York.
Russo, J., Rivera, R., and Russo, J. (1992) Influence of age and parity on the development of the human breast. Breast Cancer Res. Treat. 23, 211-218.
Russo, J., Saby, J., Isenberg, W., and Russo, I.H. (1977). Pathogenesis of mammary carcinomas induced in rats by 7,12-dimethylbenz(a)anthracene. J. Natl Cancer Inst. 59, 435-445.
Russo, I.H., and Russo, J. (1978). Developmental stage of the rat mammary gland as determinant of its susceptibility to 7,12-dimethylbenz(a)anthracene. J. Natl. Cancer Inst. 61, 1439-1449.
Russo, J., and Russo, I.H. (1978). DNA labeling index and structure of the rat mammary gland as determinants of its susceptibility to carcinogenesis. J. Natl. Cancer Inst. 61, 1451-1459.
Russo, J., Wilgus, G., and Russo, I.H. (1979). Susceptibility of the mammary gland to carcinogenesis. I. Differentiation of the mammary gland as determinant of tumor incidence and type of lesion. Am. J. Pathol. 96, 721-734.
Russo, J. (1983). Basis of cellular autonomy in the susceptibility to carcinogenesis. Toxicologic Pathology 11, 149-166.
Russo, J., Tay, L.K., and Russo, I.H. (1982). Differentiation of the mammary gland and susceptibility to carcinogenesis: A Review. Breast Cancer Res. Treat. 2, 5-73.
Russo, J., Gusterson, B.A., Rogers, A.E., Russo, I.H., Welling, S.R., and Van Zwieten, M.J. (1990). Comparative study of human and rat mammary tumorigenesis. Lab. Invest. 62, 244-278.
Russo, J., Reina, D., Frederick, J., and Russo, I.H. (1988). Expression of phenotypical changes by human breast epithelial cells treated with carcinogens in vitro. Cancer Res. 48, 2837-2857.
Russo, J., Calaf, G., and Russo, I.H. (1993). A critical approach to the malignant transformation of human breast epithelial cells. CRC Critical Reviews in Oncogenesis 4, 403-417.
Russo, J., and Russo, I.H. (1997). Role of differentiation in the pathogenesis and prevention of breast cancer. Endocrine-Related Cancer, 4-21.
Russo, J., and Russo, I.H. (1993). Developmental pattern of human breast and susceptibility to carcinogenesis. European J. Cancer Prevent. 2, 85-100.
Russo, J., and Russo, I.H. (1994). Toward a physiological approach to breast cancer prevention. Cancer Epidemiol., Biomarkers, and Prevention 3, 353-364.
Russo, J., Romero, A.L., and Russo, I.H. (1994). Architectural pattern of the normal and cancerous breast under the influence of parity. J. Cancer Epidemiol, Biomarkers, and Prevention, 3, 219-224.
Trapido, E.J. (1983). Age at first birth, parity and breast cancer risk. Cancer 51, 946-948.
Vessey, M.D., McPherson, K., Roberts, M.M., Neil, A., and Jones, L. (1985). Fertility in relation to the risk of breast cancer. Br. J. Cancer 52, 625-628.
Wellings, S.R., Jensen, M.M., and Marcum, R.G. (1975). An atlas of sub-gross pathology of the human breast with special reference to possible pre-cancerous lesions. J. Natl. Cancer Inst. 55, 231-275.

GROWTH FACTOR SIGNAL TRANSDUCTION AND HORMONE INDEPENDENCE IN BREAST CANCER

David L. Miller and Francis G. Kern

Advances in Oncobiology
Volume 2, pages 11-79.

ISBN: 0-7623-0388-3

BACKGROUND

In 1996, the breast was the most common site of newly diagnosed cancer in women, with an estimated 184,000 new cases reported. Also that year, approximately 44,000 women died as a result of the disease. Until recently, breast cancer was the leading cause of cancer deaths in women; now, it is second only to lung cancer (American Cancer Society, 1997). What is perhaps more alarming than the number of cancers, however, is the underlying trend. Over the past forty years, the incidence of breast cancer has steadily climbed by approximately 1% annually (Marshall, 1993). While some of this increase is more apparent than real, because the widespread use of mammography is detecting cancers earlier than previously possible, there is nevertheless a slow upward climb in incidence. Furthermore, the mortality rate for women diagnosed with breast cancer has remained virtually unchanged for over forty years, although there has been a recent slight decrease. In other words, a woman diagnosed with breast cancer today has roughly the same likelihood of dying of the disease as her grandmother nearly half a century ago.

The limited success of treating tumors with traditional approaches such as surgery, radiation, and chemotherapy has fueled a search for a more basic understanding of the molecular basis of the disease in the hope that more effective therapies can be devised. The molecular regulators of cell growth known as growth factors are attractive targets for novel therapies because a great deal of evidence indicates that they play a central role in the regulation of cell growth in both normal and cancerous tissues. One mechanism by which a cell may become cancerous, it is hypothesized, is through the unregulated expression of these factors. In breast tissue, the action of growth factors and their cellular receptors may mediate, at least in part, the mitogenic effects of the hormone estrogen, and the aberrant expression of these proteins may contribute to a diminished reliance on estrogen to promote cell growth. In addition to promoting cell growth, growth factors may be involved in the induction of angiogenesis, the process of new blood vessel growth within the tumor that provides both nourishment for the tumor cells and a pathway for metastatic spread of the disease. There is much interest, therefore, in gaining a better understanding of the precise role of growth factors in controlling cell growth.

Mitogenesis in breast tissue is mediated in part through the interaction of the steroid hormone estrogen, denoted E2 or 17-β-estradiol, with an intracellular protein, the estrogen receptor (ER). The E2–ER complex acts as a transcription factor to induce expression of specific genes. Hormone-dependent breast cancers express ER and are stimulated by E2, which acts in these cells at least in part through induction of genes coding for polypeptide growth factors that in turn promote cell growth. The E2-mediated growth stimulation of these cells has led to the use of antiestrogenic drugs to inhibit the growth of these tumors. Hormone-independent breast cancers are not stimulated by E2 and may have no detectable ER expression and are therefore insensitive to antiestrogen treatment. These tumors often express growth factors at high levels. This inverse correlation between ER and overexpression of growth factors suggests that the expression of such proteins in an unregulated manner by hormone-dependent breast cancer may contribute to its growth in the absence of E2, and it has led to the hypothesis that the constitutive expression of growth factors by a hormone-dependent cancer may contribute to the development of hormone independence. It is the general thesis of this review that amplification of the endogenous growth-factor receptor signaling pathway through overexpression of polypeptide growth factors or their cognate receptors is sufficient to diminish or overcome the requirement for E2 stimulation of growth. If this is true, it follows that disruption of these growth signals may present an effective target for future therapies directed against hormone-independent cancers.

The poorer prognosis associated with tumors that do not express ER also suggests that growth stimulation independent of E2 may result in more aggressive behavior by the tumor. In this sense, use of a growth factor-mediated pathway results not only in the ability to overcome dependence on estrogen but also the development of a more malignant phenotype. Therapies directed at growth factors might therefore not only restore some sensitivity to antiestrogen therapies in hormone independent cancers that retain ER expression, but also diminish the ability of the tumor cells to invade and metastasize.

BREAST CANCER, ESTROGEN, AND ESTROGEN RECEPTOR

Although environmental influences such as diet (Lipsett, 1975) are believed to be important, the primary tumor-promoting agent in breast cancer is thought to be E2 itself (Figure 1; Menedez-Botet and Schwartz, 1993). Prolonged exposure of breast tissue to E2, whether as a result of early menarche or late menopause, as well as late age at first pregnancy, which leaves the breast in a hyperproliferative, but not fully differentiated, state, increase the relative risk of developing breast cancer (Hulka, et al., 1994). A role for E2 exposure in promoting breast cancer is supported by the observations that women who undergo bilateral oophorectomy before age 40 have a significantly reduced incidence of breast cancer when compared to women who experience normal menopause (Feinleib, 1968; Hirayama and Wynder, 1962) and that the incidence of breast cancer among men is only 1% of that in women (Kinne, 1990).

The progression of mammary epithelium from normal to malignant is thought to occur in several steps. An increase in the number of cells contained within the duct is termed hyperplasia or proliferative disease, and this may be present with or without cellular atypia. Further changes in cellular morphology may progress to ductal carcinoma in situ (DCIS), a term applied to cells determined to be cancerous but contained within the myoepithelial cell layer. Evidence suggests that some, but not all, DCIS will progress to fully invasive cancer (Harris et al., 1980). Studies of allelic loss of heterozygosity (LOH) within these various cells suggest that the less advanced lesions, proliferative disease without atypia and atypical ductal hyperplasia,

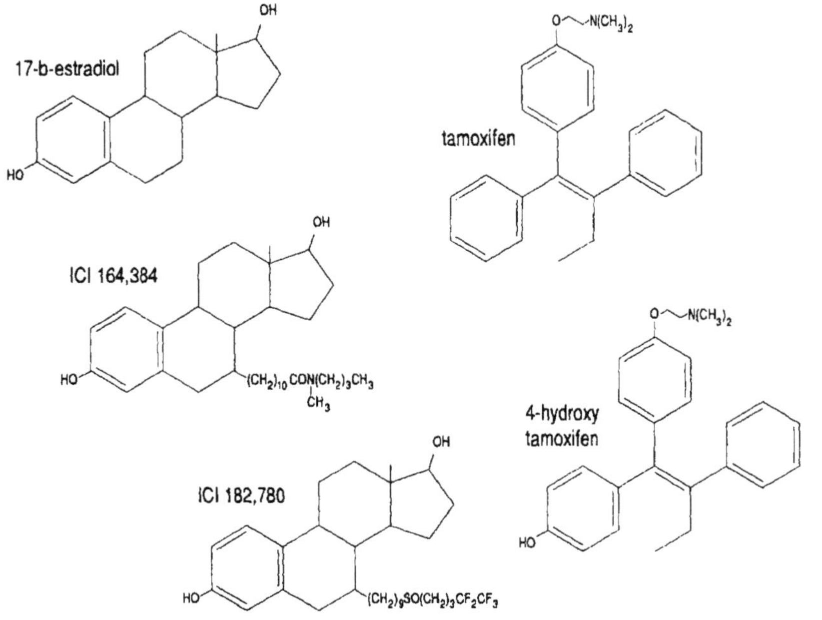

Figure 1. Molecular structure of estrogen and antiestrogens. Estrogen (E2) is an 18-carbon steroid hormone derived from cholesterol. It is synthesized primarily in the ovaries and contains an aromatized A ring. Tamoxifen and its metabolite 4-hydroxy tamoxifen are nonsteroidal antiestrogens that compete with E2 for binding to the estrogen receptor (ER). Both compounds exhibit partial agonist activity addition to antagonist activity. ICI 164, 184 and ICI 182,784 and ICI 182,789 are considered "pure" steroidal antiestrogens because they do not possess agonist activity. These drugs also compete with E2 for binding to the ER.

as well as DCIS, may give rise to later cancers because the more advanced lesions share significant patterns of LOH with the less malignant cells (Allred et al. 1993; O'Connell et al. 1994; Smith et al., 1993). Tumors in which the cells have invaded the basement membrane are classified as infiltrating ductal carcinoma, and represent the most common form of the disease at diagnosis.

Infiltrating ductal carcinomas may contain or lack estrogen receptors (ERs). E2 stimulates the growth of both normal breast tissue and a subset of breast cancers, and expression of ER is an important indicator of likelihood of response to treatment with antiestrogens (Wittliff, 1984). Not every tumor that expresses ER is inhibited by antiestrogens, however, and many tumors that do initially respond subsequently recur and are no longer sensitive to hormone ablation, possibly as a result of selection pressure in favor of cells that develop hormone-independent growth. At the time of diagnosis, approximately ½–¾ of breast cancers are considered estrogen receptor positive (ER+; Jensen, 1981). ER+ tumors respond to hormone ablative therapy 55% of the time, and this response rate is raised to 80% if only ER+/PR+ tumors are considered (Wittliff, 1984), as presence of progesterone receptor (PR) is an indicator of a functional ER. In addition to predicting response to antiestrogen therapy, estrogen receptor positive (ER+) tumors tend to be more differentiated and less malignant in behavior, whereas estrogen receptor negative (ER–) tumors typically have an undifferentiated and more malignant phenotype (Silfversward et al., 1980). ER expression has also been used along with factors such as size of the primary tumor, presence of tumor cells in surrounding lymph nodes, and histological grade of the tumor cells to predict overall survival. Patients with more advanced ER+ disease have a better prognosis than patients with less advanced ER– disease, indicating the utility of ER expression in the further classification of patients already stratified by tumor staging (Godolphin et al., 1981). This suggests that continued ER expression may be a marker for a less aggressive tumor and that loss of ER expression may indicate the development of a more malignant phenotype.

Much of our current understanding of the molecular basis of growth control in breast cancer cells has been gained from study of human breast cancer cell lines. First isolated from patients, the majority of these cell lines were established from metastases or pleural effusions and subsequently adapted to grow in culture. These cell lines display a wide range of phenotypic behavior and molecular characteristics that reflect some of the changes believed to occur during progression of breast cancer (Engel and Young, 1978). For example, MCF-7 cells are ER+ and E2 responsive in vitro and are strictly dependent on E2 for growth in vivo; however, they are poorly tumorigenic in vivo even in the presence of E2 (Seibert et al., 1983). In contrast, ER– cell lines such as MDA- MB-231 and MDA-MB-468 grow equally well in vitro in the presence or absence of E2 and many are quite aggressive in vivo. Investigation of the mechanisms underlying the failure of ER– breast cancer cell lines to express ER indicates that, while these cells retain ER DNA sequences (Yaich et al., 1992), there is no detectable mRNA transcribed or protein made from the gene (Barrett-Lee et al., 1987; Piva et al., 1990; Weigel and deConinck, 1993; Zhang et al., 1991). The 5' promoter region of the ER gene in MDA-MB-231 cells contains hypermethylated CpG dinucleotides (Ottaviano et al., 1994), suggesting methylation as a possible regulatory mechanism (Bird, 1986). Treatment of these cells with 5-azacytidine, an analog of cytosine used to generate DNA without methylation, induces ER expression, and the ER is functional based on its ability to induce expression of both PR and a reporter gene (Ferguson et al., 1995). It should be noted, however, that treatment with 5-azacytidine nonspecifically allows expression of many genes, and the possible role of other unknown gene products in this effect is not known.

There is in fact much evidence that factors other than the appropriate receptor alone are required for cellular responsiveness to E2. In experiments also done in MDA- MB-231 cells, expression of a recombinant ER results in growth inhibition in the presence of E2 (Garcia et al., 1992; Jiang and Jordan, 1992; Zajchowski et al.,

1993). Ectopic expression of ER in other ER– cells has also shown that expression of ER alone is insufficient to result in E2-responsive growth (Levenson and Jordan, 1994). In mouse fibroblasts transfected with the human ER, addition of E2 has no growth-promoting effect (Gaben and Mester, 1991). Transfection of ER into HeLa cells, a human adenocarcinoma cell line of cervical origin, results in cells whose growth is inhibited by E2 in a manner that is dependent on ER expression levels, although it is not clear whether this growth inhibition is a result of induction of growth-inhibitory genes or a more nonspecific effect of ER overexpression (Maminta et al., 1991). These results indicate that expression of ER alone does not render cells responsive to E2 or simply re- create the ER+ state in breast cancer.

The ER is a member of a family of DNA-binding proteins that act as transcription factors. Other members of the steroid hormone receptor superfamily include receptors for other steroids, thyroid hormone, retinoic acid, and vitamin D (O'Malley, 1992). Mature steroid receptors are divided into five subdomains (Figure 2), and the function of each of the domains is separable; replacing the DNA-binding domain of the glucocorticoid receptor with that of the ER results in a hybrid receptor that activates E2-responsive genes in the presence of glucocorticoids but not E2 (Berg, 1989). The three primary functions of the ER, ligand binding, sequence-specific DNA binding to an estrogen response element (ERE), and transactivation, take place in the E, C, and A/B domains of the protein, respectively. The A/B domain of the estrogen receptor is located at the N terminus and is the regulatory region of the receptor. The A/B domain also contains the transcriptional activating function 1 (AF-1) of the ER, a region that promotes transcription of genes in a ligand-independent manner. This activity, however, is dependent on both cell type and promoter context; for example, AF-1 activates transcription efficiently in chicken embryonic fibroblasts but poorly in HeLa cells (Tora et al., 1989). In addition, deletions of the A/B domain, including AF-1, have no effect on activation of a vitellogenin ERE but severely impair activation of the human pS2 promoter (Kumar et al., 1987). The C region of the ER is compossed of the DNA binding domain, two zinc finger motifs folded to form a single structural unit termed a class II zinc-binding motif (Harrison, 1991). Most of the ability of this DNA-binding region to discriminate among the DNA sequences of the various steroid response elements is determined by three amino acids within the first zinc finger (Mader et al., 1989). The DNA-binding domain also contains a weak dimerization signal (Kumar and Chambon, 1988). Region E appears to be the most complex in terms of both structure and function. This region, in addition to mediating hormone binding, contains sequences responsible for nuclear localization (Guiochon-Mante et al., 1989; Picard and Yamamoto, 1987) and interactions with heat shock proteins (Chambraud et al., 1990), as well as a stronger dimerization signal than found in the DNA-binding domain (Fawell et al., 1990a; Kumar and Chambon, 1988; Salomonsson et al., 1994). This region also contains the AF-2 function, which is similar to AF-1 in the A/B domain; however, AF-2 is active only in the presence of hormone (Webster et al., 1988). In the intact receptor, AF-1 and AF-2 act together to produce full transcriptional activity (Danielian et al., 1992).

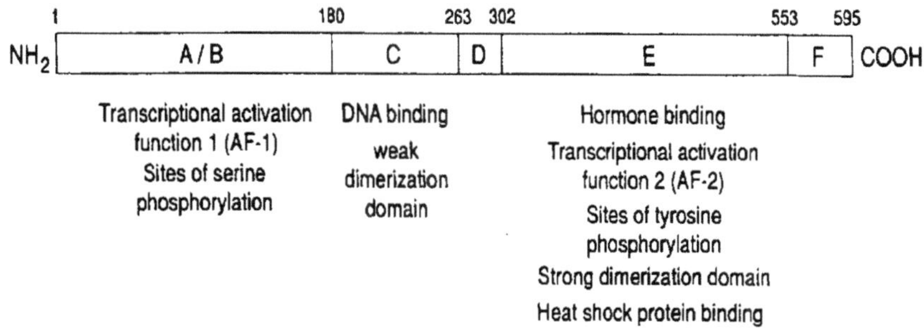

Figure 2. Functional domains of the human estrogen receptor. The ER is divided into six subdomains. The A/B domain is involved in transactivation of target genes, the C domain mediates binding to a specific DNA sequence present in the regulatory region of E2-responsive genes, and the E domain mediates hormone binding. Numbers above the receptor correspond to amino acid position of the domain boundaries.

The ER, like all steroid receptors, is a phosphoprotein, and evidence suggests that transcriptional activation may coincide with an increase in phosphorylation of the receptor. In a mouse mammary tumor model, ER isolated from hormone-independent breast tumors is hypophosphorylated compared to that in hormone-dependent tumors (Migliaccio et al., 1992), although the biological significance of this difference is not clear. A kinase purified from calf uterus phosphorylates the human ER on a tyrosine residue at a site within or near the hormone-binding domain (Migliaccio et al., 1989), and this phosphorylation promotes binding of hormone to the ER (Koffman et al., 1991; Migliaccio et al., 1991). *Src*-family tyrosine kinases can phosphorylate tyrosine 537 in the ER hormone-binding domain independently of estrogen treatment (Arnold et al., 1995b), and phosphorylation of this residue facilitates subsequent dimerization and binding of the hormone–ER complex to an ERE (Arnold et al., 1995b).

In contrast to these reports of tyrosine phosphorylation, other groups report phosphorylation of ER exclusively on serine (Denton et al., 1992; Washburn et al., 1991). Discrepancies exist, however, as to the number of phosphorylation sites and the identity of the individual residues. In both mouse uterus and MCF-7 breast cancer cells, E2 treatment increases the serine phosphorylation of ER (Denton et al., 1992; Washburn et al., 1991). In mouse ER, multiple sites of serine phosphorylation have been reported upon estrogen binding (Lahooti et al., 1994); similar patterns have been observed in humans (Ali et al., 1993). Serine residues at position 104, 106, and 118 in the A/B domain of the ER are phosphorylated in vivo (Le Goff et al., 1994). Mutation of these residues inhibits induction of a reporter gene (Le Goff et al., 1994), suggesting that phosphorylation is important for full receptor activation. Other studies, however, have found only a single site of serine phosphorylation in the ER, either at position 118, which lies within the AF-1 domain (Ali et al., 1993; Joel et al., 1995) or position 167 (Arnold et al., 1994). Kinases implicated in the phosphorylation of ER at these sites include casein kinase II (Arnold et al., 1994), a

DNA-dependent protein kinase (Arnold et al., 1995c), and MAP kinase (Arnold et al., 1995a; Bunone et al., 1996; Kato et al., 1995). Although agents that activate protein kinase A or C result in ER phosphorylation (Le Goff et al., 1994), neither of these enzymes directly phosphorylates the ER (Arnold et al., 1995c), which suggests an indirect mechanism or an intermediate kinase.

The classical model of transcriptional control by steroid hormone receptors involves dimerization of the receptor in the presence of ligand, followed by DNA-binding at a specific hormone response element and the enhancement of transcription due to the recruitment of accessory proteins to the initiation complex of RNA polymerases at the promoter. In the absence of hormone, the ER is localized to the cell nucleus but is not tightly associated with DNA (Greene et al., 1984; Welshons et al., 1988) and exists in a complex with heat shock proteins (Landel et al., 1994). Hormone binding induces a conformational change in the receptor that results in dimerization (Wang et al., 1995b) and binding of dimerized receptors to DNA at a specific sequence termed an estrogen response element (ERE). DNA-binding is mediated through interactions of the zinc fingers of the C domain and the major groove of the DNA double helix (Schwabe et al., 1993), resulting in transcriptional activation of target genes (Kumar and Chambon, 1988).

Other proteins, apart from the basal transcription factors, are also associated with activation of E2-responsive genes. A number of coactivator and corepressor molecules have recently been described that modulate the extent of transcriptional activation mediated by nuclear receptors (Horwitz et al., 1996). A protein called steroid receptor coactivator-1 or SRC-1 specifically interacts with several steroid receptors including ER in a ligand- dependent fashion and increases the degree of transcription induction (Onate et al., 1995). In ER+ MCF-7 breast cancer cells, the hormone-binding domain of ER specifically interacts with a 160 kDa protein in the presence of E2, and mutant ERs unable to activate transcription can not bind this protein (Halachmi et al., 1994). In ER+ ZR-75-1 breast cancer cells, treatment with E2 causes receptor-interacting proteins (RIPs) of 180, 160, and 80 kDa to interact with the hormone-binding domain of the ER. These proteins do not interact with an ER complexed with antiestrogens, and ER hormone-binding domains that are defective in transcription activation due to mutation of AF-2 do not associate with these RIPs (Cavailles et al., 1994). Subsequent cloning and characterization of the 140-kDa RIP reveal that this protein interacts in vitro with the AF-2 region of the ER and seems to play a dual role in ER activation. When expressed at low levels relative to ER, RIP140 stimulates ER-mediated transcription of a reporter gene two-fold; however, when expressed at high levels, RIP140 almost completely abolishes ER-mediated transcription (Cavailles et al., 1995). Because RIP140 is induced by estrogen in ZR-75-1 cells (Cavailles et al., 1994), these results suggest RIP140 may act to both promote the initial activation of the ER and subsequently act as a feedback inhibitor of the ER. These results indicate that specific interactions with other proteins coincide with ligand binding and transcriptional activation and suggest that these events may influence receptor activity in both the presence and absence of hormone.

ANTIESTROGENS, TAMOXIFEN RESISTANCE, AND HORMONE-INDEPENDENT GROWTH

As early as 1896, surgical removal of the ovaries was used to induce remission in premenopausal women with advanced breast cancer (Beatson, 1896); today, however, hormone ablation is effected chemically through the use of antiestrogenic drugs. The most commonly used antiestrogen is tamoxifen, which may be converted in vivo to 4- hydroxy tamoxifen (Figure 1), a metabolite that has a 10-fold higher affinity for the ER (Furr and Jordan, 1984). Both are nonsteroidal antiestrogens that compete with E2 for binding to ER (Jordan et al., 1977; Jordan and Koerner, 1975) and inhibit gene transcription due to the presence of a side chain (Jordan et al., 1988). Tamoxifen exhibits both weak agonist as well as antagonist activity (Jordan, 1984) that is both species- and tissue-specific; in mice, tamoxifen virtually acts as a complete agonist (Terenius, 1970) but can still inhibit the growth of ER+ xenografts (Langdon et al., 1994; Osborne et al., 1995); in humans, doses of tamoxifen that induce remission in mammary tumors produce estrogenic changes in vaginal tissues (Ferrazzi et al., 1977). The antagonist activity of tamoxifen is thought to arise from its inability to induce AF-2 function on promoters that require this region for transcription and its agonist properties from activation of promoters where AF-1 action alone is sufficient for gene transcription (Berry et al., 1990; Tzukerman et al., 1994).

Because tamoxifen possesses activity as a partial agonist, antiestrogens that do not possess this activity and therefore act as complete antagonists have been synthesized. These so-called "pure" steroidal antiestrogens include ICI 164,184 and ICI 182,780, which each contain a side chain attached to the B ring of the steroid (Figure 1); the effectiveness of these compounds as antiestrogens depends on this side chain (Bowler et al., 1989). Because the clinical usefulness of ICI 164,184 is limited due to its hydrophobicity, a fluorine-substituted analog with greater solubility was generated: ICI 182,780 (Figure 1); this drug has the further advantage of having a slightly higher affinity for the ER (Wakeling et al., 1991). Both ICI compounds are believed to inhibit ER activation by inhibiting E2 binding and subsequent dimerization of the receptor (Fawell et al., 1990b; Parker, 1993). Addition of either ICI compound to cells decreases the levels of ER by reducing the half-life of the protein (Dauvois et al., 1992).

Because the molecular mechanisms underlying the acquisition of resistance to antihormone therapy in breast cancer are poorly understood, ER+ breast cancer cell lines have been studied in vitro and in vivo in many model systems that attempt to re-create changes in tumor behavior seen in patients in response to hormone ablation. Such manipulation has successfully generated E2-independent or antiestrogen-insensitive cell lines from ER+ precursors, but most retain ER expression. Variant sublines of ER+ MCF-7 cells have been selected either by in vitro growth in E2-depleted medium (Katzenellenbogen et al., 1987; Welshons and Jordan, 1987) or by serial passage through nude mice in vivo in the absence of E2 and subsequent isolation of rare spontaneous tumors (Clarke et al., 1989b). Cells thus selected grow as well under steroid-free conditions as do normal cells in the presence of E2 but cannot be further stimulated to grow in vitro by E2; they also exhibit elevated ER levels, but retain the ability to induce E2-responsive genes upon hormone addition (Katzenellenbogen et al., 1987; Welshons and Jordan, 1987).

In the E2-independent MCF-7 subline selected in nude mice, in vivo tumor formation can still be enhanced by E2, suggesting that E2 may induce host factors that stimulate the tumor cells in a paracrine fashion (Clarke et al., 1989b). Characterization of an in vitro cell line established from these cells selected in vivo reveals many of the same changes observed in cells originally selected in vitro, including retention of both ER expression and sensitivity to antiestrogens (Brünner et al., 1993a). Because of the failure of E2 withdrawal from ER+ breast cancer cells to induce loss of ER and resistance to tamoxifen, these E2-independent cells were further selected for growth in the presence of tamoxifen in vitro. Although the subsequent tamoxifen-resistant cells continue to express ER, the ability of ER to induce PR expression is diminished; however, this cell line remains sensitive to the steroidal antiestrogen ICI 182,780 (Brünner et al., 1993b). In a reciprocal study, ER+ breast cancer cells selected for resistance to either ICI 164,384 or ICI 182,780 continue to express ER, retain sensitivity to tamoxifen, and become stimulated for in vivo tumor formation by E_2 (Lykkesfeldt et al., 1995).

Acquisition of resistance to antiestrogen therapy has also been modeled by in vitro selection of ER+ ZR-75-1 breast cancer cells in the presence of antiestrogens and 5-azacytidine. Although these cells acquire hormone independence, expression of ER is lost and an increase in expression of transmembrane tyrosine kinase growth-factor receptors is observed (van Agthoven et al., 1994b). This result suggests that, in both the absence of E2 and the presence of an agent that promotes expression of previously silent genes or prevents the silencing of active genes, cells emerge that utilize alternate mitogenic pathways that may still result in loss of ER expression.

Although some tumors that arise after adjuvant tamoxifen treatment of patients with tumors that were originally ER+ exhibit an ER- phenotype, the majority of tumors that initially respond to tamoxifen continue to express functional ER when they acquire resistance to the drug (Johnston et al., 1995; Johnston et al., 1997). The failure of most cells lines selected for E2-independent and tamoxifen-resistant growth to lose ER expression also indicates that mechanisms of hormone-independent growth and resistance to tamoxifen do not necessarily require loss of ER. Increased resistance to the antiestrogenic effects of tamoxifen without alterations in ER expression could be affected through diminished uptake or intracellular accumulation of the drug. Evidence for such a mechanism is found in nude mice bearing MCF-7-derived tumors, where 10- fold lower levels of tamoxifen are seen in tumors resistant to tamoxifen compared to levels of the drug circulating in serum (Osborne et al., 1991). Furthermore, loss of ER expression in response to tamoxifen treatment would not explain the response of some ER+ cells selected for tamoxifen resistance that are actually *stimulated* by the drug. One tamoxifen-stimulated tumor has been found to express an ER containing a point mutation in the ligand-binding domain that is presumably responsible for the tamoxifen- stimulated growth of these cells (Wolf and Jordan, 1994), and other mutations within the hormone-binding domain of the ER have been shown to convert tamoxifen and ICI 164,384 to complete agonists (Mahfoudi et al., 1995). It should be noted, however, that mutations of the ER are rare in patients with tamoxifen-resistant tumors (Karnik et al., 1994). Mitogenic effects of tamoxifen may also arise from altered metabolism of the drug, as both elevated levels of isomers of tamoxifen that are less potent antiestrogens as well as increases in estrogenic metabolites of the drug are found in some tumors (Johnston et al., 1993; Osborne et al., 1991; Wiebe et al., 1992).

Not all breast cancer therapy, however, utilizes antiestrogenic drugs directed at the ER, and consequently not all attempts to isolate variant cells have selected for hormone-independent growth. One such cell line is derived from ER+ MCF-7 cells selected for resistance to the chemotherapeutic drug adriamycin, an intercalating agent that inhibits the replication and transcription of DNA. These cells known as MCF7/Adr^R have acquired drug resistance due to amplification of the multidrug resistance gene (MDR); interestingly, these cells have also lost expression of ER and acquired the ability to form tumors in nude mice in the absence of E2 (Vickers et al., 1988). Although exposure to cytotoxic drugs can decrease ER expression (Clarke et al., 1986; Yang and Samaan, 1983), long-term passage of the cells following selection in drug-free medium does not cause ER to reappear. The 5' promoter region of the ER gene in MCF-7/Adr^R cells is hypermethylated (Ottaviano et al., 1994), a phenomenon reminiscent of the ER gene in MDA-MB-231 cells, suggesting methylation may be a common mechanism for loss of ER expression. MCF-7/Adr^R cells also possess 100-fold higher levels of the transmembrane tyrosine kinase epidermal growth factor receptor (EGFR). Recent evidence indicates that acquired overexpression of EGFR or the related transmembrane tyrosine kinase c-erbB2 is not a mechanism by which tumors acquire resistance to tamoxifen (Newby et al., 1997). However, expression of either receptor in tumors prior to treatment is predictive of a poor response to this antiestrogen (Carlomagno et al., 1996; Newby et al., 1997). This result suggests that, even in the absence of antiestrogen therapy, amplification of the signaling capacity of growth factor–activated pathways may contribute to loss of ER expression and hormone-independent growth . In a similar study, MCF-7 cells selected for adriamycin resistance in the presence of verapimil, an antagonist of the MDR protein that prevents amplification of the gene, also upregulate expression of EGFR. In these cells, EGFR expression is lower than in the MCF-7/Adr^R cells, and these cells retain ER expression (Dickstein et al., 1993).

The central limitation in current hormone ablative therapy of breast cancer is the ability of tumor cells to acquire resistance to the drugs utilized; it is not the case, however, that all hormone-independent cells lose ER expression. It must therefore be true that hormone-independent growth can occur in the presence of ER. The utility of classifying a tumor as ER+ or ER– at the time of diagnosis is in predicting prognosis and likelihood of response to antiestrogen therapy. The fact that only

55% of ER+ tumors respond to tamoxifen, however, indicates that nearly half of ER+ tumors are at least partially hormone-independent. Although ER+ and ER– cell lines are used to model hormone-dependent and hormone-independent breast cancer cells respectively, it should be emphasized that it is the ability of cells to grow in the absence of E2 and the presence of antiestrogenic drugs, not the presence or absence of ER per se, that represents the limitation on the current use of antiestrogens clinically. It is likely that ER expression can continue in a cell after hormone dependence has been overcome and that a later event leads to loss of expression altogether. It should also be noted that the ER present in breast cancer cells undergoing estrogen withdrawal or treatment with antiestrogenic drugs may not be uninvolved in mitogenic signaling, despite the absence of its natural ligand.

ESTROGEN RECEPTOR AND THE *erb*B GENE FAMILY OF GROWTH FACTOR RECEPTORS

If the use of alternate mitogenic pathways is one mechanism of hormone independent growth, what might be the molecular mechanism? Part of the difficulty in addressing this question lies in our limited understanding of the complexity of the regulation of cell growth. When ER+ and ER– breast cancer cell lines are compared, the sheer number of growth factors and receptors overexpressed in ER– cells makes identification of one or two critical factors all but impossible. The search for proteins that play a direct role in the acquisition of hormone-independent growth has therefore focused on the set of genes that correlate inversely with ER expression or that seem particularly useful in predicting prognosis.

One such gene codes for the epidermal growth factor receptor (EGFR), a transmembrane glycoprotein with intrinsic tyrosine kinase capability. EGFR is the prototypical member of the *erb*B family of growth factor receptors, and expression of EGFR is a valuable prognostic indicator in breast cancer. In ER+ breast cancer cell lines, expression of EGFR is very low, whereas some ER- cells express over one million sites per cell (Davidson et al., 1987). In breast cancers, expression of EGFR correlates inversely with both ER (Bolufer et al., 1990; Koenders et al., 1991; Lee et al., 1990; Lewis et al., 1990) and PR (Battaglia et al., 1988a; Battaglia et al., 1988b; Delarue et al., 1988; Foekens et al., 1989). In tumors where expression is heterogeneous and no inverse correlation between ER and EGFR is seen, individual cells seem to be either ER+ or EGFR+ and the two proteins rarely co-localize to the same cell (van Agthoven et al., 1994a). Overexpression of EGFR correlates with both poor response to tamoxifen treatment (Nicholson et al., 1988; Nicholson et al., 1989; Nicholson et al., 1990) and poor prognosis (Harris et al., 1992; Nicholson et al., 1991; Toi et al., 1991). Significantly, expression of EGFR predicts prognosis independent of ER status, and ER+/EGFR+ patients have a worse prognosis than ER–/EGFR- patients, suggesting that the phenotype conferred by EGFR overexpression is dominant over that conferred by ER (Harris et al., 1992; Sainsbury et al., 1987). The observation that

ER+/EGFR– primary tumors may give rise to ER–/EGFR+ metastases (Mori et al., 1991; Sainsbury et al., 1985; Toi et al., 1991) suggests overexpression of EGFR may also be involved in the acquisition of a more aggressive phenotype.

The mature EGFR protein can be divided into two primary regions separated by a short transmembrane domain that anchors the protein in the plasma membrane (Figure 3). An extracellular domain is responsible for binding various ligands. An

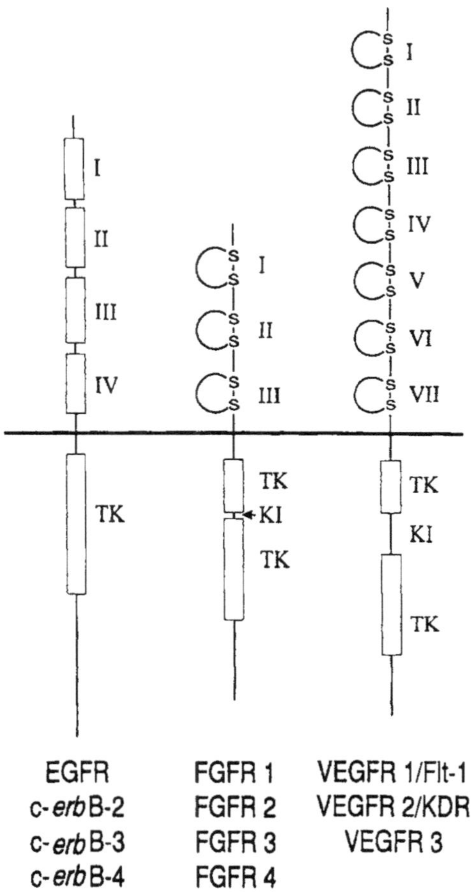

Figure 3. Prototypic receptors for EGF/erbB, FGF, and VEGF family ligands. The members of the EGFR/erbB gene family have four extracellular subdomains and a single intracellular tyrosine kinase domain. FGF receptors have either two or three extracellular domains of homology to IG molecules and an intracellular tyrosine kinase domain that is interrupted by a small kinase insert region. VEGF receptors contain seven extracellular Ig domains and an intracellular tyrosine kinase domain with a larger insert than FGF receptors. Shown is a representation of the structure of a typical receptor of each type. S–S denotes disulfide-linked cysteine residues that maintain the structure of the Ig loops.

intracellular tyrosine kinase domain binds numerous intracellular substrates and phosphorylates them on tyrosine residues. The extracellular and intracellular domains are further subdivided into subdomains, and the boundaries of these subdomains roughly correspond to intron/exon boundaries (Callaghan et al., 1993). Binding of ligand to the extracellular domain results in receptor dimerization and activation of the intracellular kinase activity thereby transducing the signals generated outside the cell across the membrane to the interior of the cell.

The extracellular region of EGFR consists of four subdomains of 150 amino acids each contained within two direct 300-amino-acid repeats. Subdomains I and III are involved in ligand binding, whereas subdomains II and IV are cysteine-rich and presumably involved in maintaining the secondary and tertiary structure of the extracellular region (Lax et al., 1988; Ullrich and Schlessinger, 1990; Wu et al., 1990). A soluble form of the EGFR extracellular domain binds ligand normally and is capable of oligomerization, indicating that this region possesses an inherent ability to mediate both functions (Hurwitz et al., 1991). Dimerization of two receptors is sufficient to activate tyrosine kinase activity because EGFR molecules engineered to oligomerize in the absence of ligand due to introduction of an intermolecular disulfide bond are constitutively active (Sorokin et al., 1994).

Whereas EGFR tyrosine kinase activity is induced by dimerization, enzymatic activity is modulated by at least two covalent modifications to the intracellular domain of the receptor. Just within the inner surface of the plasma membrane is a threonine at position 654 that is a substrate for protein kinase C; phosphorylation of this residue results in inhibition of EGFR kinase activity (Davis, 1988) and loss of the high-affinity ligand-binding state of the receptor (Livenh et al., 1987). Increases in intracellular cyclic AMP, an activator of protein kinase A, inhibit EGF-induced mitogenesis, although this does not involve direct phosphorylation of the EGFR itself (Cook and McCormick, 1993; Wu et al., 1993). Additionally, two serine residues at positions 1046 and 1047 are substrates for phosphorylation by CAM kinase II, a kinase that is activated by the calcium-binding protein calmodulin (George et al., 1990); modification of these residues results in an inhibition of receptor tyrosine kinase activity (Countaway et al., 1992). These observations suggest that negative feedback may play a role in EGFR self- modulation as well as cross-talk between different classes of mitogenic agents that activate intracellular proteins kinases.

The carboxy terminal region, which contains the receptor autophosphorylation sites, mediates signal transduction of extracellular stimuli through interaction with specific intracellular second-messenger proteins. The autophosphorylation sites within this region are thought to act as competitive inhibitors of intracellular substrates for the tyrosine kinase activity because mutation of these tyrosines to phenylalanines has little effect on the kinetics of ligand-binding or phosphorylation of cellular targets (Honegger et al., 1988b), yet it seems to increase the affinity of the receptor for peptide substrates (Honegger et al., 1988a). Elimination of these tyrosines through progressive truncation of the intracellular domain of the receptor, however, results in decreased affinity for ligand (Livneh et al., 1986), altered substrate specificity (Decker et al., 1992), and ultimately receptors that are deficient in internalization and degradation (Decker et al., 1992; Livneh et al., 1986).

The epidermal growth factor receptor is encoded by the c-*erb*B gene locus on chromosome 7p12 (Shimizu and Kondo, 1982) and consists of 26 exons that span 110 kb (Haley et al., 1987). EGFR is a class I tyrosine kinase receptor of approximately 170 kDa (Ullrich et al., 1984) and belongs to a family of receptors that includes c-*erb*B-2, c-*erb*B-3, and c-*erb*B-4 (Figure 3) (Plowman et al., 1993a; Plowman et al., 1990; Ullrich and Schlessinger, 1990). Expression of EGFR is transiently induced by E2 treatment of ER+ MCF-7 cells (Yarden et al., 1994), although it remains far below the levels seen in some ER– cell lines. In breast cancer, the level of EGFR protein correlates with the level of mRNA, and when EGFR is overexpressed, it almost always occurs at the level of transcription (Davidson et al., 1987); rarely is the gene rearranged or amplified (King et al., 1985). Despite this fact, two naturally occurring mutants of EGFR have been reported. The v-*erb*B gene, which is responsible for the diseases induced by the avian erythroblastosis virus in chickens, is an oncogenic form of the chicken EGFR (Downward et al., 1984). It contains an N-terminal deletion of almost the entire extracellular domain and a smaller C-terminal deletion as well as several point mutations throughout the molecule (Hayman and Enrietto, 1991). As a result of these mutations, the receptor is constitutively active (Kris et al., 1985), although its ability to transform different cell types depends on the extent of mutations in the intracellular domain (Pelley et al., 1989; Shu et al., 1990). In addition to the v-*erb*B oncogene, a second naturally occurring, constitutively active mutant of EGFR has been isolated. First found in human gliomas (Nishikawa et al., 1994) and subsequently detected in tumors from various tissues including breast (Moscatello et al., 1995; Widstrand et al., 1996), the EGFRvIII mutant contains an in-frame deletion of exons two through seven that gives rise to a receptor with a truncated extracellular domain. Although this mutant receptor no longer binds ligand, it displays low levels of constitutive kinase activity (Batra et al., 1995) and modest, but constitutive, activation of downstream signaling molecules (Montgomery et al., 1995). Expression of this mutant in NIH 3T3 cells results in weak transformation independent of ligand (Yamazaki et al., 1990). Although no physiologic role has yet been shown for EGFRvIII in breast cancer, the observation that expression of this mutant in a glioblastoma cell line greatly enhances in vivo tumor formation (Nishikawa et al., 1994) suggests that this deletion may, through activation of the receptor kinase signal transduction pathway, promote tumor growth.

The c-*erb*B-2 gene is the human homologue of the rat *neu* gene that was originally isolated from ethylnitrosourea-induced neuro- and glioblastomas in rodents (Bargmann et al., 1986). Cloning and sequencing of c-*erb*B-2 revealed it to be highly homologous to EGFR, having a similar tyrosine kinase domain and transmembrane and extracellular domains (Coussens et al., 1985). The activated *neu* found in tumors differs from the normal rat gene, c-*neu*, by a single point mutation in the transmembrane region that results in a constitutively active kinase (Stern et al., 1988; Weiner et al., 1989). Interestingly, no one has yet reported a similar mutation in the human homologue; if c- *erb*B-2 is involved in human carcinogenesis or tumor progression, it is most likely through overexpression, not mutation of the transmembrane region (Lemoine et al., 1990). In human breast cancer, c- *erb*B-2 overexpression has been noted in 20–30% of cases (Slamon et al., 1987; Slamon et al., 1989). The prognostic significance of c-*erb*B-2 overexpression in breast cancer has been extensively investigated and the vast majority of studies have found a prognostic effect in at least a subgroup of patients (Gullick et al., 1991; Perren, 1991). In a number of clinical studies, c-*erb*B-2 overexpression is less frequently observed in ER+ breast tumors (Berns et al., 1992; Perren, 1991). Overexpression of c-*erb*B-2 in MCF-7 cells results in increased tumorigenicity in vivo in the absence of E2 (Liu et al., 1995b; Pietras et al., 1995) and, in one report, a 50% reduction in expression of ER (Pietras et al., 1995). On a molecular level, EGFR and c-*erb*B-2 are capable of heterodimerization and cross-phosphorylation in the presence of EGFR ligands, a process known as transmodulation (King et al., 1988; Stern and Kamps, 1988; Wada et al., 1990). Co- expression of a kinase-negative c-*erb*B-2 mutant with EGFR in murine fibroblasts is sufficient to block the EGF-dependent transformation of these cells (Qian et al., 1994), suggesting that transmodulation is an important component of both the signaling capacity and transforming ability of these receptors.

The gene encoding c-*erb*B-3 was isolated from human cDNA library using an EGFR probe and low stringency screening (Plowman et al., 1990); the c-*erb*B-4 gene was cloned using degenerate oligonucleotides to amplify a conserved region of *erb*B family tyrosine kinase domains and subsequent library screening (Plowman et al., 1993a). Little is currently known about the biologic roles of either c-*erb*B-3 or c-*erb*B-4, although increased expression and constitutive activation of c-*erb*B-3 have been noted in some breast cancer cell lines (Kraus et al., 1993). Of the four family members known, c-*erb*B-3 is the least related, and two changes in the cytoplasmic domain may profoundly affect its function. Firstly, there are seven repeats of an autophospwhorylation site within the carboxy terminus that are absent in the other family members, possibly conferring signaling specificity on c-*erb*B-3 that is lacking in the other receptors. Secondly, there are four amino acid changes in the tyrosine kinase domain, and one has been shown to abolish enzymatic activity in other kinases (Carraway and Cantley, 1994). Not surprisingly, c-*erb*B-3 exhibits diminished tyrosine kinase ability (Guy et al., 1994) raising the possibility that c-*erb*B-3 might require accessory proteins for full kinase activity or that its primary role is modulation of signals generated by other *erb*B family members through heterodimer formation. The cytoplasmic domain of c-*erb*B-3 also interacts with intracellular second messenger proteins different than those activated by other *erb*B family members, supporting the hypothesis that c-*erb*B-3 may be involved in mediating differential signaling events (Fedi et al., 1994).

LIGANDS FOR EGFR AND *erb*B FAMILY RECEPTORS

In ER+ breast cancer cells, E2 regulates expression of a number of growth factors, including several ligands for the *erb*B family of transmembrane tyrosine kinase receptors (Bates et al., 1988; Bronzert et al., 1987; Dickson et al., 1990; Lippman and Dickson, 1989; Martinez-Lacaci et al., 1995; Osborne and Arteaga, 1990; Rochefort et al., 1989; Yee et al., 1988). Many of these same factors are constitutively expressed at high levels by ER– breast cancer cell lines (Lippman et al., 1987) and have mitogenic effects on E2-deprived ER+ MCF-7 breast cancer cells (Arteaga et al., 1988; Cullen et al., 1990; Freter et al., 1988; Karey and Sirbasku, 1988; Manni et al., 1990; Ogasawara and Sirbasku, 1988; Vignon et al., 1986). Support for the hypothesis that E2 acts in part through the induction of growth factors is suggested by the observation that infusion of conditioned medium from E2-treated MCF-7 cells is sufficient to induce transient tumor formation by these cells in ovariectomized nude mice in the absence of E2 (Dickson et al., 1986b). These results indicate that factors induced by E2 are sufficient to partially mediate its effects, and they further suggest that constitutive expression of such factors might diminish the requirement of E2 for growth in ER+ breast cancer cells and possibly promote E2-independent growth.

A number of polypeptide growth factors have been shown to interact specifically with EGFR, including epidermal growth factor (EGF), transforming growth factor alpha (TGF-α; Massague, 1993), and amphiregulin (AR; Shoyab et al., 1989). The first to be isolated was EGF, a 53-amino acid polypeptide containing six regularly spaced cysteine residues that form three disulfide bonds characteristic of this family of ligands (Fisher and Lakshmanan, 1990). The mature form of the protein is derived from a 1207–amino acid transmembrane precursor that contains seven "EGF-like" repeats (Bell et al., 1995). EGF is expressed in a number of tissues including the mammary gland and can stimulate mammary epithelial cell proliferation in vitro (Richards et al., 1982; Tonelli and Sorof, 1980). In breast cancer, expression of EGF correlates positively with expression of ER, suggesting that EGF expression is a reflection of the original cell type from which the tumor is derived (Dotzlaw et al., 1990). In breast cancer cell lines, expression of EGF is seen primarily in ER+ cell lines, although it is induced by progestins and not E2 (Murphy et al., 1988).

Another polypeptide growth factor shown to bind and activate EGFR is TGF-α, which, like EGF, is synthesized as a transmembrane molecule and subsequently cleaved; the mature form of TGF-α is a 50–amino acid protein derived from a 160–amino acid transmembrane prohormone (Derynck et al., 1984). TGF-α shares only about 30% sequence homology with EGF but retains the three disulfide bonds and has a similar three-dimensional structure (Derynck, 1988). TGF-α was initially isolated from conditioned media of retrovirus-transformed rodent fibroblasts and human tumor cells based on its ability to bind EGFR (Pike et al., 1982; Todaro et al., 1980). Although TGF-α and EGF bind the same receptor, there are distinguishable differences between them. TGF-α is a more potent angiogenic factor than EGF (Schreiber et al., 1986). Although chemical cross-linking shows that both EGF and TGF-α bind to subdomain III of the EGFR extracellular domain (Lax et al., 1988; Wu et al., 1990), an anti-EGFR antibody that inhibits TGF-α binding has no effect on EGF binding, suggesting the two ligands recognize different specific binding sites or different conformations of the receptor (Winkler et al., 1989). Furthermore, a mutation in the EGFR extracellular subdomain IV, which is not believed to be involved directly in ligand binding, has no effect on EGF binding but eliminates the high affinity binding site for TGF-α (Moriai et al., 1994).

TGF-α is induced by E2 in MCF-7 breast cancer cell lines (Dickson et al., 1986a) and is overexpressed by a number of ER– cell lines (Bates et al., 1988). In breast cancer, the role of TGF-α in the acquisition of an E2-independent phenotype has been investigated by transfection of TGF-α into ER+ MCF-7 cells. Cells that constitutively overexpress the protein remain E2 responsive in vitro and do not form tumors in ovariectomized nude mice (Clarke et al., 1989a). Similarly, transfection of TGF-α into rodent fibroblast cell lines results in a more transformed phenotype (Finzi et al., 1987; Watanabe et al., 1987), but increased tumorigenicity requires overexpression of both ligand and EGFR (Di Marco et al., 1989). Conversely, overexpression of EGFR in NIH 3T3 cells is transforming only in the presence of EGF (DiFiore et al., 1987; Riedel, et al., 1988; Velu, et al., 1987). These results indicate that overexpression of TGF-α alone is unlikely to be sufficient for transformation in the absence of an increase in cell surface EGFR levels.

AR was initially isolated from the conditioned medium of phorbol ester-treated MCF-7 cells (Kimura et al., 1990). Like EGF and TGF-α, AR is synthesized as a transmembrane precursor of 252 amino acids that is then cleaved to yield an 84–amino acid form; this can be further processed by removal of six N-terminal amino acids to yield a 78–amino acid form (Salomon et al., 1995). AR mRNA and protein are expressed in a number of ER+ and ER– breast cancer cell lines, although slightly more so in ER+ lines (Normanno et al., 1993). Although AR expression correlates with ER expression in breast tumors, tumors that overexpress EGFR also express AR in 35% of cases. This suggests a possible role for AR and EGFR in an autocrine loop in a subset of tumors (LeJeune et al., 1993). AR differs from EGF and TGF-α in that it contains two putative N-terminal nuclear localization signals and is sometimes found in the cell nucleus, although its function there remains unclear (Salomon et al., 1995).

Other members of the EGF family of proteins include a heparin-binding form of EGF, HB-EGF (Higashiyama et al., 1991), β-cellulin (Shing et al., 1993), and cripto-1 (Ciccodicola, et al., 1989). All three proteins, like EGF, TGF-α, and AR, are synthesized as transmembrane precursors that are proteolytically cleaved to yield a soluble form. Mature HB-EGF, a protein of approximately 20 kDa, was isolated from a human lymphoma cell line; it binds EGFR and induces tyrosine phosphorylation (Higashiyama et al., 1992). HB-EGF is mitogenic for both smooth muscle cells and A431 epidermal carcinoma cells, which overexpress EGFR (Higashiyama et al., 1991). β-cellulin was initially purified from the conditioned medium of a mouse pancreatic β-cell tumor cell line (Shing et al., 1993). It is an 80–amino acid protein that is thought to interact primarily with EGFR (Watanabe et al., 1994), although it appears to also bind c-*erb*B-4 (Riese et al., 1996). Cripto-1 is a 188–amino acid protein that was identified in an embryonal carcinoma cell line (Ciccodicola et al., 1989). Although cripto-1 is a member of the EGF family of growth factors based on the presence of the six conserved cysteine residues (Ciccodicola et al., 1989) and is transforming in a number of cell types (Ciardiello et al., 1991; Ciccodicola et al., 1989), it does not seem to bind EGFR (Brandt et al., 1994); the natural receptor for cripto-1 remains unidentified.

In contrast to the multiple growth factors shown to bind and activate EGFR specifically, only two ligands for c-*erb*B-2 have been described. One, termed *neu* activating factor or NAF, was purified from the conditioned medium of a transformed human T cell line (Dobashi et al., 1991). NAF is a protein of approximately 15–17 kDa that both activates c-*erb*B-2 in vivo (Dobashi et al., 1991) and binds to purified receptor protein in vitro (Samanta, et al., 1994). Binding of NAF to c-*erb*B-2 is associated with dimerization and tyrosine phosphorylation of the receptor as well as phosphorylation of several small nuclear proteins; these proteins are also phosphorylated in response to EGF in cells expressing EGFR (Samanta and Greene, 1995). This suggests that EGFR and c-*erb*B-2 may act through some of the same downstream effectors.

The second candidate ligand for c-*erb*B-2 is an integral membrane glycoprotein termed ascites sialoglycoprotein-2 or ASGP-2 (Sheng et al., 1992). ASGP-2 is synthesized as part of a larger mRNA; this precursor is post-translationally cleaved into a heterodimeric complex of two proteins, ASGP-1 and ASGP-2 (Sherblom and Carraway, 1980). Expression of RNA encoding the ASGP-1/ASGP-2 precursor is observed in some human breast tumors, suggesting that an increase in expression of this protein may play a role in tumor progression (Wu et al., 1994). ASGP-1 is a highly glycosylated mucin-like protein whose expression has been implicated in an increase in metastatic potential (Steck and Nicolson, 1983). ASGP-2, which contains two EGF-like domains (Sheng et al., 1992), can be co-immunoprecipitated with c-*erb*B-2 from rat mammary ascites cells, suggesting that these two proteins exist in a complex that may maintain the receptor in a constitutively active form (Sheng et al., 1992).

The family of proteins known alternatively as heregulins (Holmes et al., 1992), *neu* differentiation factors (Peles et al., 1992; Wen, et al., 1992), glial growth factors (Marchionni et al., 1993), and acetylcholine receptor-inducing activity (Battaglia et al., 1993; Falls et al., 1993) includes ligands for c-*erb*B-3 and c-*erb*B-4. Many of these factors were initially isolated based on their ability to induce tyrosine phosphorylation of c-*erb*B-2 and were thought to be ligands for this receptor; subsequently, the phosphorylation of c-*erb*B-2 was found to result from heterodimerization between c-*erb*B-2 and either c-*erb*B-3 or c-*erb*B-4 driven by ligand binding to the non-c-*erb*B-2 portion of the dimer (Plowman et al., 1993b; Sliwkowski et al., 1994). Heregulins are a family of related proteins that each contain one of two types of EGF-like domains, designated as α and β (Wen et al., 1994). These proteins arise from tissue-specific alternative splicing of a single mRNA (Holmes et al., 1992). The presence of 12 cysteine residues within these growth factors makes them members of the EGF family of proteins, although they do not bind EGFR directly (Riese et al., 1995). Exogenous addition of recombinant heregulin-β-1 to wild-type MCF-7 cells or transfection of MCF-7 cells with heregulin-β-1 results in hormone-independent growth in vivo and ligand-independent activation of ER (Pietras et al., 1995), suggesting that these proteins may play a role in the growth of ER+ breast cancer cells in the absence of E2. MCF-7 cells transfected with heregulin-β-2 exhibit a more aggressive, hormone-independent phenotype and lose the ability to induce expression of an ERE-reporter construct, suggesting that ER signaling is disrupted by activation of the c-*erb*B-2, -3, and -4 signaling cascade (Tang et al., 1996).

The cellular response to EGF family ligands, including heregulins, is complex and appears to be regulated not only by the presence of an appropriate receptor but also the presence of other *erb*B family members that can act as co-receptors. A model system has been described that utilizes a hematopoietic cell line devoid of endogenous *erb*B receptor expression except for a low level of c-*erb*B-3, which is transfected with various pairwise combinations of *erb*B family members. In response to EGF, EGFR forms not only homodimers with itself but also heterodimers with the three other *erb*B family members (Riese, et al., 1996). Addition of β-cellulin to these same transfectants results in transmodulation of c-*erb*B-2 and c-*erb*B-3 in cells coexpressing either EGFR or c-*erb*B-4 (Riese et al., 1996). These results indicate that EGF and β-cellulin may act not only through EGFR/EGFR homodimers but also through heterodimers that consist of EGFR and another *erb*B receptor. In this same system, addition of the EGF-like domain of heregulin is sufficient to induce tyrosine phosphorylation of both c-*erb*B-2 and c-*erb*B-4, but not EGFR or c-*erb*B-3, when each receptor is expressed singly. The failure to induce tyrosine phosphorylation of c-*erb*B-3 is likely due to the impaired kinase activity of this receptor (Guy et al., 1994). When expressed in pairwise combination, however, addition of heregulin to cells expressing either c-*erb*B-3 or c-*erb*B-4 induces tyrosine phosphorylation of all other *erb*B family members (Riese et al., 1995). These results indicate that the pattern of *erb*B receptor expression may influence the qualitative response of a cell to a particular ligand and raise the possibility that ligands may bind differentially to homo- versus heterodimers of *erb*B family receptors. This additional level of interaction provides a further degree of complexity and specificity of cellular responses to a peptide growth factor.

EGFR AND SIGNAL TRANSDUCTION

Activation of the epidermal growth factor receptor begins by ligand binding to the extracellular domain subsequently inducing a conformational change in the receptor and results in receptor dimerization. This oligomerization induces the dimerized receptor to cross- phosphorylate the C-terminal tail of its partner (Honegger et al., 1989; Kashles et al., 1988). Mutations that abolish tyrosine kinase activity also abolish the ability of the receptor to phosphorylate cellular substrates and to transduce signals (Chen et al., 1987); these mutations also prevent proper degradation of receptors following ligand-induced internalization (Felder et al., 1990). There is the suggestion that EGFR molecules that lack tyrosine kinase activity due to either a point mutation or deletion of the intracellular region may be able to suppress the function of wild type EGFR molecules through heterodimerization and formation of hemiphosphorylated dimers that are believed to be deficient in signal transduction (Kashles et al., 1991). Whether such mutants act in a dominant–negative fashion in all cell types, however, is not clear (Campos-Gonzales and Glenney, 1992; Hack et al., 1993; Selva et al., 1993).

The signals generated at the plasma membrane by cross-phosphorylation of two EGFR molecules are transduced within the cell via numerous proteins that contain SH2 domains (Pawson and Gish, 1992). SH2 domains are regions of homology to the oncogene *src* and mediate protein–protein interactions by facilitating binding to phosphorylated tyrosine residues (Marais et al., 1995). Whereas the presence of phosphotyrosine creates a binding site for a protein containing an SH2 domain, the specificity of the protein binding is conferred by the amino acids that surround the tyrosine on the target protein (Pawson and Gish, 1992). The first step in the intracellular signal transduction pathway activated by EGFR is autophosphorylation of C-terminal tyrosine residues of the receptor, thereby creating docking sites for SH2-containing proteins (McCormick, 1993; Figure 4).

Two proteins immediately responsible for the transduction of the EGFR signal are Grb2 and Shc. Grb2 is a member of a family of proteins initially cloned based on the ability to bind phosphotyrosine residues (Skolnik et al., 1991). Grb2 is a 23-kDa protein that possesses no intrinsic enzymatic activity and consists almost entirely of a central SH2 domain flanked by two SH3 domains (Lowenstein et al., 1992). Both SH2 and SH3 domains are involved in protein–protein interactions; SH2 domains bind phosphotyrosine,whereas SH3 domains mediate binding to proline-rich sequences (Pawson and Gish, 1992). The *shc* gene, which was cloned using an SH2 domain as a probe, encodes three overlapping proteins of 46, 52, and 66 kDa that contain a single SH2 domain and, like Grb2, possess no enzymatic activity (Pelicci et al., 1992). Shc proteins can also interact with tyrosine-phosphorylated EGFR molecules via their SH2 domains (Batzer et al., 1995).

In a resting cell, Grb2 is found in the cytoplasm in a complex with the human homologue of the *Drosophila* Son of Sevenless gene, *Sos* (Li et al., 1993). Tyrosine phosphorylation of the cytoplasmic tail of EGFR as well as other receptor tyrosine kinases creates docking sites for SH2-containing proteins and promotes the recruitment of both Shc and Grb2–Sos to the plasma membrane where the Shc proteins are substrates for EGFR kinase activity (Okada et al., 1995; Pelicci et al., 1992). Although both Shc and Grb2 contain SH2 domains and are therefore capable of binding EGFR, the predominant interaction in EGF-stimulated cells is between EGFR and Shc, and the interaction of Grb2–Sos with EGFR occurs indirectly through the binding of the Grb2 SH2 domain to phosphorylated tyrosines on Shc (Sasaoka et al., 1994). Recruitment of Sos to the inner surface of the plasma membrane results in the activation of Ras through the Sos-induced promotion of GTP binding (Li et al., 1993). Ras is a 21-kDa protein that is anchored to the inner surface of the plasma

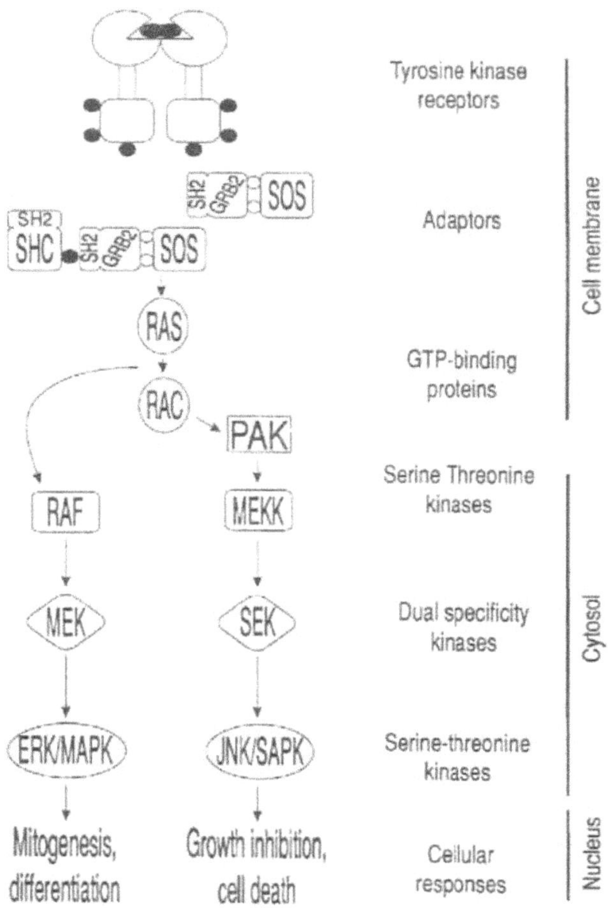

Figure 4. Schematic representation of the MAPK cascades. Activation of the cascade begins with ligand binding to transmembrane receptors, which results in phosphorylation on tyrosine residues (black circles) of the receptors and intracellular substrates. The signals generated through tyrosine phosphorylation are transduced through numerous proteins acting through different mechanisms. Adapters containing SH2 domains bind to phosphorylated tyrosines and recruit factors that result in the activation of GTP-binding proteins. These proteins, Ras and Rac, are in turn responsible for activation of the initial kinases in the MAP kinase cascade, Raf and MEKK. Raf and MEKK are serine/threonine kinases, which phosphorylate and activate MEK and SEK, respectively. These proteins are dual specificity kinases, capable of phosphorylating the ERK/MAPKs and JNK/SAPKs on both tyrosine and threonine residues. This dual phosphorylation promotes activation of the serine/threonine kinase activity of the ERK/MAPKs and JNK/SAPKs, and results in the phosphorylation of nuclear transcription factors and modulation of gene transcription. Details of the activation and regulation of these cascades, including proteins not shown, are in Sections 1.6 and 1.9.

membrane; it is inactive when bound to GDP, but in its GTP-bound form activates a pathway central to transduction of growth factor–induced signals (Grand and Owen, 1991). Translocation of Sos from the cytosol to the plasma membrane is sufficient to activate Ras because a Sos protein engineered to localize to the plasma membrane results in constitutive activation of Ras (Aronheim et al., 1994). The activation of Ras by Sos is balanced by the GTPase-activating protein GAP, which promotes hydrolysis of GTP to GDP and the return of Ras to its inactive state (Bourne et al., 1991).

Ras-GTP is in turn responsible for activation of the Raf-1 serine/threonine kinase. Raf genes consist of three known members: A-*raf*-1 and B-*raf*-1, which are expressed in a tissue-specific manner, and c-*raf*-1, which is expressed ubiquitously (Rapp, 1991). The c-*raf*-1 protein, Raf-1, is a 73-kDa phosphoprotein with intrinsic kinase activity toward serine and threonine residues (Li et al., 1991). The Raf proteins consist of an N- terminal regulatory domain and a C-terminal catalytic domain; truncation of the N- terminal regulatory domain yields a constitutively active oncogenic form of the protein that results in cellular transformation (Heidecker et al., 1990). Ras activates Raf-1 through the formation of a multiprotein complex at the plasma membrane that includes Ras, Raf-1 and MEK (Moodie et al., 1993; Van Aelst et al., 1993), a substrate for Raf-1 kinase activity. The precise mechanism of Raf-1 activation by Ras is unclear, but is in part due to recruitment of Raf-1 to the plasma membrane because a Raf-1 protein engineered to localize to the plasma membrane is constitutively active independent of Ras (Leevers et al., 1994; Marais et al., 1995).

The mitogen-activated protein kinases (MAPKs) are a large family of serine/threonine protein kinases that mediate cellular responses to a wide variety of mitogens and external signals (Seger and Kerbs, 1995). The most widely studied and best characterized MAPK pathway involves the extracellular-regulated kinases (ERKs), ERK-1 and ERK-2, and is primarily involved in the transmission of mitogenic signals. This pathway is activated by Raf-1 phosphorylation of a protein known as mitogen-activated protein kinase kinase (MAPKK or MEK, MAPK or ERK kinase; Dent et al., 1992; Kyriakis et al., 1992). MEK is responsible for activation of ERK1 and ERK2, also known as p42–p44 MAPKs (Boulton et al., 1991; Chen et al., 1992; Howe et al., 1992; Macdonald et al., 1993). The activated MEK phosphorylates the ERK-1 and ERK-2 MAP kinases on both tyrosine and threonine in a threonine–glutamate–tyrosine motif (Kosako et al., 1992). The ERKs are a related family of protein kinases of 42 and 44 kDa that phosphorylate and activate the nuclear transcription factors Myc, Fos, and Elk- 1 (Chen et al., 1992; Whitmarsh et al., 1995; Zinck et al., 1995) thereby completing the transmission of the tyrosine kinase receptor signal to the nucleus and resulting in the transcriptional activation of target genes. A central role for this pathway in the control of cell growth is suggested by the observation that expression of constitutively active mutants of Raf-1 or MEK results in cellular transformation (Heidecker et al., 1990; Mansour et al., 1994). In addition to nuclear transcription factors, another substrate for the ERKs is the Sos protein (Cherniach et al., 1994). Phosphorylation of Sos results in dissociation of Sos from Grb2 and inhibition of further stimulation of Ras activity thereby attenuating the activation of the cascade (Waters et al., 1995).

A second, related intracellular MAPK cascade, in contrast to the mitogenic one above, is primarily associated with the response of a cell to stress such as UV irradiation (Derijard et al., 1994). This pathway is mediated through proteins related to ERKs but initially characterized based on their ability to phosphorylate the amino terminus of the c- Jun transcription factor (Derijard et al., 1994) as well as in the response of cells to stress (Kyriakis et al., 1994), and this has lead to the dual nomenclature of Jun N-terminal kinase/Stress- activated protein kinases (JNK/SAPKs or p46–p54 JNKs). This cascade begins with activation of MEK kinase (MEKK), which is responsible for the phosphorylation and activation of the p46–p54 JNKs (Lin et al., 1995). Downstream control of this pathway is analogous to that of the ERK pathway, with MEKK occupying a position similar to Raf-1 in the ERK pathway. Although MEKK is capable of phosphorylating MEK (of the ERK pathway) when overexpressed, under physiologic conditions, the preferred substrate is the JNK kinase SEK(Yan et al., 1994). MEKK is the direct activator of SEK, which in turn is the direct activator of JNK (Derijard et al., 1994; Lin et al., 1995; Minden et al., 1994a). Activation of the JNKs leads to phosphorylation of the transactivation domain of c-Jun (Angel et al., 1988; Minden et al., 1994b), and phosphorylated c-Jun homodimers have potent AP-1 activity (Angel et al., 1988). Activation of the JNK pathway may occur via EGFR in a Ras-dependent manner in rat pheochromocytoma PC12 cells, although JNK activation by tumor necrosis factor α in these cells occurs independent of Ras (Minden et al., 1994a). The fundamental difference between the JNK pathway and the ERK pathway is illustrated by the observation that, unlike expression of a constitutively active mutant of Raf-1 or MEK, expression of a constitutively active MEKK results in growth inhibition (Lange-Carter and Johnson, 1994; Yan et al., 1994).

The small GTP-binding protein Rac1 is also involved in JNK activation by growth factors, v-*src*, or H-*ras*. Rac(N17), a dominant–negative form of the protein, attenuates JNK activation while having no effect on the activation of ERKs (Coso et al., 1995; Minden et al., 1995). Rac1(N17) also inhibits foci formation induced by v-*src* or *H-ras* (Qiu et al., 1995; Khosravi-far et al., 1995) suggesting that JNK activation may be important in eliciting this transformed phenotype, although there is also evidence a distinct rac effector pathway may be involved (Joneson et al., 1996). This finding places this member of the rho family of GTP-binding proteins downstream of Ras in a pathway that branches off from Ras and is separate from the Raf/ERK pathway. Rac1 therefore functions as an intermediate between Ras and MEKK (Vojtek and Cooper 1995). The Rac(N17) inhibition of JNK activity is partial, however, suggesting that more than a single sequential cascade is involved. Input from trimeric G proteins and phosphatidylinositide 3-kinase pathways are also likely (Minden et al., 1995).

ANGIOGENESIS: VASCULAR GROWTH FACTORS

Although the previous discussion has focused on factors that produce mitogenic effects on the tumor cells themselves, it must be remembered that tumor formation occurs in the context of a whole organism, and that interactions between the tumor cells and host cells are likely to be as important as interactions among the tumor cells themselves. A large body of evidence indicates that tumors require the formation of new blood vessels to grow beyond a size of 1–2 mm^3 (reviewed in Folkman, 1990). In addition to this regulatory role of angiogenesis in tumor growth, recent evidence also supports the notion that neovascularization is important in the development of metastasis (Weidner et al., 1991). The metastatic phenotype is likely to involve a combination of genetic and epigenetic alterations that allow a tumor cell to leave the site of the primary lesion, intravasate into the venous capillaries, escape immune surveillance, attach to the capillary endothelial cell lining, extravasate, colonize, and eventually proliferate in normal tissue at a distant site in the host (Liotta and Stracke, 1988; Liotta et al., 1991). The association of neovascularization with metastasis is not surprising because the invasive processes associated with new blood vessel formation would increase the opportunity for shed tumor cells to enter the circulation. An immunohistochemical stain for factor VIII, used to identify endothelial cells and quantitate the extent of microvessel formation within breast carcinoma tissues, reveals that tumors from patients with metastases have a mean number of vessels approximately twice as high as tumors from matched patients without metastases. The prevalence of metastatic disease increases as the microvessel count increases (Weidner et al., 1992; Weidner et al., 1991). This correlation is also seen in several studies examining a number of tumors of different origin (reviewed in Gaben and Mester, 1991; Gasparini and Harris, 1995). The correlation of neoangiogenesis with metastasis points out the importance of new blood vessel growth and the need to gain a better understanding of the mechanism underlying the acquisition of an angiogenic phenotype.

Regulation of the extent of angiogenesis within tumors is likely to be a complex process. One contribution may be the downregulation of a tumor suppressor gene that codes for the angiogenesis inhibitor thrombospondin (Rastinejad et al., 1989). A second possibility might involve production by the tumor cells of angiostatin or other angiogenesis inhibitors (Callaghan et al., 1993; Chen et al., 1995). A third possibility that is not mutually exclusive involves increased production by the tumor of an angiogenic growth factor or the increased infiltration of the tumor by monocytic cells that produce angiogenic cytokines. One of the difficulties facing the approach of inhibiting metastasis through the inhibition of angiogenesis is the large number of factors that have been shown in various in vitro or in vivo systems to be capable of either direct or indirect stimulation of this process (reviewed in Falls et al., 1993; Fidler and Ellis, 1994; Folkman, 1995). The most well studied mitogens for endothelial cells include several members of the fibroblast growth factor (FGF) family of heparin-binding growth factors (see discussion below); however, transforming growth factor α and β, epidermal growth factor, tumor necrosis factor-a, pleiotrophin, hepatocyte growth factor, interleukin 4, interleukin 8, platelet-derived endothelial cell growth factor, and vascular endothelial growth factors (VEGF) all have been reported to act as mediators of tumor angiogenesis (reviewed in Kageyama et al., 1988; Kim et al., 1993).

Given the plethora of growth factors with angiogenic potential, essentially all of which have been found to be expressed in tumors or tumor cell lines, it is impossible at this point to determine which factor or combination of factors is driving the process in tumors. A number of studies, however, have pointed out the potential importance of VEGF-A, an extremely potent endothelial cell growth factor (Kageyama et al., 1988; Keck et al., 1989; Leung et al., 1989). In addition to being an endothelial cell mitogen, this factor also allows neovascularization by promoting remodeling of the extracellular matrix and was originally described as Vascular Permeability Factor (VPF). VEGF-A is structurally related to PDGF but, unlike PDGF, is not mitogenic for fibroblasts and appears to be stimulatory only for endothelial cells (Takahashi et al., 1989; Tischer et al., 1991), a quality that makes it unique among the angiogenic factors described thus far. Secretion of this factor may therefore be an important element controlling angiogenesis. Three major isoforms of 189, 165, and 121 amino acids have been described for VEGF- A, and a fourth VEGF-A peptide of 206 amino acids is found in placental tissue (Houck et al., 1991). VEGF-A contains a secretory signal peptide and three of the four isoforms bind to heparin (Conn et al., 1990; Folkman and Haudenschild, 1981). These isoforms result from alternative splicing of a single mRNA (22). VEGF-A facilitates the development of a matrix important for the subsequent establishment of the supporting stromal network of fibroblasts, endothelial cells, and connective tissue. Three other members of the VEGF/PDGF family of growth factors have also been identified. VEGF-B is abundant in the heart and skeletal muscle and stimulates DNA synthesis in endothelial cells (Olofsson et al., 1996a). Two isoforms of VEGF-B result from alternative splicing to yield proteins of 167 and 186 amino acids have been described (Olofsson et al., 1996b). The third member, VEGF-C, was identified through screening of conditioned medium from a prostatic cancer cell line for the ability to induce autophosphorylation of a transmembrane tyrosine kinase VEGF receptor (Joukov et al., 1996). VEGF- D was isolated using a computer-based homology search (Achen et al., 1998).

Three structurally related transmembrane tyrosine kinase receptors for VEGFs exhibit a structure characterized by seven immunoglobulinlike loops in the extracellular domain (Figure 3). Two of these three receptors bind VEGF-A with high affinity. The two receptors for VEGF-A were originally described as Flt-1 (*fms*-like-tyrosine kinase, based on its homology to c-*fms*, the receptor for colony-stimulating factor 1) and KDR (kinase insert domain-containing receptor; de Vries et al., 1992; Terman et al., 1992); for clarity, these receptors have been renamed VEGFR-1 (Flt-1) and VEGFR-2 (KDR). The third VEGF receptor, VEGFR-3, was originally cloned as Flt-4 and was the receptor used to isolate VEGF-C. Endothelial cell mitogenesis appears to be meditated through signaling by VEGFR-2; signaling via VEGFR-1 does not have this capability (Waltenberger et al., 1994). Neither VEGF-A nor VEGF-B bind to VEGFR-3 with high affinity, although VEGF-C binds to both VEGFR-2 and VEGFR-3 and stimulates capillary endothelial cell tube formation in collagen gels (Joukov et al., 1996). VEGF-B does not bind any VEGFR yet described, but it is still capable of stimulating the proliferation of endothelial cells. VEGF-D binds both VEGFR-2 and VEGFR-3. Because VEGFR-3 is expressed in lymphatic endothelial cells, the binding specificities of VEGF-C and VEGF-D suggest that they may regulate lymphatic angiogenesis (Achen et al., 1998). In addition to the three transmembrane tyrosine kinase receptors, the 165-amino form of VEGF-A, but not the 121 amino acid form, binds to lower affinity/lower molecular weight sites present on human umbilical vein endothelial cells and MDA-MB-231 breast carcinoma cells. These sites appear to be unrelated to either VEGFR-1 or -2, do not appear to be phosphorylated or induce mitogenesis in response to VEGF treatment, and presently have unknown structure and function (Soker et al., 1996).

An important role for VEGF receptor-mediated signal transduction as a major regulator of neoangiogenesis is supported by work demonstrating a correlation of the temporal and spatial expression patterns of VEGF-A and VEGFR-2 in areas of new blood vessel formation in the developing mouse embryo (Millauer et al., 1993; Quinn et al., 1993). Furthermore, specific gene knockouts in mice of VEGF-A, VEGFR-1, or VEGFR-2 generated by targeted recombination in embryonic stem cells, result in lethality early in embryonic development, with mice exhibiting evidence of impaired vasculogenesis (Carmeliet et al., 1996; Ferrara et al., 1996; Fong et al., 1995; Shalaby et al., 1995).

The signaling pathways of the VEGF receptors are not well described. VEGF is only biologically active when covalently dimerized by disulfide bridges (Potgens et al., 1994) and there is evidence that the ligand is bivalent (Keyt et al., 1996) supporting the generally held belief that, like other transmembrane tyrosine kinase receptors, there is a requirement of receptor dimerization for subsequent tyrosine phosphorylation and kinase activation. Heterodimer formation among the three VEGFRs is hypothesized but has yet to be demonstrated conclusively. Downstream mediators in the signal transduction pathway are likely to include MAP kinases, which are tyrosine phosphorylated in response to VEGF stimulation of VEGFR-2 in bovine brain capillary endothelial cells (D'Angelo et al., 1995) or VEGFR-1-expressing rat liver sinusoidal endothelial cells (Seetharam, et al., 1995); PLC-γ is also tyrosine phosphorylated in these VEGF-treated cells (D'Angelo et al., 1995; Seetharam et al., 1995). There is no evidence of any association of PI-3 kinase activity with either receptor in any cell lines tested (Waltenberger et al., 1994), a finding consistent with the absence of a recognition motif for the PI-3 kinase in either VEGF receptor. In porcine aortic endothelial cells, which do not normally express either receptor, transfection with VEGFR-1 or -2 results in association of Src tyrosine kinase family members with both receptors, although this association appears to be weaker than that seen with other transmembrane tyrosine kinase receptors. In these cells, VEGF addition does not result in PLC-γ phosphorylation in response to VEGF treatment. In contrast to the VEGFR-1- expressing sinusoidal endothelial cells (Seetharam et al., 1995), tyrosine phosphorylation of GAP is only weakly induced by VEGF treatment of porcine endothelial cells, and GAP is not found to be associated directly with either receptor (Waltenberger et al., 1994). These results suggest that factors other than the receptors alone are necessary for efficient coupling of receptor tyrosine kinase activity to downstream effectors.

In some tumor models, VEGF enhances tumor growth by promoting neoangiogenesis without altering the in vitro growth rates of the tumor cells themselves. VEGF overexpression confers an in vivo growth advantage to CHO cells in the absence of any other increase in the in vitro parameters of transformation (Ferrara et al., 1993). Inhibition of VEGF-mediated angiogenesis using an anti-VEGF antibody suppresses the in vivo growth of two sarcoma cell lines and one glioblastoma cell line, even though no effects of the antibody are observed on the in vitro growth of these cells (Kim et al., 1993). VEGF neutralizing antibodies reduce the number of spontaneous micrometastases formed by A431 epidermoid carcinoma cells in a manner independent of the reduction in primary tumor size (Melnyk et al., 1996). Anti-VEGF antibody administration also reduces the number and size of liver metastases formed by a human colon cancer cell line after splenic-portal injection (Warren et al., 1995) and the number of lung metastases formed by human fibrosarcoma cells after tail vein injection (Asano et al., 1995). In a different approach aimed at inhibiting VEGF expression, transfection of a rat glioblastoma cell line with a vector directing VEGF antisense RNA expression greatly impairs tumor formation in vivo, and the small tumors that do form exhibit reduced levels of vascularization and increased amounts of necrosis (Saleh et al., 1996). Transfection of a poorly metastatic human melanoma cell line that naturally expresses a low level of the protein with sense VEGF increases and antisense VEGF decreases, the number of lung nodules found after tail vein injection (Claffey, et al., 1996). Inhibition of VEGF-mediated signaling in endothelial cells within a tumor by infection with a recombinant retrovirus directing the expression of a dominant–negative mutant of VEGFR-2 also inhibits the growth of glioblastoma cells in nude mice (Millauer et al., 1994). A role for VEGF in the facilitation of metastasis is consistent with its angiogenic capabilities, the correlation of extent of microvessel density within a tumor, and the likelihood of metastatic spread. In addition, VEGF induces the expression of proteases and collagenases presumed to be mediators of extracellular matrix invasion (Mandriota et al., 1995; Unemori et al., 1992).

VEGF-A is expressed in breast cancer (Anan et al., 1996; Brown et al., 1995; Toi et al., 1994; Toi et al., 1995; Toi et al., 1996; Yoshiji et al., 1996) and the extent of expression appears to correlate with the number of microvessels within a tumor (Toi et al., 1995). In a univariate analysis, VEGF expression is a prognostic indicator for relapse-free survival, but this is not the case in a multivariate analysis (Toi et al., 1995). VEGFR-1 mRNA and protein are strongly expressed in the endothelial cells of small vessels adjacent to infiltrating ductal carcinoma cells or metastatic ductal carcinoma in sections examined by in situ hybridization. VEGFR-1 and VEGFR-2 mRNA are also found in endothelial cells adjacent to comedo type DCIS as well, but there is no evidence of expression in endothelial cells in the normal areas of the breast (Brown et al., 1995). Although relapse-free survival is significantly reduced in tamoxifen-treated patients with highly angiogenic tumors (Macaulay et al., 1997), a study to determine whether high angiogenesis is predictive for tamoxifen resistance has not been reported. Nevertheless, there is a considerable amount of evidence suggesting a role for VEGF as an important determinant in establishing a neoangiogenic phenotype in breast cancer.

Results of a transfection study using an expression vector for the 121–amino acid form of VEGF and MCF-7 breast cancer cells as recipients lend support to the notion that VEGF-mediated angiogenesis is important in facilitating the growth of breast tumors. Whereas $VEGF_{121}$ overexpression increases the size of estrogen-dependent tumors that form in nude mice, no effect is seen on estrogen dependence, tamoxifen sensitivity, or formation of macrometastasis (Zhang et al., 1995). A similar study performed by our laboratory using the 165–amino acid isoform found that overexpression does facilitate tumor formation in tamoxifen-treated mice and tumor cell dissemination (Bullocks et al., 1997). This apparent difference between studies may reflect different expression levels of the transfected gene or the use of different VEGF isoforms because the heparin-binding ability of the 165–amino acid form may protect it from inactivation by binding to $\alpha2$ macroglobulin (Soker et al., 1993). There is recent evidence that $VEGF_{165}$ and $VEGF_{121}$ exhibit different binding specificities for VEGFR-1 and -2, and that this difference is related to the heparin-binding domain present in $VEGF_{165}$ (Gitay-Goren et al., 1996; Soker et al., 1996); however, the biological significance of this difference in binding to these receptors is not known. The difference in effects on the metastatic phenotype may also be more apparent than real. The recipient MCF-7 cells in our study express the bacterial β-galactosidase gene, which allows staining of tissue sections with a chromogenic substrate to allow detection of micrometastases (McLeskey et al., 1993), but we do not observe macrometastases in animals bearing $VEGF_{165}$-overexpressing tumors.

The pattern of specific expression of VEGF receptors in areas of tumor-induced neoangiogenesis in breast tumors is also seen with other tumor types (Brown et al., 1993; Warren et al., 1995). This raises the possibility that therapies targeted to the VEGF receptors might be particularly useful in treating a wide range of tumors. VEGF- toxin conjugates and targeted expression of truncated VEGFR-2 proteins that act in a dominant–negative fashion to inhibit signaling in endothelial cells expressing wild type VEGFR-2 receptors have been used successfully to inhibit neovascularization as has a VEGFR-2 tyrosine kinase inhibitor (Millauer et al., 1996; Ramakrishnan et al., 1996; Strawn et al., 1996). There is some evidence of VEGFR expression in nonendothelial cells, however, including placental trophoblasts, embryonic kidney, choriocarcinoma, ovarian carcinoma, melanoma, and leukemia cells (Boocock et al., 1995; Charnock-Jones et al., 1994; Katoh et al., 1995; Liu et al., 1995a). In the melanoma cell lines examined, evidence for an autocrine loop was established (Liu et al., 1995a) but no such response is seen in the ovarian cells, and the functional significance of the receptor expression is therefore uncertain (Boocock et al., 1995). Recent studies also show VEGFR-1 expression in normal monocytes (Barleon et al., 1996; Clauss et al., 1996) and VEGFR- 2 expression in pancreatic duct cells (Oberg et al., 1994), normal hematopoetic stem cells, megakaryoctes, and platelets (Katoh et al., 1995). They also provide evidence for a functional role for such expression in monocyte chemotaxis and inhibition of hematopoetic stem cell apoptosis. These recent findings of VEGFR expression in sites other than endothelial cells suggest that the enthusiasm centered on the potential tumor- specific nature of these VEGFR-directed therapies should be tempered with some degree of caution.

FIBROBLAST GROWTH FACTORS

Ten members of the fibroblast growth factor (FGF) family of heparin-binding growth factors and CDNAs encoding four additional FGF–related proteins have been described (Baird and Klagsbrun, 1991; Basilico and Moscatelli, 1992; Miyamoto et al., 1993; Tanaka et al., 1992; Emoto et al., 1997; Smallwood et al., 1996). Many of these factors, including FGF-1, FGF-2, and FGF-4, stimulate endothelial cell proliferation in vitro and angiogenesis in vivo (Brustle et al., 1992; Delli-Bovi et al., 1988; Folkman and Klagsbrun, 1987; Folkman and Shing, 1992; Moscatelli et al., 1986; Thomas et al., 1985). FGFs are also mitogenic for other mesodermally and neuroectodermally derived cells (Burgess and Maciag, 1989) and, under some circumstances, for epithelial cells (Takahashi et al., 1989). The potential ability of FGFs to stimulate the growth of both tumor cells and endothelial cells has made them the subject of much research. The identification of four of the ten members of the FGF family by in vivo or in vitro transformation assays suggests that stimulation of preneoplastic cells by these growth factors via autocrine or paracrine mechanisms may be a step in a progression pathway. A possible role for members of the FGF family of ligands in breast cancer tumorigenesis was originally suggested by the finding that the FGF-3 and FGF-4 loci are frequent sites of proviral activation by the mouse mammary tumor virus (Dickson et al., 1984; Peters et al., 1989). In a number of human malignancies including breast cancer, the 11q13 chromosomal region that contains these two loci is frequently amplified (Brison, 1993; Lammie et al., 1991; Theillet et al., 1989); however, transcripts for FGF-3 or FGF-4 are not consistently present in human breast cancer tissues even when the corresponding chromosomal region is amplified (Ding et al., 1992; Fantl et al., 1990; Lammie et al., 1991; Liscia et al., 1989; Penault-Llorca et al., 1995; Theillet et al., 1989).

In contrast, both FGF-2 and FGF-1 are expressed in a significant proportion of human breast tumor specimens (Anandappa et al., 1994; Ding et al., 1992; Ke et al., 1993; Penault-Llorca et al., 1995; Smith et al., 1994). FGF-2 mRNA is detectable in a number of immortalized or tumor-derived mammary epithelial cell lines in culture (Anandappa et al., 1994; Ke et al., 1993; Li and Shipley, 1991; Luqmani et al., 1992; Souttou et al., 1994). An earlier immunohistochemical analysis of formalin-fixed tissue sections suggested that the source of FGF-2 expression in breast tumors was the normal stromal and myoepithelial component in the samples (Gomm et al., 1991; Ke et al., 1993; Smith et al., 1994). A second, more recent report using a different antibody and frozen tissue sections demonstrates FGF-2 immunostaining in neoplastic epithelial cells in 38% of tumors analyzed, with stromal FGF-2 staining observed in 37% of the tumors. In this study, disease recurrence is associated with strong FGF-2 staining (Visscher et al., 1995). The difference between studies may arise from the two antibodies used or the differences in sample preparation. This explanation seems reasonable in light of a second report from the first group in which FGF-2 expression was re-examined using frozen sections. Whereas FGF-2 immunostaining is seen in normal breast epithelial cells, myoepithelial cells, and breast tumor epithelial cells, stronger staining is seen in normal cells. FGF-2 protein levels correlate with mRNA levels and low expression of FGF-2 mRNA correlates with decreased disease-free and overall survival (Yiangou et al., 1997). Definitive conclusions about the relationship of FGF-2 expression and prognosis are difficult to make, however, due to the relatively small number of samples with follow-up information available in these studies. It should also be emphasized that immunohistochemical studies demonstrate where FGF-2 is localized, not necessarily the source of the expression; in situ hybridization studies will be required to determine the origin of the protein.

An increase in the synthesis or deposition of FGF-1 may also be an early feature in the development of breast cancer. One study finds FGF-1 in the majority of carcinoma tissues and adjacent normal tissues but not in the benign breast disease or normal reduction mammoplasty tissues (Smith et al., 1994). Although another study reports higher levels of FGF-1 mRNA in "normal breast" compared to tumor tissue, the normal sections in many cases were adjacent to the tumor tissue and many of the tumor tissues examined had high levels of FGF-1 mRNA present (Bansal et al., 1995). Thus these two reports may not be contradictory. A more recent immunohistochemical study using frozen sections found FGF-1 in the stroma surrounding breast carcinoma cells in all of the invasive cancers that were studied if the sections were first incubated with protease inhibitors (Coope et al., 1997). In a study performed in our laboratory, we found readily detectable levels of FGF-1 mRNA in approximately one-third of the tumor samples analyzed by RNAse protection assay, and approximately an equal proportion showed FGF-1-positive immunostained tumor cells using an antibody for FGF-1 (Ding et al., 1992; Zhang et al., 1997).

FGF-2 and FGF-1 do not have a signal peptide and a number of studies have indicated that extracellular localization is required for either factor to fully elicit the transforming effects of overexpression in immortalized rodent fibroblast cell lines (Blam et al., 1988; Bunnag et al., 1991; Forough et al., 1993; Jaye et al., 1988; Neufeld et al., 1988; Rogelj et al., 1988; Sasada et al., 1988). Despite the lack of a signal peptide, both proteins are often found outside of cells and a number of suggestions for the release or export of FGF-2 and FGF-1 via alternate mechanisms have been proposed (Jackson et al., 1992; Jackson et al., 1995; Kandel et al., 1991; Mignatti et al., 1991; Mignatti and Rifkin, 1991; Muthukrishnan et al., 1991). Both positive and negative effects of exogenous FGFs on breast cancer cell lines have been described. FGF-2, but not FGF-1, stimulates the growth of early passage breast epithelial cells derived from either nonmalignant or carcinoma tissues (Takahashi et al., 1989). FGF-2 also stimulates the anchorage-independent growth of immortalized breast epithelial cells transduced with c- *myc* or SV40 T antigen retroviral vectors (Valverius et al., 1990). Endogenously produced FGF-2 functions as an autocrine growth factor for HBL-100, an SV-40 immortalized breast epithelial cell line, and the same cell line exhibits a mitogenic response to exogenous addition of FGF-1 (Souttou et al., 1994). In this same cell line, however, transfection of FGF-4 results in increased growth in soft agar and serum-free media and in acquisition of a tumorigenic phenotype (Souttou et al., 1996). We and others have found FGFs to stimulate anchorage-dependent growth of MCF-7 cells (Briozzo et al., 1991; McLeskey et al., 1994; Wellstein et al., 1990). We have also observed that overexpression of either FGF-1 or FGF-4 has important effects on the estrogen dependence, antiestrogen sensitivity, and metastatic potential of MCF-7 cells. Transfection with either FGF results in cell lines that give rise to progressively growing tumors in ovariectomized mice without estrogen supplementation and in mice treated with tamoxifen or ICI 182,780 (McLeskey et al., 1993; McLeskey et al., 1998; Zhang et al., 1997). FGF-transfected cells also acquire the ability to form spontaneous metastases in the lymph nodes, lungs, and other organs with high frequency and short latency (Kurebayashi et al., 1993; McLeskey et al., 1993; Zhang et al., 1997).

Although stimulatory effects of FGF-2 on MCF-7 cells are seen under restrictive growth conditions (Karey and Sirbasku, 1988; Stewart et al., 1992), exogenous FGF-2 inhibits MCF-7 cell growth in the presence of estradiol and insulin (Fenig et al., 1997). We have observed that exogenous FGF-2 or FGF-1 inhibits the growth of the MDA-MB- 134 cell line, which overexpresses FGF receptors (McLeskey et al., 1994). In other systems, the inhibitory effects of endogenous FGFs may be related to nuclear uptake via nuclear localization signals present in alternatively translated forms of the protein (Kiefer et al., 1997; Kiefer and Dickson, 1995; Quarto et al., 1991a; Quarto et al., 1991b). This is not the underlying cause for the FGF-2 inhibition of MCF-7 cell growth, however, because the nuclear localization signal is not present in the exogenous recombinant FGF-2 used in this study (Fenig et al., 1997). Treatment of Swiss 3T3 cells with FGF-2 stimulates DNA synthesis and results in localization of FGF receptor 1 with the peripheral nuclear matrix where it appears to retain tyrosine kinase activity (Maher, 1996). Nuclear uptake of exogenous FGF-1 may also occur, but in this case, a receptor- mediated mechanism is involved leading to perinuclear localization of ligand-occupied receptor and occuring over a longer time course (Prudovsky et al., 1996). Although subsequent utilization of a nuclear localization signal present at the amino terminus of the FGF-1 protein (Friedman et al., 1994) may then allow the ligand to enter the nucleus, the biological consequences of this event are currently unknown.

FGF RECEPTORS AND SIGNAL TRANSDUCTION

The various FGFs interact with two distinct classes of receptors: (1) low affinity heparan sulfate proteoglycans on the cell surface and basement membrane and (2) four high affinity transmembrane tyrosine kinase receptors (Figure 3; Dionne et al., 1990; Givol and Yayon, 1992; Johnson and Williams, 1993; Keegan et al., 1991; Partanen et al., 1991; Partanen et al., 1992). The primary transcripts of the FGFRs undergo differential splicing that yields numerous secreted or cell-bound isoforms as a result of different exon usage (Givol and Yayon, 1992; Hou et al., 1991; Johnson et al., 1991; Johnson and Williams, 1993; Partanen et al., 1992). Although multiple ligands bind multiple receptors with similar affinities, some degree of specificity is conferred by alternate usage of exons encoding segments of the immunoglobulinlike domains present in the extracellular regions (Ornitz et al., 1996; Zimmer et al., 1993).

Receptor isoforms with three and two immunoglobulinlike domains exist for FGFR-1 and FGFR-2, but only the three immunoglobulinlike domain forms of FGFR-3 and FGFR-4 are known (Jaye et al., 1992). The two-Ig loop form of FGFR-1 (FGFR- 1β) has a higher affinity for FGF-1 and heparin than the three-Ig loop form (FGFR-1α; Wang et al., 1995a) and it is the FGFR-1α isoform that appears to be involved in perinuclear trafficking (Prudovsky et al., 1996). In a number of systems, a switch in the ratio of the FGFR-1β to FGFR-1α is observed and is associated with an increase in malignant potential. Higher levels of FGFR-1β compared to FGFR-1α are observed in astrocytomas as they progress from a benign to malignant phenotype (Yamaguchi et al., 1994). Similarly, although FGFR-1β and FGFR-1α are present in equal proportions in normal pancreatic acinar cells, FGFR-1 expression is elevated and the two-Ig loop FGFR-1β form predominates in pancreatic ductal carcinoma cells (Kobrin et al., 1993). In breast tumors, normal breast tissue expresses the two forms in equal proportion, but a higher ratio of FGFR-1β to FGFR-1α is seen in cancerous breast tissues and furthermore, for individual tumors, a higher ratio of FGFR-1β to FGFR-1α is associated with reduced relapse-free survival (Luqmani et al., 1995).

FGFR-1, FGFR-2, and FGFR-3 also undergo differential splicing of an exon that encodes the second half of the third Ig-like domain, a process that can alter the affinity for the various FGF family members (Ornitz et al., 1996). For FGFR-2, the choice of the exon is particularly important in determining whether the receptor binds either FGF-2 or FGF-7 (a.k.a. keratinocyte growth factor or KGF) with high affinity. These two splicing options for FGFR-2 appear to be mutually exclusive: epithelial cells primarily utilize exon IIIb, which confers specificity for FGF-7/KGF, whereas mesenchymal cells use exon IIIc, which results in a receptor with high affinity for FGF-2 (Miki et al., 1992; Yan et al., 1993). A change in exon utilization from IIIb to IIIc is associated with the epithelial- to-mesenchymal transition that occurs in a rat bladder carcinoma cell line in response to FGF-1 treatment (Savagner et al., 1994) suggesting that exon switching may be associated with a more motile or invasive phenotype. A similar change in utilization from IIIb to IIIc also occurs in a rat prostate cancer progression model and coincides with a transition to an undifferentiated and more malignant state (Yan et al., 1993). In breast tissues, both variants are expressed and there is no difference in the levels of mRNA for either form between noncancerous and malignant tissues; however, patients with advanced clinical staging have a higher ratio of mRNA of the IIIc form compared to the IIIb form (Luqmani et al., 1996) further suggesting an association between utilization of the IIIb form and acquisition of a more aggressive phenotype.

Given this complexity, it is likely that the functional specificity and extent of the biological response elicited by expression of a particular FGF family member will to a large extent be dependent upon the repertoire and level of high affinity receptors expressed by a particular cell type as well as the levels and varieties of the receptor isoforms expressed. The low affinity heparan sulfate proteoglycans present on the cell surface and within the extracellular matrix are likely to function both as a reservoir for heparin-binding growth factors and in the formation of a ternary complex required for high affinity receptor signaling (Kan et al., 1993; Klagsbrun and Baird, 1991; Murgue et al., 1994; Spivak-Kroizman et al., 1994; Yayon et al., 1991), and the affinities and levels of the various heparin-binding growth factors bound to these sites are also likely to be important (Guimond et al., 1993).

Messenger RNA for each of the four receptors can be detected in breast cancer cell lines and primary breast cancer tissues (Ding et al., 1992; Luqmani et al., 1992; McLeskey et al., 1994). Amplification of FGFR-1 and FGFR-2 occurs in 10–15% of human breast cancers (Adnane et al., 1991; Jacquemier et al., 1994). Two-to four-fold amplification of FGFR-4 is found in approximately 10% of the breast tumor tissues (Jaakkola et al., 1993), and appreciable levels of FGFR-3 mRNA are observed in human breast tumor tissues (Ding et al., 1992). Taken together, these observations provide further support for a role of FGF signal transduction in breast cancer tumorigenesis or progression. Co-expression of more than one receptor is typically observed in breast cancer cell lines and tissues (Ding et al., 1992; Lehtola et al., 1992; McLeskey et al., 1994). Therefore, additional complexity in FGF signaling in breast cancer cells may be generated by the participation of receptor heterodimers in FGF signal transduction (Bellot et al., 1991). Other splice variants that encode secreted, kinase-deficient, or cell-bound receptor isoforms that lack a putative juxtamembrane regulatory domain may also modulate the extent of downstream signaling. A fifth, nontyrosine kinase receptor has also been identified (Burrus et al., 1992) as well as a possible role for an intracrine loop involving a cytoplasmically localized receptor isoform, but the consequences of these proteins with regard to transmission of a growth signal have not been elucidated (Imamura et al., 1990; Maciag et al., 1994; Yan et al., 1992).

Finally, recent results raise the possibility that FGF receptor activation may occur independent of ligand. Interaction with heparin (Gao and Goldfarb, 1995) or cell adhesion molecules (Doherty et al., 1994) can activate FGF receptors in the absence of FGF (Green et al., 1996). Mutations in FGFR-2 and FGFR-3 that result in constitutive tyrosine kinase activity have been identified in several skeletal disorders in humans when inherited through the germline (Bellus et al., 1995; Neilson and Friesel, 1995; Tavormina et al., 1995; Webster et al., 1996; Webster and Donoghue, 1996); FGFR-3 is thought to play a role in the negative regulation of bone growth (Deng et al., 1996). Nonetheless, it remains to be determined whether constitutive activation can lead to oncogenic transformation. Thus, when one considers the consequences of operative FGF signal transduction in the process of breast cancer progression, it does not necessarily have to be thought of in the context of production of a particular FGF family member.

Like other transmembrane tyrosine kinase growth factor receptors, dimerization of FGF receptors appears to be required for signal transduction resulting from ligand binding. Presumably this dimerization facilitates transmodulation, the intermolecular phosphorylation of the dimerized receptors on tyrosine residues. This process increases the intrinsic tyrosine kinase activity of the receptors and is required for the subsequent binding and phosphorylation of substrates important in downstream signaling events (Ullrich and Schlessinger, 1990). For FGFR-1, seven intracellular tyrosine residues have been definitively identified as phosphorylation sites (Mohammadi et al., 1996). Phosphorylation of tyrosine 653 is strictly dependent on the *trans* intermolecular mechanism, and phosphorylation of tyrosines 653 and 654, which lie within the tyrosine kinase domain and are conserved in other FGFRs, is required for phosphorylation and activation of substrates (Shi et al., 1993). Mutational analysis indicates the five other tyrosine residues (Y-463, Y-583, Y-585, Y-730, and Y-766) are dispensible for mitogenesis or induction of differentiation (Mohammadi et al., 1996). Tyrosine 766 is a binding site for the SH2 domain of PLC-γ. Mutation of this residue to phenylalanine results in a receptor that fails to induce phosphotidylinositol hydolysis and Ca^{++} flux without affecting its ability to induce mitogenesis (Mohammadi et al., 1992; Peters et al., 1992), suggesting that activation of substrates other than PLC-γ is important for growth stimulation.

Despite the complexity of the FGF signal transduction system, there are indications that the signals leading to control of growth or differentiation share at least one common pathway: use of the Ras/Raf/ERK cascade. The Shc adaptor protein and MAP kinases are phosphorylated in response to FGF treatment of FGFR-1-transfected BaF3 murine lymphoid cells or L6 rat myoblast cells (Wang et al., 1994; Vainikka et al., 1994; Shaoul et al., 1995), but there is no in vivo evidence of a direct association of Shc with FGFRs. Ras appears to play a role in FGF signal transduction because expression of a dominant inhibitory mutant of Ras in PC-12 cells antagonizes FGF-1-induced hyperphosphorylation of Raf-1 and activation of MAP kinases (Wood et al., 1992), but the mechanism by which it is placed into its active GTP-bound form following ligand activation of FGFRs remains to be elucidated. Addition of FGF-1 to quiescent mouse 3T3 cells induces changes in electrophoretic mobility of Raf-1 protein due to phosphorylation (Morrison et al., 1988), and injection of RNA encoding a dominant-negative Raf-1 mutant into *Xenopus* embryos blocks the ability of FGF to induce mesoderm (MacNichol et al., 1993).

The available data clearly suggest that the mechanism of FGF signal transduction is highly conserved among different cell types, but characterization of the pathway has yet to be described in breast cancer cells. It is also probable that FGF signaling involves multiple interacting pathways. Aside from PLC-γ, other proteins frequently associated with other transmembrane tyrosine kinases such as Grb-2, PI-3K, and GAP do not interact with FGFRs. There is evidence to suggest the involvement of other signal transduction pathways as well. Pretreatment of membranes from FGFR-1-transfected NIH 3T3 cells with pertussis toxin will reduce the binding affinity of FGF-1 in a heparin-dependent manner (Jarvis et al., 1992) and also block the mitogenic effect of FGF on Balb/c 3T3 cells, suggesting the coupling of FGF receptors to a trimeric G protein (Logan and Logan, 1991).

Src tyrosine kinases may represent a parallel signaling pathway to the Ras/Raf/ERK pathway and there are indications that this family of cytoplasmic tyrosine kinases is involved in FGF signal transduction. FGF-stimulated NIH3T3 cells, murine lung endothelial cells, and human foreskin fibroblasts show evidence of Src family kinase activation in response to FGF-2 treatment (Landgren et al., 1995). The interaction of FGFR-1 with Src correlates with the association of Src with cortactin and the subsequent tyrosine phosphorylation of this protein (Zhan et al., 1994). Cortactin is a F-actin-binding protein that is present in focal adhesion sites, suggesting a link between FGF receptor activation and cytoskeleton changes associated with cell division or motility. These observations may partially explain why prolonged exposure to FGF-1 is required for the maximal induction of DNA synthesis in BALB/c 3T3 cells (Zhan et al., 1993). Induction of immediate early genes and increased p90 tyrosine phosphorylation (see below) require only short-term exposure to the factor, but increased phosphorylation of Src and association with cortactin are maximal in mid-to-late G1 (Zhan et al., 1994). This suggests that the interplay of FGFR activation, Src activation, and cortactin phosphorylation has functional significance in the induction of DNA synthesis.

Two other proteins that may be involved in FGF signal transduction are an 85-kDa serine kinase and a phosphoprotein of approximately 90 kDa. The 85-kDa protein is associated with FGFR-4 in transfected L6 rat myoblast cells, but the functional significance of this association is not currently known (Vainikka et al., 1996). Stronger evidence suggests that the 90-kDa protein may play a role in signal transduction. This 90-kDa protein is related to the protein SNT (Ong et al., 1996), which was originally identified as a protein in PC12 cells phosphorylated in response to FGF or NGF treatment. An SNT-like protein (SLP) has since been found in many cell types. It is rapidly phosphorylated on tyrosine, serine, and threonine residues following FGF addition and appears to interact weakly with the FGFR-1. Interestingly, none of the FGFR autophosphorylation sites other than those located in the kinase domain appear to be required for the protein to be phosphorylated (Klint et al., 1995; Mohammadi et al., 1996). The phosphorylated SLP is membrane-bound and associated with Grb2 and Sos in FGF-stimulated cells and may therefore function in Ras activation (Kouhara et al., 1997); in contrast, Sos is not associated with Shc–Grb2 complexes in these cells (Wang et al., 1996). It would therefore appear that the Shc–Grb2 complex may not be be responsible for delivering the Sos exchange factor to membrane-bound Ras as it does in the EGFR-activated Ras activation pathway. Shc is also associated with an unidentified 145-kDa phosphotyrosine-containing protein in FGF- treated cells (Kavanaugh and Williams, 1994), but the significance of this association is unknown.

Various FGFRs also differ in their mitogenic and differentiation-inducing potential. Studies of BaF3 murine lymphoid cells and L6 mouse myoblast cells show that FGFR-4 induces weaker phosphorylation of Raf-1 and MAP kinases than does FGFR-1. FGFR-1 also induces stronger phosphorylation of MAP kinases than does the IIIb isoform of FGFR-2 having high affinity for FGF-7/KGF (Shaoul et al., 1995). FGF signaling may also be mediated through the JNK family of MAP kinases. EGFR signaling has been shown to include the use of this pathway and it is reasonable to suspect that FGF signaling may utilize this pathway as well; however, this remains to be established. There is evidence that a particular splice variant of the FGFR-2/IIIb, which lacks a carboxy terminal PLC - 1 association site at Tyr 769, may have a greater transforming potential than the spliced form that contains this site (Ishii et al., 1995). The signaling and biological responses resulting from FGF activation of receptors may therefore differ depending not only on the identity of the ligand presented to the breast tumor cell but also on the combination of receptors expressed by the target cell.

NONCLASSICAL MECHANISMS OF ESTROGEN RECEPTOR ACTIVATION

Although some ER+ cell lines selected for hormone independence lose ER expression, the majority do not. The continued expression of a receptor that no longer appears to promote cell growth, even as it retains its ability to induce expression of a reporter gene, suggests that the mitogenic function ER in these hormone-independent sublines has been superseded by the emergence of other mitogenic pathways that do not necessarily require loss of ER expression. In fact, there is the suggestion that some of the actions of growth factor–activated pathways may be mediated at least in part through the ER. The ability of the growth factor–induced kinase cascade to activate nuclear transcription factors such as Myc and Jun suggests that the unliganded ER, itself a transcription factor, might also be a substrate for phosphorylation. Activation of the ER by phosphorylation-induced by the action of growth factors may therefore explain why some hormone-independent breast cancer cells retain ER expression.

Steroid receptors are phosphoproteins, and although all the effects of phosphorylation are probably not known, there is evidence that one effect of phosphorylation is to enhance transcriptional activity in the presence of ligand or to allow activation of the receptor in the absence of ligand. Phosphorylation of chicken PR, induced by activators of protein kinase A, induces transcription of a reporter gene in a manner similar to treatment with progesterone (Denner et al.,

1990). Treatment of cells with dopamine, an activator of adenyl cyclase that results in activation of protein kinase A, increases transcriptional activity of several steroid receptors in the absence of ligand, including ER and PR (Power et al., 1991). Similar induction of transcription is seen in cells treated with okadaic acid, an inhibitor of protein phosphatases, suggesting hyperphosphorylation as an important factor in steroid receptor transcriptional activity. In MCF-7 breast cancer cells, expression of a reporter plasmid downstream of an ERE is induced synergistically by treatment with E2 and activators of either protein kinase A or protein kinase C (Cho and Katzenellenbogen, 1993). Similarly, in rat uterus, treatment of cells with either E2, IGF-I, or agents that increase intracellular cyclic AMP (and thereby activate protein kinase A) increases transcription from an ERE reporter construct, and this increase in transcription can be blocked by treatment with either ICI 164,384 or inhibitors of protein kinases (Aronica and Katzenellenbogen, 1993). In an ER+ ovarian adenocarcinoma cell line, addition of either E2 or EGF induces expression of an ERE reporter construct, and this induction is inhibited by either ICI 164,384 or an anti-EGFR antibody (Ignar-Trowbridge et al., 1993). In mouse uterus, either E2 or EGF induces DNA synthesis and localization of the ER to the nucleus, and both events are inhibited by antiestrogen treatment (Ignar-Trowbridge, et al., 1992) and mice lacking ER due to targeted disruption of the gene do not exhibit estrogenic changes in response to EGF (Curtis et al., 1996). Deletion of the A/B domain of the ER, which contains the AF-1 function, inhibits the ability of growth factors to activate transcription of an ERE reporter construct in an ER+ ovarian carcinoma cell line suggesting this region of the ER as critical in the transduction of growth factor–generated signals (Ignar-Trowbridge et al., 1996). Interestingly, whereas combinations of different growth factors generate additive effects, addition of growth factors and E2 together results in synergistic activation of the ER (Ignar-Trowbridge et al., 1996; Ignar-Trowbridge et al., 1993) suggesting that each set of mitogens is acting through a distinct mechanism to activate the ER and transcription.

Recent studies demonstrate that the ER is a substrate for direct phosphorylation by MAP kinase. In vitro, a truncated human ER containing the A/B and C domains is phosphorylated by MAP kinase; mutation of serine 118, which lies within the AF-1 region in the A/B domain, abolished this phosphorylation strongly suggesting this residue as a target for phosphorylation. In vivo, mutation of this serine to alanine eliminates the ability of EGF to induce transcription of an ERE reporter plasmid (Bunone et al., 1996; Kato et al., 1995). Other growth factors also activate ER in the absence of E2, including heregulin (Pietras et al., 1995) and IGF-1 (Aronica and Katzenellenbogen, 1993; Ignar-Trowbridge et al., 1996). Mutation of serine 118 to alanine abolishes phosphorylation and ligand-independent ER activation mediated through three independent mechanisms of ERK activation: overexpression of wild type Ras (Kato et al., 1995), a constitutively active mutant of Ras (Kato et al., 1995), or a constitutively active mutant of MEK (Bunone et al., 1996; Kato et al., 1995). The ability of Ras overexpression to activate ER in the absence of ligand is enhanced by a constitutively active MEK and is abrogated by expression of a MAPK-specific phosphatase (Kato et al., 1995) indicating that the Ras-generated signals are transmitted through the MAPK cascade. The ability of EGF to induce ligand-independent activation of ER is inhibited by expression of dominant–negative Ras, a dominant–negative MEK, and an inhibitor of protein kinase C (Bunone et al., 1996) suggesting these proteins as downstream mediators of the EGF-induced activation of ER.

Interestingly, the agonist activity of a tamoxifen–ER complex is significantly enhanced by phosphorylation, whereas ICI 164,384-ER complexes retain full antagonistic function (Kato et al., 1995). The fact that tamoxifen possesses partial agonist activity while the pure antiestrogen does not suggests that one component of ligand-independent activation of ER may be enhancement by phosphorylation of AF-1 activity in ER molecules bound to tamoxifen. The consequence of phosphorylation is not solely to activate AF-1, however. In HeLa cells, where AF-1 function alone induces transcription poorly from an ERE (Tora et al., 1989), phosphorylation of the ER by ERKs nevertheless results in activation of transcription (Bunone et al., 1996) suggesting a possible role for AF-2. Other studies, however, suggest that activation in the absence of ligand is not mediated through AF-2. In studies of ER– SK-Br-3 breast cancer cells and HeLa cells, EGF induces ligand-independent activation of a transiently transfected ER containing a mutated AF-2 region (Bunone et al., 1996). Interestingly, this mutant retains full transcription activity in the presence of EGF when complexed to either tamoxifen or ICI 164,384 (Bunone et al., 1996), although ICI 164,384 retains its antagonistic properties when complexed to the wild type ER (Bunone et al., 1996; Kato et al., 1995). These results suggest that ligand-independent activation of the ER by growth factors is likely to involve at least two distinct mechanisms: activation of AF-1 through phosphorylation of serine 118, and at least one other, unknown mechanism independent of AF-2.

Despite the uncertainty surrounding the precise mechanism of growth factor action, the significance of these findings is that they offer evidence for a direct interaction between mediators of the E2 and growth factor–signaling pathways. This provides both a molecular basis through which growth factors might mimic the presence of E2 in an ER+ cell, and suggests that growth factor-promoted phosphorylation of ER may result in enhancement of the agonistic properties of tamoxifen. The interplay between the ER bound to tamoxifen and the MAPK cascade might have particularly important clinical implications for patients undergoing hormone ablative therapy whose tumors overexpress growth factors. The finding that phosphorylation of the ER complexed to tamoxifen by MAPK enhances the agonist activity of tamoxifen (Kato et al., 1995) suggests that tumors with activated MAPK cascades as a result of growth factor overproduction may not only be resistant to the growth inhibitory effects of the drug but may actually be stimulated by it. These results suggest that one consequence of growth factor overexpression may be to convert tamoxifen from an antagonist to an agonist in ER+ breast cancer cells thereby not only overcoming its activity as a growth suppressor but also turning it into a mitogen.

CONCLUSIONS

One of the central questions that is the focus of much current research into growth factors and their receptors in breast cancer is whether such agents act to mimic the effect of E2 through ligand-independent activation of the ER or, alternatively, to bypass the ER signaling pathway altogether. Although different growth factors, both alone and in combination, are likely to evoke different effects depending on the specific nature of each tumor, the mechanism of hormone-independent growth utilized by a particular tumor will ultimately determine whether patients who fail on endocrine therapy due to amplification of growth factor-mediated signaling are likely to respond to a second round of antiestrogen therapy. The failure of most ER+ cell lines selected in vitro or in vivo for hormone independence or resistance to one class of antiestrogens to develop cross-resistance to other antiestrogens is corroborated by clinical observation that patients who relapse following tamoxifen treatment sometimes respond to a second round hormone ablation using a pure antiestrogen (Howell et al., 1995). Patients who progress following first-line hormonal therapy may also respond to treatment with inhibitors of the aromatase involved in E2 synthesis suggesting that, despite relapse following initial hormone ablation, their tumors are still responsive to the circulating levels of E2 (Brodie and Njar, 1996; Buzdar et al., 1996). This observation suggests that, even in some patients who have relapsed following tamoxifen treatment, estrogenic stimulation of tumor growth may still occur. The actions of E2 and growth factors need not be mutually exclusive, therefore; additive or synergistic effects on growth resulting from the simultaneous activation of both pathways are also possible.

Because patients who are initially ER+ and subsequently acquire resistance to tamoxifen do not necessarily lose ER expression, it is possible that growth factor–mediated alteration of the cellular response to the tamoxifen–ER complex may play a role in the acquisition of tamoxifen resistance. If mitogenic signaling in such a tumor is mediated through the ER, it is likely that a second round of antiestrogen therapy using an ICI compound might prove useful in patients who have relapsed after tamoxifen therapy. There is much evidence to suggest that some of the effects of EGFR signaling involve activation of the ER. As discussed previously, EGF addition can mimic estrogenic activation of the ER in numerous cell types both in tissue culture and in vivo. That the ER is a target for the EGFR-activated Ras/Raf/ERK cascade lends further support to the hypothesis that one mechanism of growth factor action is to mimic the effect of E2 by activating the ER in the absence of E2 or the presence of tamoxifen. Because the agonist activity of tamoxifen is enhanced by phosphorylation of the ER by ERKs (Kato et al., 1995), ER+ breast cancer cells undergoing hormone ablation due to tamoxifen treatment may actually be stimulated as a result of the interaction between the tamoxifen–ER complex and the signal transduction cascade activated by growth factors. ICI 164,384, however, retains its properties as a pure antiestrogen even when the ER is phosphorylated. This observation suggests that some ER+ patients who fail on tamoxifen therapy, even those whose tumors overexpress EGFR and may have an activated MAPK pathway and phosphorylated ER, might respond to a second treatment with a pure antiestrogen.

An alternate mechanism through which growth factors may contribute to E2 independence is through activation of a mitogenic pathway that does not intersect with the ER. Overexpression of growth factors, therefore, would not necessarily have to lead to ligand-independent activation of ER; it may bypass the ER entirely. Rather than mimicking the presence of E2, this mechanism would eliminate the need for E2. There is experimental evidence to support the hypothesis that, in contrast to the studies described earlier in which growth factor overexpression results in ligand-independent ER activation, other growth factors may stimulate growth in a manner that is completely independent of ER or that is antagonistic to the ER. Heregulin-β-2--transfected MCF-7 cells acquire in vitro E2 independence and the ability to form tumors in nude mice in the absence of E2 (Tang et al., 1996). In these cells the ER is downregulated but unable to induce expression of PR. Whereas transfection of MCF-7 cells with heregulin-β-1 also results in hormone-independent growth and downregulation of ER, overexpression of this isoform leads to constitutive activation of ER (Pietras et al., 1995). The ability of two different isoforms of the same protein transfected into the same breast cancer cell line to induce opposite phenotypes suggests that the response of a cell to a growth factor is likely to result from a complex interplay of the specific factor or factors present and the combination and levels of receptors expressed by the target cell, as well as many other variables.

An ER-independent mechanism for hormone independence is also supported by work in our lab using FGF-transfected MCF-7 cells. In addition to effects on tumorigenicity or metastasis, overexpression of FGF-1 or FGF-4 in MCF-7 cells increases the growth rate in E2-depleted media or media containing the antiestrogen 4-hydroxy tamoxifen (McLeskey et al., 1993; Zhang et al., 1997). Thus, FGF signaling appears to act in an autocrine or intracrine fashion to provide a mitogenic stimulus to ER+ breast cancer cells that allows these cells to circumvent their requirement for E2 for optimal growth. We have also found that FGF-1 and FGF-4 overexpressing cells have a much higher cloning efficiency than do control transfected cells when plated in soft agar containing ICI 182,780 and are still able to form tumors in mice treated with compound (McLeskey et al., 1998). FGF-transfected cells do not have downregulated ER or constitutively elevated levels of E2-responsive genes as might be expected if ER were activated (Pietras et al., 1995). These results suggest that FGF signaling bypasses, rather than mimics, ER-mediated growth-signaling mechanisms.

These results raise the specter that patients with ER-positive breast tumors that are nonresponsive or resistant to tamoxifen as a result of activation of FGF signal transduction pathways would be unlikely to respond to a second line therapy with pure antiestrogens. However, there is also the possibility that elucidation of FGF signal transduction pathways operating in ER-positive breast cancer cells will identify targets for intervention at points upstream of where the two growth-signaling pathways converge to allow antiestrogens to again become effective. Even if interference with growth factor–mediated mitogenesis fails to restore sensitivity to endocrine therapies in breast tumors that have progressed following tamoxifen treatment, however, the disruption of the autocrine or paracrine loops operating within a tumor may nevertheless result in more effective treatments. Whereas targeting of any one growth factor may prove useful, the frequent overexpression of numerous growth factors by a single tumor implies that a strategy of targeting several pathways simultaneously, as well as inhibiting angiogenesis within a tumor, may show the most promise for ultimately curing breast cancer.

ABBREVIATIONS

E2, estrogen; ER, estrogen receptor; EGFR, epidermal growth factor receptor; TGF-α, transforming growth factor alpha; AR, amphiregulin; VEGFR, vascular endothelial growth factor receptor; FGFR, fibroblast growth factor receptor; PLC-γ, phospholipase C gamma; PI-3K, phosphoinositol-3 kinase

ACKNOWLEDGMENTS

This work was supported by NIH grants CA50376 and CA53185 and Department of Defense grant DAMD17-94-J4167.

REFERENCES

Achen, M.G., Jeltsch, M., Kukk, E., Makinen, T., Vitali, A., Wilks, A.F., Alitalo, K., and Stacker, S.A. (1998). Vascular endothelial growth factor D (VEGF-D) is a ligand for the tyrosine kinases VEGF receptor 2 (Flk1) and VEGF receptor 3 (Flt4). Proc. Nat. Acad. Sci. USA. 95, 548-553.

Adnane, J., Gaudray, P., Dionne, C.A., Crumley, G., Jaye, M., Schlessinger, J., Jeanteur, P., Birnbaum, D., and Theillet, C. (1991). BEK and FLG, two receptors to members of the FGF family, are amplified in subsets of human breast cancers. Oncogene 6, 659-663.

Ali, S., Metzger, D., Bornert, J.M., and Chambon, P. (1993). Modulation of transcriptional activation by ligand-dependent phosphorylation of the human oestrogen receptor A/B region. EMBO J. 12, 1153-1160.

Allred, D.C., O'Connell, P., and Fuqua, S.A. (1993). Biomarkers in early breast neoplasia. J. Cell. Biochem. (suppl.) 17G, 125-131.

American Cancer Society. (1997). Breast cancer facts and figures.

Anan, K., Morisaki, T., Katano, M., Ikubo, A., Kitsuki, H., Uchiyama, A., Kuroki, S., Tanaka, M., and Torisu, M. (1996). Vascular endothelial growth factor and platelet- derived growth factor are potential angiogenic and metastatic factors in human breast cancer. Surgery 119, 333-339.

Anandappa, S.Y., Winstanley, J.H., Leinster, S., Green, B., Rudland, P.S., and Barraclough, R. (1994). Comparative expression of fibroblast growth factor mRNAs in benign and malignant breast disease. Br. J. Cancer 69, 772-776.

Angel, P., Hattori, K., Smeal, T., and Karin, M. (1988). The jun proto-oncogene is positively autoregulated by its product, Jun/AP-1. Cell 55, 875-885.

Arnold, S.F., Obourn, J.D., Jaffe, H., Notides, A.C. (1995a). Phosphorylation of the human estrogen receptor by mitogen-activated protein kinase and casein kinase II: consequence on DNA binding. J. Steroid Biochem. Mol. Biol. 55, 163-172.

Arnold, S.F., Obourn, J.D., Jaffe, H., and Notides, A.C. (1995b). Phosphorylation of the human estrogen receptor on tyrosine 537 in vivo and by *src* family tyrosine kinases in vitro. Mol. Endocrinol. 9, 24-33.

Arnold, S.F., Obourn, J.D., Jaffe, H., and Notides, A.C. (1994). Serine 167 is the major estradiol-induced phosphorylation site on the human estrogen receptor. Mol. Endocrinol. 8, 1208-1214.

Arnold, S.F., Obourn, J.D., Yudt, M.R., Carter, T.H., and Notides, A.C. (1995c). In vivo and in vitro phosphorylation of the human estrogen receptor. J. Steroid Biochem. Mol. Biol. 52, 159-171.

Aronheim, A., Engelberg, D., Li, N., Al-Alawi, N., Schlessinger, J., and Karin, M. (1994). Membrane targeting of the nucleotide exchange factor Sos is sufficient for activating the Ras signaling pathway. Cell 78, 949-961.

Aronica, S.M. and Katzenellenbogen, B.S. (1993). Stimulation of estrogen receptor- mediated transcription and alteration in the phosphorylation state of the rat uterine estrogen receptor by estrogen, cyclic adenosine monophosphate, and insulinlike growth factor-1. Mol. Endocrinol. 7, 743-752.

Arteaga, C.L., Coronado, E., and Osborne, C.K. (1988). Blockade of the epidermal growth factor receptor inhibits transforming growth factor α-induced but not estrogen-induced growth of hormone-dependent human breast cancer. Mol. Endocrinol. 2, 1064-1069.

Asano, M., Yukita, A., Matsumoto, T., Kondo, S., and Suzuki, H. (1995). Inhibition of tumor growth and metastasis by an immunoneutralizing monoclonal antibody to human vascular endothelial growth factor/vascular permeability factor 121. Cancer Res. 55, 5296- 5301.

Baird, A. and Klagsbrun, M. (1991). The Fibroblast Growth Factor family. Cancer Cells 3, 239-243.

Bansal, G.S., Yiangou, C., Coope, R.C., Gomm, J.J., Luqmani, Y.A., Coombes, R.C., and Johnston, C.L. (1995). Expression of fibroblast growth factor 1 is lower in breast cancer than in normal human breast. Br. J. Cancer 72, 1420-1426.

Bargmann, C.L., Hung, M., and Weinberg, R.A. (1986). The *neu* oncogene encodes an epidermal growth factor receptor-related protein. Nature 319, 226-230.

Barleon, B., Sozzani, S., Zhou, D., Weich, H.A., Mantovani, A., and Marme, D. (1996). Migration of human monocytes in response to vascular endothelial growth factor (VEGF) is mediated via the VEGF receptor Flt-1. Blood 87, 3336-3343.

Barrett-Lee, P.J., Travers, M.T., McClelland, R.A., Luqmani, Y., and Coombes, R.C. (1987). Characterization of estrogen receptor messenger RNA in human breast cancer. Cancer Res. 47, 6653-6659.

Basilico, C. and Moscatelli, D. (1992). The FGF family of growth factors and oncogenes. Adv. Cancer Res. 59, 115-165.

Bates, S.E., Davidson, N.E., Valverius, E.M., Freter, C.E., Dickson, R.B., Tam, J.P., Kudlow, J.E., Lippman, M.E., and Salomon, D.S. (1988). Expression of transforming growth factor α and its messenger ribonucleic acid in human breast cancer: Its regulation by estrogen and its possible functional significance. Mol. Endocrinol. 2, 543-555.

Batra, S.K., Castelino-Prabhu, S., Wikstrand, C.J., Zhu, X., Humphrey, P.A., Friedman, H.S., Bigner, D.D. (1995). Epidermal growth factor ligand-independent, unregulated, cell-transforming potential of a naturally occurring human mutant EGFRvIII gene. Cell Growth Differ. 6, 1251-1259.

Battaglia, C., Berlingieri, M.T., Martelli, M.L., Trapasso, F., Delli Bovi, P., and Fusco, A. (1993). Mitogenic and dedifferentiating effect of the *K-fgf/hst* oncogene on rat thyroid PC clone 3 epithelial cells. Cell Growth Differ. 4, 185-192.

Battaglia, F., Polizzi, G., Scambia, G., Rossi, S., Panici, P.B., Iacobelli, S., Crucitti, F., and Mancuso, S. (1988a). Receptors for epidermal growth factor and steroid hormones in human breast cancer. Oncology 45, 424-427.

Battaglia, F., Scambia, G., Rossi, S., Panici, P.B., Bellantone, R., Polizzi, G., Querzoli, P., Negrini, R., Iacobelli, S., Crucitti, F., et al. (1988b). Epidermal growth factor receptor in human breast cancer: Correlation with steroid hormone receptors and axillary lymph node involvement. Eur. J. Cancer Clin. Oncol. 24, 1685-1690.

Batzer, A.G., Blaikie, P., Nelson, K., Schlessinger, J., Margolis, B. (1995). The phosphotyrosine interaction domain of Shc binds an LXNPXY motif on the epidermal growth factor receptor. Mol. Cell. Biol. 15, 4403-4409.

Beatson, G.T. (1896). On the treatment of inoperable cases of carcinoma of the mammae: Suggestions for a new method of treatment with illustrative cases. Lancet 2, 104-107.

Bell, G.I., Fong, N.M., Stempien, M.M., Wormsted, M.A., Caput, D., Ku, L., Urdea, M.S., Rall, L.B., and Sanchez-Pescador, R. (1995). Human epidermal growth factor receptor precursor: cDNA sequence, expression in vitro and gene organization. J. Biol. Chem. 256, 21393-21396.

Bellot, F., Crumley, G., Kaplow, J.M., Schlessinger, J., Jaye, M., and Dionne, C.A. (1991). Ligand-induced transphosphorylation between different FGF receptors. EMBO J. 10, 2849-2854.

Bellus, G.A., McIntosh, I., Smith, E.A., Aylsworth, A.S., Kaitila, I., Horton, W.A., Greenhaw, G.A., Hecht, J.T., and Francomano, C.A. (1995). A recurrent mutation in the tyrosine kinase domain of fibroblast growth factor receptor 3 causes hypochondroplasia. Nat. Genet. 10, 357-359.

Berg, J.M. (1989). DNA binding specificity of steroid receptors. Cell 57, 1065-1068.

Berns, E.M., Klijn, J.G., van Staveren, I.L., Portengen, H., Noordegraaf, E., and Foekens, J.A. (1992). Prevalence of amplification of the oncogenes c-myc, HER2/*neu*, and int-2 in one thousand human breast tumors: Correlation with steroid receptors. Eur. J. Cancer 28, 697-700.

Berry, M., Metzger, D., and Chambon, P. (1990). Role of the two activating domains of the oestrogen receptor in the cell-type and promoter-context-dependent agonistic activity of the anti-oestrogen 4-hydroxytamoxifen. EMBO J. 9, 2811-2818.

Bird, A.P. (1986). CpG-rich islands and the function of DNA methylation. Nature 321, 209-213.

Blam, S.B., Mitchell, R., Tischer, E., Rubin, J.S., Silva, M., Silver, S., Fiddes, J.C., Abraham, J.A., and Aaronson, S.A. (1988). Addition of growth hormone secretion signal to basic fibroblast growth factor results in cell transformation and secretion of aberrant forms of the protein. Oncogene 3, 129-136.

Bolufer, P., Miralles, F., Rodriguez, A., Vazquez, C., Lluch, A., Garcia-Conde, J., and Olmos, T. (1990). Epidermal growth factor receptor in human breast cancer: Correlation with cytosolic and nuclear ER receptors and with biological and histological tumor characteristics. Eur. J. Cancer 26, 283-290.

Boocock, C.A., Charnock-Jones, D.S., Sharkey, A.M., McLaren, J., Barker, P.J., Wright, K.A., Twentyman, P.R., and Smith, S.K. (1995). Expression of vascular endothelial growth factor and its receptors flt and kdr in ovarian carcinoma. J. Natl. Cancer Inst. 81, 506-516.

Boulton, T.G., Nye, S.H., Robbins, D.J., Ip, N.Y., Redziejewska, E., Morgenbesser, S.D., DePinho, R.A., Panayotatos, N., Cobb, M.H., and Yancopoulos, G.D. (1991). ERKs: a family of protein-serine/threonine kinases that are activated and tyrosine phosphorylated in response to insulin and NGF. Cell 65, 663-675.

Bourne, H.R., Sanders, D.A., and McCormick, F. (1991). The GTPase superfamily: conserved structure and molecular mechanism. Nature 349, 117-127.

Bowler, J., Lilley, T.J., Pittam, J.D., Wakeling, A.E. (1989). Novel steroidal pure antiestrogens. Steroids 54, 71-99.

Brandt, R., Normanno, N., Gullick, W.J., Harkins, R., Schneider, D., Jones, B.-W., Ciardiello, F., Persico, M.G., Kim, N., and Salomon, D.S. (1994). Identification and biological characterization of an epidermal growth factor-related protein: cripto-1. J. Biol. Chem. 269, 17320-17328.

Briozzo, P., Badet, J., Capony, F., Pieri, I., Montcourrier, P., Barritault, D., and Rochefort, H. (1991). MCF7 mammary cancer cells respond to bFGF and internalize it following its release from extracellular matrix: A permissive role of cathepsin D. Exp. Cell Res. 194, 252-259.

Brison, O. (1993). Gene amplification and tumor progression. Biochim. Biophys. Acta. 1155, 25-41.

Brodie, A.M. and Njar, V.C. (1996). Aromatase inhibitors and breast cancer. Sem. Oncol. 23, 10-20.

Bronzert, D.A., Pantazis, P., Antoniades, H.N., Kasid, A., Davidson, N., Dickson, R.B., and Lippman, M.E. (1987). Synthesis and secretion of platelet-derived growth factor by human breast cancer cell lines. Proc. Natl. Acad. Sci. USA 84, 5763-5767.

Brown, L.F., Berse, B., Jackman, R.W., Tognazzi, K., Guidi, A.J., Dvorak, H.F., Senger, D.R., Connolly, J.L., and Schnitt, S.J. (1995). Expression of vascular permeability factor (vascular endothelial growth factor) and its receptors in breast cancer. Hum. Pathol. 26, 86-91.

Brown, L.F., Berse, B., Jackman, R.W., Tognazzi, K., Manseau, E.J., Senger, D.R., and Dvorak, H.F. (1993). Expression of vascular permeability factor (vascular endothelial growth factor) and its receptors in adenocarcinomas of the gastrointestinal tract. Cancer Res. 53, 4727-4735.

Brünner, N., Boulay, V., Fojo, A., Freter, C.E., Lippman, M.E., and Clarke, R. (1993a). Acquisition of hormone-independent growth in MCF-7 cells is accompanied by increased expression of estrogen-regulated genes but without detectable DNA amplifications. Cancer Res. 53, 283-290.

Brünner, N., Frandsen, T.L., Holst-Hansen, C., Bei, M., Thompson, E.W., Wakeling, A.E., Lippman, M.E., and Clarke, R. (1993b). MCF7-LCC2: A 4-hydroxytamoxifen resistant human breast cancer variant that retains sensitivity to the steriodal antiestrogen ICI 182,780. Cancer Res. 53, 3229-3232.

Brustle, O., Aguzzi, A., Talarico, D., Basilico, C., Kleihues, P., and Wiestler, O.D. (1992). Angiogenic activity of the K-fgf/hst oncogene in neural transplants. Oncogene 7, 1177- 1183.

Bullocks, J., Zhang, L., Ding, I.Y.F., McLeskey, S.W., Tobias, C.A., Miller, D.L., Kern, F.G. (1997). Overexpression of vascular endothelial growth factor (VEGF) in MCF-7 breast carcinoma cells facilitates growth in tamoxifen-treated nude mice and tumor cell dissemination. Proc. Am. Assoc. Cancer Res. (38, A3521.)

Bunnag, P., Waddell, K.S., Varban, M.L., and Chiu, I.M. (1991). Transformed phenotype conferred to NIH/3T3 cells by ectopic expression of heparin-binding growth factor 1/acidic fibroblast growth factor. Cell. Dev. Biol. 27, 89-96.

Bunone, G., Briand, P.-A., Miksicek, R.J., and Picard, D. (1996). Activation of the unliganded estrogen receptor by EGF involves the MAP kinase pathway and direct phosphorylation. EMBO J. 15, 2174-2183.

Burgess, W.H. and Maciag, T. (1989). The heparin-binding (fibroblast) growth factor family of proteins. Annu. Rev. Biochem. 58, 575-606.

Burrus, L.W., Zuber, M.E., Lueddecke, B.A., and Olwin, B.B. (1992). Identification of a cysteine-rich receptor for fibroblast growth factors. Mol. Cell. Biol. 12, 5600-5609.

Buzdar, A.U., Smith, R., Vogel, C., Bonomi, P., Keller, A.M., Favis, G., Mulaga, M., and Cooper, J. (1996). Fadrozole HCL (CGS-16949A) versus megesterol acetate treatment of postmenopausal patients with metastatic breast cancer. Results of two randomized double blind cotrolled multiinstitutional trials. Cancer 77, 2503-2513.

Callaghan, T., Antczak, M., Flickinger, T., Raines, M., Myers, M., and Kung, H.-J. (1993). A complete description of the EGF-receptor exon structure: Implication in oncogenic activation and domain evolution. Oncogene 8, 2939-2948.

Campos-Gonzales, R., and Glenney, J.R., Jr. (1992). Tyrosine phosphorylation of mitogen-activated protein kinase in cells with tyrosine kinase-negative epidermal growth factor receptors. J. Biol. Chem. 267, 14535-14538.

Carlomagno, C., Perrone, F., Gallo, C., De Laurentiis, M., Lauria, R., Morabito, A., Pettinato, G., Panico, L., D'Antonio, A., Bianco, A.R., and De Placido, S. (1996). c-erb B2 overexpression decreases the benefit of adjuvant tamoxifen in early-stage breast cancer without axillary lymph node metastases. J. Clin. Oncol. 14, 2702-2708.

Carmeliet, P., Ferreira, V., Breier, G., Pollefeyt, S., Kieckens, L., Gertsenstein, M., Fahrig, M., Vandenhoeck, A., Harpal, K., Eberhardt, C., Declercq, C., Pawling, J., Moons, L., Collen, D., Risau, W., and Nagy, A. (1996). Abnormal blood vessel development and lethality in embryos lacking a single VEGF allele. Nature 380, 435-439.

Carraway, K.L.I., Cantley, L.C. (1994). A in vivoneu acquaintance for *erb*B3 and *erb*B4: A role for receptor heterodimerization in growth signaling. Cell 78, 5-8.

Cavailles, V., Dauvois, S., Danielian, P.S., and Parker, M.G. (1994). Interaction of proteins with transcriptionally active estrogen receptors. Proc. Natl. Acad. Sci. USA 91, 10009-10013.

Cavailles, V., Dauvois, S., L'Horset, F., Lopez, G., Hoare, S., Kushner, P.J., and Parker, M.B. (1995). Nuclear factor RIP140 modulates transcriptional activation by the estrogen receptor. EMBO J. 14, 3741-3751.

Chambraud, B., Berry, M., Redeuilh, G., Chambon, P., and Baulieu, E.E. (1990). Several regions of the human estrogen receptor are involved in the formation of receptor-heat shock protein 90 complexes. J. Biol. Chem. 265, 20686-20691.

Charnock-Jones, D.S., Sharkey, A.M., Boocock, C.A., Ahmed, A., Plevin, R., Ferrara, N., and Smith, S.K. (1994). Vascular endothelial growth factor receptor localization and activation in human trophoblast and choriocarcinoma cells. Biol. Reprod. 51, 524-530.

Chen, C., Parangi, S., Tolentino, M.J., and Folkman, J. (1995). A strategy to discover circulating angiogenesis inhibitors generated by human tumors. Cancer Res. 55, 4230- 4233.

Chen, R.-H., Sarnecki, C., and Blenis, J. (1992). Nuclear localization and regulation of *erk*- and *rsk*-encoded protein kinases. Mol. Cell. Biol. 12, 915-927.

Chen, W.S., Lazar, C.S., Poenie, M., Tsien, R.Y., Gill, G., and Rosenfeld, M.G. (1987). Requirement for intrinsic protein tyrosine kinase in the immediate and late actions of the EGF receptor. Nature 328, 820-823.

Cherniach, A.D., Klarlund, J.K., and Czech, M.P. (1994). Phosphorylation of the Ras nucleotide exchange factor Son of sevenless by mitogen-activated protein kinase. J. Biol. Chem. 269, 4717-4720.

Cho, H. and Katzenellenbogen, B.S. (1993). Synergistic activation of estrogen receptor-mediated transcription by estradiol and protein kinase activators. Mol. Endocrinol. 7, 441-452.

Ciardiello, F., Dono, R., Kim, N., Persico, M.G., and Salomon, D.S. (1991). Expression of cripto, a novel gene of the epidermal growth factor gene family, leads to in vitro transformation of a normal mouse mammary epithelial cell line. Cancer Res. 51, 1051- 1054.

Ciccodicola, A., Dono, R., Obici, S., Simeone, A., Zollo, M., and Persico, M.G. (1989). Molecular characterization of a gene of the "EGF family" expressed in undifferentiated human NTERA2 teratocarcinoma cells. EMBO J. 8, 1987-1991.

Claffey, K.P., Brown, L.F., del Aguila, L.F., Tognazzi, K., Yeo, K.T., Manseau, E.J., and Dvorak, H.F. (1996). Expression of vascular permeability factor/vascular endothelial growth factor by melanoma cells increases tumor growth, angiogenesis, and experimental metastasis. Cancer Res. 56, 172-181.

Clarke, R., Brünner, N., Katz, D., Glanz, P., Dickson, R.B., Lippman, M.E., Kern, F.G. (1989a). The effects of a constitutive expression of transforming growth factor-alpha on the growth of MCF-7 human breast cancer cells in vitro and in vivo. Mol. Endocrinol. 3, 372-380.

Clarke, R., Brünner, N., Katzenellenbogen, B.S., Thompson, E.W., Norman, M.J., Koppi, C., Paik, S., Lippman, M.E., and Dickson, R.B. (1989b). Progression of human breast cancer cells from hormone-dependent to hormone-independent growth both in vitro and in vivo. Proc. Natl. Acad. Sci. USA 86, 3649-3653.

Clarke, R., Morwood, J., van den Berg, H.W., Nelson, J., and Murphy, R.F. (1986). Effect of cytotoxic drugs on estrogen receptor expression and response to tamoxifen in MCF-7 cells. Cancer Res. 46, 6116-6119.

Clauss, M., Weich, H., Breier, G., Knies, U., Rockl, W., Waltenberger, J., and Risau, W. (1996). The vascular endothelial growth factor receptor Flt-1 mediates biological activities. Implications for a functional role of placenta growth factor in monocyte activation and chemotaxis. J. Biol. Chem. 271, 17629-17634.

Conn, G., Bayne, M.L., Soderman, D.D., Kwok, P.W., Sullivan, K.A., Palisi, T.M., Hope, D.A., and Thomas, K.A. (1990). Amino acid and cDNA sequences of a vascular endothelial cell mitogen that is homologous to platelet-derived growth factor. Proc. Natl. Acad. Sci. USA 87, 2628-2632.

Cook, S.J. and McCormick, F. (1993). Inhibition by cAMP of Ras-dependent activation of Raf. Science 262, 1069-1072.

Coope, R.C., Browne, P.J., Yiangou, C., Bansal, G.S., Walters, J., Groome, N., Shousha, S., Johnston, C.L., Coombes, R.C., and Gomm, J.J. (1997). The location of acidic fibroblast growth factor in the breast is dependent on the activity of proteases present in breast cancer tissue. Br. J. Cancer 75, 1621-1630.

Coso, O.A., Chiariello, M., Yu, J.-C., Teramoto, H., Crespo, P., Xu, N., Miki, T., and Gutkind, J.S. (1995). The small GTP-binding proteins Rac1 and Cdc42 regulate the activity of the JNK/SAPK signaling pathway. Cell 81, 1137-1146.

Countaway, J.L., Nairn, A.C., and Davis, R.J. (1992). Mechanism of desensitization of the epidermal growth factor receptor protein-tyrosine kinase. J. Biol. Chem. 267, 1129-1140.

Coussens, L., Yang-Feng, T., Liao, Y., Chen, L., Gray, A., McGrath, J., Seeburg, P.H., Lieberman, T.A., Schlessinger, J., Francke, U., Levinson, A., and Ullrich, A. (1985). Tyrosine kinase receptor with extensive homology to EGF receptor shares chromosomal location with *neu* oncogene. Science 230, 1132-1139.

Cullen, K.J., Yee, D., Sly, W.S., Perdue, J., Hampton, B., Lippman, M.E., and Rosen, N. (1990). Insulinlike growth factor receptor expression and function in human breast cancer. Cancer Res. 50, 48-53.

Curtis, S.W., Washburn, T., Sewall, C., DiAugustine, R., Lindzey, J., Couse, J.F., and Korach, K.S. (1996). Physiological coupling of growth factor and steroid receptor signaling pathways: Estrogen receptor knockout mice lack estrogenlike response to epidermal growth factor. Proc. Natl. Acad. Sci. USA 93, 12626-12630.

D'Angelo, G., Struman, I., Martial, J., and Weiner, R.I. (1995). Activation of mitogen- activated protein kinases by vascular endothelial growth factor and basic fibroblast growth factor in capillary endothelial cells is inhibited by the antiangiogenic factor 16- kDa N-terminal fragment of prolactin. Proc. Natl. Acad. Sci. USA 92, 6374-6378.

Danielian, P.S., White, R., Lees, J.A., and Parker, M.G. (1992). Identification of a conserved region required for hormone-dependent transcriptional activation by steroid hormone receptors. EMBO J. 11, 1025-1033.

Dauvois, S., Danielian, P.S., White, R., and Parker, M.G. (1992). Antiestrogen ICI 164,384 reduces cellular estrogen receptor content by increasing its turnover. Proc. Natl. Acad. Sci. USA 89, 4037-4041.

Davidson, N.E., Gelmann, E.P., Lippman, M.E., and Dickson, R.B. (1987). Epidermal growth factor receptor gene expression in estrogen receptor-postitive and negative human breast cancer cell lines. Mol. Endocrinol. 1, 216-223.

Davis, R.J. (1988). Independent mechanisms account for the regulaion by protein kinase- C of the EGF-receptor affinity and tyrosine-protein kinase activity. J. Biol. Chem. 263, 9462-9469.

de Vries, C., Escobedo, J.A., Ueno, H., Houck, K., Ferrara, N., and Williams, L.T. (1992). The fms-like tyrosine kinase, a receptor for vascular endothelial growth factor. Science 255, 989-991.

Decker, S.J., Alexander, C., and Habib, T. (1992). Epidermal growth factor (EGF)- stimulated tyrosine phosphorylation and EGF receptor degradation in cells expressing EGF receptors truncated at residue 973. J. Biol. Chem. 267, 1104-1108.

Delarue, J.C., Friedman, S., Mouriesse, H., May-Levin, F., Sancho-Garnier, H., and Contesso, G. (1988). Epidermal growth factor receptor in human breast cancers: Correlation with estrogen and progesterone receptors. Breast Cancer Res. Treat. 11, 173- 178.

Delli-Bovi, P., Curatola, A.M., Newman, K.M., Sato, Y., Moscatelli, D., Hewick, R.M., Rifkin, D.B., Basilico, C. (1988). Processing, secretion, and biological properties of a novel growth factor of the fibroblast growth factor family with oncogenic potential. Mol. Cell. Biol. 8, 2933-2941.

Deng, C., Wynshaw-Boris, A., Zhou, F., Kuo, A., and Leder, P. (1996). Fibroblast growth factor receptor 3 is a negative regulator of bone growth. Cell 84, 911-921.

Denner, L.A., Weigel, N.L., Maxwell, B.L., Schrader, W.T., and O'Malley, B.W. (1990). Regulation of progesterone receptor-mediated transcription by phosphorylation. Science 250, 1740-1743.

Dent, P., Haser, W., Haystead, T.A.J., Vincent, L.A., Roberts, T.M., and Sturgill, T.W. (1992). Activation of mitogen-activated protein kinase kinase by *v-raf* in NIH 3T3 cells and in vitro. Science 257, 1404-1407.

Denton, R.R., Koszewski, N.J., and Notides, A.C. (1992). Estrogen receptor phosphorylation. Hormonal dependence and consequence on specific DNA binding. J. Biol. Chem. 267, 7263-7268.

Derijard, B., Hibi, M., Wu, I.H., Barrett, T., Su, B., Deng, T., Karin, M., and Davis, R.J. (1994). JNK1: A protein kinase stimulated by UV light and Ha-Ras that binds and phosphorylates the c-Jun activation domain. Cell 76, 1025-1037.

Derynck, R. (1988). Transforming growth factor α. Cell 54, 593-595.

Derynck, R., Roberts, A.B., Winkler, M.E., Chen, E.Y., and Goeddel, D.V. (1984). Human transforming growth factor α: Precursor structure and expression in *E. coli*. Cell 38, 287-297.

Di Marco, E., Pierce, J.H., Fleming, T.P., Kraus, M.H., Molloy, C.J., Aaronson, S.A., Di and Fiore, P.P. (1989). Autocrine interaction between TGF α and the EGF-receptor: quantitative requirements for induction of the malignant phenotype. Oncogene 4, 831- 838.

Dickson, C., Smith, R., Brookes, S.; and Peters, G. (1984). Tumorigenesis by mouse mammary tumor virus: Proviral activation of a cellular gene in the common integration region int-2. Cell 37, 529-536.

Dickson, R.B., Huff, K.K., Spencer, E.M., and Lippman, M.E. (1986a). Induction of epidermal growth factor-related polypeptides by 17 β-estradiol in MCF-7 human breast cancer cells. Endocrinology 118, 138-142.

Dickson, R.B., McManaway, M.E., and Lippman, M.E. (1986b). Estrogen-induced factors of breast cancer cells partially replace estrogen to promote tumor growth. Science 232, 1540-1543.

Dickson, R.B., Thompson, E.W., and Lippman, M.E. (1990). Regulation of proliferation, invasion and growth factor synthesis in breast cancer by steroids. J. Steroid Biochem. Mol. Biol. 37, 305-316.

Dickstein, B., Valverius, E.M., Wosikowski, K., Saceda, M., Pearson, J.W., Martin, M.B., Bates, S.E. (1993). Increased epidermal growth factor receptor in an estrogen-responsive, adriamycin-resistant MCF-7 cell line. J. Cell. Physiol. 157, 110-118.

DiFiore, P., Pierce, J., Fleming, T., Hazan, R., Ullrich, A., King, C.R., Schlessinger, J., and Aaronson, S. (1987). Overexpression of the human EGF receptor confers an EGF- dependent transformed phenotype to NIH 3T3 cells. Cell 51, 1063-1070.

Ding, I.Y., McLeskey, S.W., Chang, K., Fu, Y.M., Acol, J.C., Shou, M.T., Alitalo, K., and Kern, F.G. (1992). Expression of fibroblast growth factors (FGFs) and receptors (FGFRs) in human breast carcinomas. Proc. Annu. Meet. Am. Assoc. Cancer Res. 33, A1610.

Dionne, C.A., Crumley, G., Bellot, F., Kaplow, J.M., Searfoss, G., Ruta, M., Burgess, W.H., Jaye, M., and Schlessinger, J. (1990). Cloning and expression of two distinct high- affinity receptors cross-reacting with acidic and basic fibroblast growth factors. EMBO J. 9, 2685-2692.

Dobashi, K., Davis, J.G., Mikami, Y., Freeman, J.K., Hamuro, J., Greene, M.I. (1991). Characterization of a *neu/c-erb*B-2 protein-specific activating factor. Proc. Natl. Acad. Sci. USA 88, 8582-8586.

Doherty, P., Williams, E., and Walsh, F.S. (1994). Activation of the FGF receptor underlies neurite outgrowth stimulated by L1, N-CAM, and N-cadherin. Neuron 13, 583-594.

Dotzlaw, H., Miller, T., Karvelas, J., and Murphy, L.C. (1990). Epidermal growth factor gene expression in human breast cancer biopsy samples: Relationship to estrogen and progesterone receptor gene expression. Cancer Res. 50, 4204-4208.

Downward, J., Yarden, Y., Mayes, E., Scrace, G., Totty, N., Stockwell, P., Ullrich, A., Schlessinger, J., and Waterfield, M.D. (1984). Close similarity of epidermal growth factor receptor and v-*erb*-B oncogene protein sequences. Nature 307, 521-527.

Emoto, H., Tagashira, S., Mattei, M.G., Yamasaki, M., Hashimoto, G., Katsumata, T., Negoro, T., Nakatsuka, M., Birnbaum, D., Coulier, F., and Itoh, N. (1997). Structure and expression of human fibroblast growth factor-10. J. Biol. Chem. 272, 23191-23194.

Engel, L.W. and Young, N.A. (1978). Human breast carcinoma cells in continuous culture: A review. Cancer Res. 38, 4327-4339.

Falls, D.L., Rosen, K.M., Corfas, G., Lane, W.S., and Fischbach, G.D. (1993). ARIA, a protein that stimulates acetylcholine receptor synthesis, is a member of the *neu* ligand family. Cell 72, 801-815.

Fantl, V., Richards, M.A., Smith, R., Lammie, G.A., Johnstone, G., Allen, D., Gregory, W., Peters, G., Dickson, C., and Barnes, D.M. (1990). Gene amplification on chromosome band 11q13 and oestrogen receptor status in breast cancer. Eur. J. Cancer 26, 423-429.

Fawell, S.E., Lees, J.A., White, R., and Parker, M.G. (1990a). Characterization and colocalization of steroid binding and dimerization activities in the mouse estrogen receptor. Cell 60, 953-962.

Fawell, S.E., White, R., Hoare, S., Sydenham, M., Page, M., and Parker, M.G. (1990b). Inhibition of estrogen receptor-DNA binding by the "pure" antiestrogen ICI 164,384 appears to be mediated by impaired receptor dimerization. Proc. Natl. Acad. Sci. USA 87, 6883-6887.

Fedi, P., Pierce, J.H., DiFiore, P.P., and Kraus, M.H. (1994). Efficient coupling with phosphatidylinositol 3-kinase, but not phospholipase Cg or GTPase-activating protein, distinguishes ErbB-3 signaling from that of other ErbB/EGFR family members. Mol. Cell. Biol. 14, 492-500.

Feinleib, M. (1968). Breast cancer and artificial menopause: A cohort study. J. Natl. Cancer Inst. 41, 315-329.

Felder, S., Miller, K., Moehren, G., Ullrich, A., Schlessinger, J., Hopkins, C.R. (1990). Kinase activity controls the sorting of the epidermal growth factor receptor within the multivesicular body. Cell 61, 623-634.

Fenig, E., Wieder, R., Paglin, S., Wang, H., Resaud, R., Haimovitz-Friedman, A., Fuks, Z., and Yahalom, J. (1997). Basic fibroblast growth factor confers growth inhibition and mitogen-activated protein kinase activation in human breast cancer cells. Clin. Cancer Res. 3, 135-142.

Ferguson, A.T., Lapidus, R.G., Baylin, S.B., and Davidson, N.E. (1995). Demethylation of the estrogen receptor gene in estrogen receptor-negative breast cancer cells can reactivate estrogen receptor gene expression. Cancer Res. 55, 2279-2283.

Ferrara, N., Carver-Moore, K., Chen, H., Dowd, M., Lu, L., O'Shea, K.S., Powell- Braxton, L., Hillan, K.J., and Moore, M.W. (1996). Heterozygous embryonic lethality induced by targeted inactivation of the VEGF gene. Nature 380, 439-442.

Ferrara, N., Winer, J., Burton, T., Rowland, A., Siegel, M., Phillips, H.S., Terrell, T., Keller, G.A., and Levinson, A.D. (1993). Expression of vascular endothelial growth factor does not promote transformation but confers a growth advantage in vivo to Chinese hamster ovary cells. J. Clin. Invest. 91, 160-170.

Ferrazzi, E., Cartei, G., Mattarazzo, R., and Fiorentino, M. (1977). Oestrogen-like effect of tamoxifen on vaginal epithelium [letter]. Br. Med. J. 1, 1351-1352.

Fidler, I.J. and Ellis, L.M. (1994). The implications of angiogenesis for the biology and therapy of cancer metastasis. Cell 79, 185-188.

Finzi, E., Fleming, T., Segatto, O., Pennington, C.Y., Bringman, T.S., Derynck, R., and Aaronson, S.A. (1987). The human transforming type α coding sequence is not a direct-acting oncogene when overexpressed in NIH 3T3 cells. Proc. Natl. Acad. Sci. USA 84, 3733-3737.

Fisher, D.A. and Lakshmanan, J. (1990). Metabolism and effects of epidermal growth factor and related growth factors in mammals. Endocrine Rev. 11, 418-442.

Foekens, J.A., Portengen, H., Janssen, M., and Klijn, J.G. (1989). Insulinlike growth factor-1 receptors and insulinlike growth factor-1-like activity in human primary breast cancer. Cancer 63, 2139-2147.

Folkman, J. (1995). Clinical applications of research on angiogenesis. N. Engl. J. Med. 333, 1757-1763.

Folkman, J. (1990). What is the evidence that tumors are angiogenesis dependent? J. Natl. Cancer Inst. 82, 1-5.

Folkman, J. and Haudenschild, C. (1981). Angiogenesis in vitro. Nature 288, 551-556.

Folkman, J. and Klagsbrun, M. (1987). Vascular physiology. A family of angiogenic peptides [news]. Nature 329, 671-672.

Folkman, J. and Shing, Y. (1992). Angiogenesis. J. Biol. Chem. 267, 10931-10934.

Fong, G.H., Rossant, J., Gertsenstein, M., Breitman, M.L. (1995). Role of the flt-1 receptor tyrosine kinase in regulating the assembly of vascular endothelium. Nature 376, 66-70.

Forough, R., Xi, Z., MacPhee, M., Friedman, S., Engleka, K.A., Sayers, T., Wiltrout, R.H., and Maciag, T. (1993). Differential transforming abilities of nonsecreted and secreted forms of human fibroblast growth factor-1. J. Biol. Chem. 268, 2960-2968.

Freter, C.E., Lippman, M.E., Cheville, A., Zinn, S., and Gelmann, E.P. (1988). Alterations in phosphoinositide metabolism associated with 17 β-estradiol and growth factor treatment of MCF-7 breast cancer cells. Mol. Endocrinol. 2, 159-166.

Friedman, S., Zhan, X., and Maciag, T. (1994). Mutagenesis of the nuclear localization sequence in FGF-1 alters protein stability but not mitogenic activity. Biochem. Biophys. Res. Commun. 198, 1203-1208.

Furr, B.J. and Jordan, V.C. (1984). The pharmacology and clinical uses of tamoxifen. Pharmacol. Ther. 25, 127-205.

Gaben, A.-M. and Mester, J. (1991). BALB/C mouse 3T3 fibroblasts expressing human estrogen receptor: Effect of estradiol on cell growth. Biochem. Biophys. Res. Commun. 176, 1473-1481.

Gao, G. and Goldfarb, M. (1995). Heparin can activate a receptor tyrosine kinase. EMBO J. 14, 2183-2190.

Garcia, M., Derocq, D., Freiss, G., and Rochefort, H. (1992). Activation of estrogen receptor transfected into a receptor-negative breast cancer cell line decreases the metastatic and invasive potential of the cells. Proc. Natl. Acad. Sci. USA 89, 11538- 11542.

Gasparini, G. and Harris, A.L. (1995). Clinical importance of the determination of tumor angiogenesis in breast carcinoma: Much more than a new prognostic tool. J. Clin. Oncol. 13, 765-782.

George, S.E., VanBerkun, M.F., Ono, T., Cook, R., Hanley, R.M., Putkey, J.A., and Means, A.R. (1990). Chimeric calmodulin-cardiac troponin C proteins differentially activate calmodulin target enzymes. J. Biol. Chem. 265, 9228-9235.

Gitay-Goren, H., Cohen, T., Tessler, S., Soker, S., Gengrinovitch, S., Rockwell, P., Klagsbrun, M., Levi, B.Z., and Neufeld, G. (1996). Selective binding of $VEGF_{121}$ to one of the three vascular endothelial growth factor receptors of vascular endothelial cells. J. Biol. Chem. 271, 5519-5523.

Givol, D. and Yayon, A. (1992). Complexity of FGF receptors: Genetic basis for structural diversity and functional specificity. FASEB J. 6, 3362-3369.

Godolphin, W., Elwood, J.M., and Spinelli, J.J. (1981). Estrogen receptor quantitation and staging as complementary prognostic indicators in breast cancer: A study of 583 patients. Int. J. Cancer 677-683.

Gomm, J.J., Smith, J., Ryall, G.K., Baillie, R., Turnbull, L., and Coombes, R.C. (1991). Localization of basic fibroblast growth factor and transforming growth factor beta 1 in the human mammary gland. Cancer Res. 51, 4685-4692.

Grand, R.J.A. and Owen, D. (1991). The biochemistry of Ras p21. Biochem. J. 279, 609- 631.

Green, P.J., Walsh, F.S., and Doherty, P. (1996). Promiscuity of fibroblast growth factor receptors. Bioessays 18, 639-646.

Greene, G.L., Sobel, N.B., King, W.J., and Jensen, E.V. (1984). Immunochemical studies of estrogen receptors. J. Steroid Biochem. 20, 51-56.

Guimond, S., Maccarana, M., Olwin, B.B., Lindahl, U., and Rapraeger, A.C. (1993). Activating and inhibitory heparin sequences for FGF-2 (basic FGF). Distinct requirements for FGF-1, FGF-2, and FGF-4. J. Biol. Chem. 268, 23906-23914.

Guiochon-Mante, I.A., Loosfelt, H., Lescop, P., Sa, r.S., Atger, M., and Perrot-Applanat, M., E., M. (1989). Mechanisms of nuclear localization of the progesterone receptor: evidence for interaction between monomers. Cell 57, 1147-1154.

Gullick, W.J., Love, S.B., Wright, C., Barnes, D.M., Gusterson, B., Harris, A.L., and Altman, D.G. (1991). c-*erb*B-2 protein overexpression in breast cancer is a risk factor in patients with involved and uninvolved lymph nodes. Br. J. Cancer 63, 434-438.

Guy, P.M., Platko, J.V., Cantley, L.C., Cerione, R.A., and Carraway, K.L.I. (1994). Insect cell-expressed p180^{erbB3} possesses an impaired tyrosine kinase activity. Proc. Natl. Acad. Sci. USA 91, 8132-8136.

Hack, N., Sue-A-Quan, A., Mills, G.B., and Skorecki, K.L. (1993). Expression of hyman tyrosine kinase-negative epidermal growth factor receptor amplifies signaling through endogenous murine epidermal growth factor receptor. J. Biol. Chem. 268, 26441-26446.

Halachmi, S., Marden, E., Martin, G., HacKay, H., Abbondanza, C., and Brown, M. (1994). Estrogen receptor-associated proteins: Possible mediators of hormone-induced transcription. Science 264, 1455-1458.

Haley, J., Whittle, N., Bennett, P., Kinchington, D., Ullrich, A., and Waterfield, M. (1987). The human EGF receptor gene: Structure of the 110 kb locus and identification of sequences regulating its transcription. Oncogene Res 1, 375-396.

Harris, A.L., Nicholson, S., Sainsbury, R., Wright, C., and Farndon, J.R. (1992). Epidermal growth factor receptor and other oncogenes as prognostic markers. J. Natl. Cancer Inst. Monogr. 11, 181-187.

Harris, J.R., Lippman, M.E., Veronesi, U., and Willett, W. (1980). Breast cancer (2). N. Eng. J. Med. 327, 390-398.

Harrison, S.C. (1991). A structural taxonomy of DNA-binding domains. Nature 353, 715- 719.

Hayman, M.J. and Enrietto, P.J. (1991). Cell transformation by the epidermal growth factor receptor and v-erbB. Cancer Cells 3, 302-307.

Heidecker, G., Huleihel, M., Cleveland, J.L., Kolch, W., Beck, T.W., Lloyd, P., Pawson, T., and Rapp, U.R. (1990). Mutational activation of c-*raf*-1 and definition of the minimal transforming sequence. Mol. Cell. Biol. 10, 2503-2512.

Higashiyama, S., Abraham, J.A., Miller, J., Fiddes, J.C., and Klagsbrun, M. (1991). A heparin-binding growth factor secreted by macrophagelike cells that is related to EGF. Science 251, 936-939.

Higashiyama, S., Lau, K., Besner, G.E., Abraham, J.A., and Klagsbrun, M. (1992). Structure of heparin-binding EGF-like growth factor: Multiple forms, primary structure, and glycosylation of the mature protein. J. Biol. Chem. 267, 6205-6212.

Hirayama, T. and Wynder, E.L. (1962). A study of the epidemiology of cancer of the breast. Cancer 15, 28-38.

Holmes, W.E., Sliwkowski, M.X., Akita, R.W., Henzel, W.J., Lee, J., Park, J.W., Yansura, D., Abadi, N., Raab, H., Lewis, G.D., Shepard, H.M., Kuang, W., Wood, W.I., Goeddel, D.V., and Vandlen, R.L. (1992). Identification of heregulin, a specific activator of p185*erb*B2. Science 256, 1205-1210.

Honegger, A., Dull, T.J., Bellot, F., Van Obberghen, E., Szapary, D., Schmidt, A., Ullrich, A., and Schlessinger, J. (1988a). Biological activities of EGF-receptor mutants with individually altered autophosphorylation sites. EMBO. J. 7, 3045-3052.

Honegger, A., Dull, T.J., Szapary, D., Komoriya, A., Kris, R., Ullrich, A., and Schlessinger, J. (1988b). Kinetic parameters of the protein tyrosine kinase activity of EGF-receptor mutants with individually altered autophosphorylation sites. EMBO. J. 7, 3053-3060.

Honegger, A.M., Kris, R.M., Ullrich, A., and Schlessinger, J. (1989). Evidence that autophosphorylation of solubilized receptors for epidermal growth factor is mediated by intermolecular cross-phosphorylation. Proc. Natl. Acad. Sci. USA 86, 925-929.

Horwitz, K.B., Jackson, T.A., Bain, D.L., Richer, J.K., Takimoto, G.S., and Tung, L. (1996). Nuclear receptor coactivators and corepressors. Mol. Endocrinol. 10, 1167-1177.

Hou, J., Kan, M., McKeehan, K., McBride, G., Adams, P., and McKeehan, W.L. (1991). Fibroblast growth factors receptors from liver vary in three structural domains. Science 251, 665-668.

Houck, K.A., Ferrara, N., Winer, J., Cachianes, G., Li, B., and Leung, D.W. (1991). The vascular endothelial growth factor family: Identification of a fourth molecular species and characterization of alternative splicing of RNA. Mol. Endocrinol. 5, 1806-1814.

Howe, L.R., Leevers, S.J., Gomez, N., Nakielny, S.l., Cohen, P., and Marshall, C.J. (1992). Activation of the MAP kinase pathway by the protein kinase Raf. Cell 71, 335- 342.

Howell, A., DeFriend, D., Robertson, J., Blamey, R., and Walton, P. (1995). Response to a specific antioestrogen (ICI 182,780) in tamoxifen-resisitant breast cancer. Lancet 345, 29-30.

Hulka, B.S., Liu, E.T., and Lininger, R.A. (1994). Steroid hormones and risk of breast cancer. Cancer 74, 1111-1124.

Hurwitz, D.R., Emanuel, S.L., Nathan, M.H., Sarver, N., Ullrich, A., Felder, S., Lax, I., and Schlessinger, J. (1991). EGF induces increased ligand binding affinity and dimerization of soluble epidermal growth factor (EGF) receptor extracellular domain. J. Biol. Chem. 266, 22035-22043.

Ignar-Trowbridge, D.M., Nelson, K.G., Bidwell, M.C., Curtis, S.W., Washburn, T.F., Mclachlan, J.A., and Korach, K.S. (1992). Coupling of dual signaling pathways: Epidermal growth factor action involves the estrogen receptor. Proc. Natl. Acad. Sci. USA 89, 4658-4662.

Ignar-Trowbridge, D.M., Pimentel, M., Parker, M.G., McLachlan, J.A., and Korach, K.S. (1996). Peptide growth factor cross-talk with the estrogen receptor requires the A/B domain and occurs independently of protein kinase C or estradiol. Endocrinology 137, 1735-1744.

Ignar-Trowbridge, D.M., Teng, C.T., Ross, K.A., Parker, M.G., Korach, K.S., and Mclachlan, J.A. (1993). Peptide growth factors elicit estrogen receptor-dependent transcriptional activation of an estrogen-responsive element. Mol. Endocrinol. 7, 992-998.

Imamura, T., Engleka, K., Zhan, X., Tokita, Y., Forough, R., Roeder, D., Jackson, A., Maier, J.A., Hla, T., and Maciag, T. (1990). Recovery of mitogenic activity of a growth factor mutant with a nuclear translocation sequence. Science 249, 1567-1570.

Ishii, H., Yoshida, T., Oh, H., Yoshida, S., and Terada, M. (1995). A truncated K-sam product lacking the distal carboxyl-terminal portion provides a reduced level of autophosphorylation and greater resistance against induction of differentiation. Mol. Cell. Biol. 15, 3664-3671.

Jaakkola, S., Salmikangas, P., Nylund, S., Partanen, J., Armstrong, E., Pyrhonen, S., Lehtovirta, P., and Nevanlinna, H. (1993). Amplification of FGFR4 gene in human breast and gynecological cancers. Int. J. Cancer 54, 378-382.

Jackson, A., Friedman, S., Zhan, X., Engleka, K.A., Forough, R., and Maciag, T. (1992). Heat shock induces the release of fibroblast growth factor 1 from NIH 3T3 cells. Proc. Natl. Acad. Sci. USA 89, 10691-10695.

Jackson, A., Tarantini, F., Gamble, S., Friedman, S., and Maciag, T. (1995). The release of fibroblast growth factor-1 from NIH 3T3 cells in response to temperature involves the function of cysteine residues. J. Biol. Chem. 270, 33-36.

Jacquemier, J., Adelaide, J., Parc, P., Penault-Llorca, F., Planche, J., deLapeyriere, O., and Birnbaum, D. (1994). Expression of the FGFR1 gene in human breast-carcinoma cells. Int. J. Cancer 59, 373-378.

Jarvis, M.F., Gessner, G.W., Martin, G.E., Jaye, M., Ravera, M.W., and Dionne, C.A. (1992). Characterization of [125I]acidic fibroblast growth factor binding to the cloned human fibroblast growth factor receptor, FGF-flg, on NIH 3T3 cell membranes: Inhibitory effects of heparin, pertussis toxin and guanine nucleotides. J. Pharmacol. Exp. Ther. 263, 253-263.

Jaye, M., Lyall, R.M., Mudd, R., Schlessinger, J., and Sarver, N. (1988). Expression of acidic fibroblast growth factor cDNA confers growth advantage and tumorigenesis to Swiss 3T3 cells. EMBO J. 7, 963-969.

Jaye, M., Schlessinger, J., and Dionne, C.A. (1992). Fibroblast growth factor receptor tyrosine kinases: Molecular analysis and signal transduction. Biochim. Biophys. Acta. 1135, 185-199.

Jensen, E.V. (1981). Hormone dependency of breast cancer. Cancer 47, 2319-2326.

Jiang, S.-Y. and Jordan, V.C. (1992). Growth regulation of estrogen receptor-negative breast cancer cells transfected with complementary DNAs for estrogen receptor. J. Natl. Cancer Inst. 84, 580-591.

Joel, P.B., Traish, A.M., and Lannigan, D.A. (1995). Estradiol and phorbol ester cause phosphorylation of serine 118 in the human estrogen receptor. Mol. Endocrinol. 9, 1041- 1052.

Johnson, D.E., Lu, J., Chen, H., Werner, S., and Williams, L.T. (1991). The human fibroblast growth factor receptor genes: A common structural arrangement underlies the mechanisms for generating receptor forms that differ in their third immunoglobulin domain. Mol. Cell. Biol. 11, 4627-4634.

Johnson, D.E., and Williams, L.T. (1993). Structural and functional diversity in the FGF receptor multigene family. Adv. Cancer Res. 60, 1-41 .

Johnston, S.R., Haynes, B.P., Smith, I.E., Jarman, M., Sacks, N.P., Ebbs, S.R., and Dowsett, M. (1993). Acquired tamoxifen resistance in human breast cancer and reduced intratumoral drug concentration. Lancet 342, 1521-1522.

Johnston, S.R., Saccani-Jotti, G., Smith, I.E., Salter, J., Newby, J., Coppen, M., Ebbs, S.R., and Dowsett, M. (1995). Changes in estrogen receptor, progesterone receptor, and pS2 expression in tamoxifen-resistant human breast cancer. Cancer Res. 55, 3331-3338.

Johnston, S.R., Lu, B., Dowsett, M., Liang, X., Kaufmann, M., Scott, G.K., Osborne, C.K., and Benz, C.C. (1997). Comparison of estrogen receptor DNA binding in untreated and acquired antiestrogen-resistant human breast tumors. Cancer Res. 57, 3723-3727.

Joneson, T., McDonough, M., Bar-Sagi, D., and Van Aelst, L. (1996). RAC regulation of actin polymerization and proliferation by a pathway distinct from Jun kinase. Science 274, 1374-1376.

Jordan, V.C. (1984). Biochemical pharmacology of antiestrogen action. Pharmacol. Rev. 36, 245-276.

Jordan, V.C., Collins, M.M., Rowsby, L., and Prestwich, G. (1977). A monohydroxylated metabolite of tamoxifen with potent antiestrogenic activity. J. Endocrinology 75, 305-316.

Jordan, V.C., Koch, R., Langan, S., and McCague, R. (1988). Ligand interaction at the estrogen receptor to program antiestrogen action: A study with nonsteroidal compounds in vitro. Endocrinology 122, 1449-1454.

Jordan, V.C. and Koerner, S. (1975). Tamoxifen (ICI 46,474) and the human carcinoma 8S oestrogen receptor. Eur. J. Cancer 11, 205-206.

Joukov, V., Pajusola, K., Kaipainen, A., Chilov, D., Lahtinen, I., Kukk, E., Saksela, O., Kalkkinen, N., and Alitalo, K. (1996). A novel vascular endothelial growth factor, VEGF- C, is a ligand for the Flt4 (VEGFR-3) and KDR (VEGFR-2) receptor tyrosine kinases. EMBO J. 15, 290-298

Kageyama, R., Merlino, G.T., Pastan, I. (1988). Epidermal growth factor (EGF) receptor gene transcription: Requirement for Sp1 and an EGF receptor specific factor. J. Biol. Chem. 263, 6329-6336.

Kan, M., Wang, F., Xu, J., Crabb, J.W., Hou, J., and McKeehan, W.L. (1993). An essential heparin-binding domain in the fibroblast growth factor receptor kinase. Science 259, 1918-1921.

Kandel, J., Bossy-Wetzel, E., Radvanyi, F., Klagsbrun, M., Folkman, J., and Hanahan, D. (1991). Neovascularization is associated with a switch to the export of bFGF in the multistep development of fibrosarcoma. Cell 66, 1095-1104.

Karey, K.P. and Sirbasku, D.A. (1988). Differential responsiveness of human breast cancer cell lines MCF-7 and T47D to growth factors and 17 β-estradiol. Cancer Res. 48, 4083-4092.

Karnik, P.S., Kulkarni, S., Liu, X.P., Budd, G.T., and Bukowski, R.M. (1994). Estrogen receptor mutations in tamoxifen-resistant breast cancer. Cancer Res. 54, 349-353.

Kashles, O., Szapary, D., Bellot, F., Ullrich, A., Schlessinger, J., and Schmidt, A. (1988). Ligand-induced stimulation of epidermal growth factor receptor mutants with altered transmembrane regions. Proc. Natl. Acad. Sci. USA 85, 9567-9571.

Kashles, O., Yarden, Y., Fischer, R., Ullrich, A., and Schlessinger, J. (1991). A dominant negative mutation suppresses the funciton of normal epidermal growth factor receptors by heterodimerization. Mol. Cell. Biol. 11, 1454-1463.

Kato, S., Endoh, H., Masuhiro, H., Kitamoto, T., Uchiyama, S., Sasaki, H., Mashushige, S., Gotoh, Y., Nishida, E., Kawashima, H., Metzger, D., and Chambon, P. (1995). Activation of the estrogen receptor through phosphorylation by mitogen-activated protein kinase. Science 270, 1491-1494.

Katoh, O., Tauchi, H., Kawaishi, K., Kimura, A., and Satow, Y. (1995). Expression of the vascular endothelial growth factor (VEGF) receptor gene, kdr, in hematopoietic cells and inhibitory effect of VEGF on apoptotic cell death caused by ionizing radiation. Cancer Res. 55, 5687-5692.

Katzenellenbogen, B.S., Kendra, K.L., Norman, M.J., and Berthois, Y. (1987). Proliferation, hormonal responsiveness, and estrogen receptor content of MCF-7 human breast cancer cells grown in the short-term and long-term absence of estrogens. Cancer Res. 47, 4355-4360.

Kavanaugh, W.M. and Williams, L.T. (1994). An alternative to SH2 domains for binding tyrosine-phosphorylated proteins. Science 266, 1862-1865.

Ke, Y., Fernig, D.G., Wilkinson, M.C., Winstanley, J.H., Smith, J.A., Rudland, P.S., and Barraclough, R. (1993). The expression of basic fibroblast growth factor and its receptor in cell lines derived from normal human mammary gland and a benign mammary lesion. J. Cell Sci. 106, 135-143.

Keck, P.J., Hauser, S.D., Krivi, G., Sanzo, K., Warren, T., Feder, J., and Connolly, D.T. (1989). Vascular permeability factor, an endothelial cell mitogen related to PDGF. Science 246, 1309-1312.

Keegan, K., Johnson, D.E., Williams, L.T., and Hayman, M.J. (1991). Isolation of an additional member of the fibroblast growth factor receptor family, FGFR-3. Proc. Natl. Acad. Sci. USA 88, 1095-1099.

Keyt, B.A., Nguyen, H.V., Berleau, L.T., Duarte, C.M., Park, J., Chen, H., and Ferrara, N. (1996). Identification of vascular endothelial growth factor determinants for binding KDR and FLT-1 receptors. Generation of receptor-selective VEGF variants by site- directed mutagenesis. J. Biol. Chem. 271, 5638-5646.

Khosravi-Far, R., Solski, P.A., Clark, G.J., Kinch, M.S., and Der, C.J. (1995). Activation of Rac1, RhoA, and mitogen-activated protein kinases is required for Ras transformation. Mol. Cell Biol. 15, 6443-6453.

Kiefer, P., Acland, P., Pappin, D., Peters, G., and Dickson, C. (1997). Competition between nuclear localization and secretory signals determines the subcellular fate of a single CUG-initiated form of FGF3. EMBO J. 13, 4126-4136.

Kiefer, P. and Dickson, C. (1995). Nucleolar association of fibroblast growth factor 3 via specific sequence motifs has inhibitory effects on cell growth. Mol. Cell. Biol. 15, 4364- 4374.

Kim, K.J., Li, B., Winer, J., Armanini, M., Gillett, N., Phillips, H.S., and Ferrara, N. (1993). Inhibition of vascular endothelial growth factor-induced angiogenesis suppresses tumor growth in vivo. Nature 382, 841-844 .

Kimura, H., Fischer, W.H., and Schubert, D. (1990). Structure, expression and function of a schwannoma-derived growth factor. Nature 348, 257-260.

King, C.R., Borrello, I., Bellot, F., Comoglio, P., and Schlessinger, J. (1988). EGF binding to its receptor triggers a rapid tyrosine phosphorylation of the *erb*B-2 protein in the mammary tumor cell line SK-BR-3. EMBO J. 7, 1647-1651.

King, C.R., Kraus, M.H., Williams, L.T., Merlino, G.T., and Pastan, I.H. (1985). Human tumor cell lines with EGF receptor gene amplification in the absence of aberrantly sized mRNA. Nucleic Acids Res. 8477-8486.

Kinne, D.W. (1990). Management of male breast cancer. Oncology 5, 45-47.

Klagsbrun, M. and Baird, A. (1991). A dual receptor system is required for basic fibroblast growth factor activity. Cell 67, 229-231.

Klint, P., Kanda, S., and Claesson-Welsh, L. (1995). Shc and a novel 89-kda component couple to the Grb2-Sos complex in fibroblast growth factor-2-stimulated cells. J. Biol. Chem. 270, 23337-23344.

Kobrin, M.S., Yamanaka, Y., Friess, H., Lopez, M.E., and Korc, M. (1993). Aberrant expression of type I fibroblast growth factor receptor in human pancreatic adenocarcinomas. Cancer Res. 53, 4741-4744.

Koenders, P.G., Beex, L.V.A.M., Geurts-Moespot, A., Heuvel, J.J.T.M., Kienhuis, C.B.M., and Benraad, T.J. (1991). Epidermal growth factor receptor negative tumors are predominantly confined to the subgroup of estradiol receptor positive human primary breast cancers. Cancer Res. 51, 4544-4548.

Koffman, B., Modarress, K.J., Beckerman, T., and Bashirelahi, N. (1991). Evidence for involvement of tyrosine in estradiol binding by rat uterus estrogen receptor. J. Steroid Biochem. Mol. Biol. 38, 135-139.

Kouhara, H., Hadari, Y.R., Spivak-Kroizman, T., Schilling, J., Bar-Sagi, D., and Schlessinger, J. (1997). A lipid-anchored Grb2-binding protein that links FGF-receptor activation to the Ras-MAPK signaling pathway. Cell 89, 693-702.

Kosako, H., Gotoh, Y., Matsuda, S., Ishikawa, M., and Nishida, E. (1992). *Xenopus* MAP kinase activator is a serine/threonine/tyrosine kinase activated by threonine phosphorylation. EMBO J. 11, 2903-2908.

Kraus, M.H., Fedi, P., Starks, V., Muraro, R., and Aaronson, S.A. (1993). Demonstration of ligand-dependent signaling by the *erb*B-3 tyrosine kinase and its constitutive activation in human breast tumor cells. Proc. Natl. Acad. Sci. USA 90, 2900-2904.

Kris, R.M., Lax, I., Gullick, W., Waterfield, M.D., Ullrich, A., Fridkin, M., and Schlessinger, J. (1985). Antibodies against a synthetic peptide as a probe for the kinase activity of the avian EGF receptor and v-erbB protein. Cell 40, 619-625.

Kumar, V. and Chambon, P. (1988). The estrogen receptor binds tightly to its responsive element as a ligand-induced homodimer. Cell 55, 145-156.

Kumar, V., Green, S., Stack, G., Berry, M., Jun, J.-R., and Chambon, P. (1987). Functional domains of the human estrogen receptor. Cell 51, 941-951.

Kurebayashi, J., McLeskey, S.W., Johnson, M.D., Lippman, M.E., Dickson, R.B., and Kern, F.G. (1993). Quantitative demonstration of spontaneous metastasis by MCF-7 human breast cancer cells cotransfected with fibroblast growth factor 4 and LacZ. Cancer Res 53, 2178-2187.

Kyriakis, J.M., App, H., Zhang, X.-F., Banerjee, P., Brautigan, D.L., Rapp, U.R., and Avruch, J. (1992). Raf-1 activates MAP kinase-kinase. Nature 358, 417-421.

Kyriakis, J.M., Banerjee, P., Nikolakaki, E., Dai, T., Rubie, E.A., Ahmad, M.F., Avruch, J., and Woodgett, J.R. (1994). The stress-activated protein kinase subfamily of c-*jun* kinases. Nature 369, 156-160.

Lahooti, H., White, R., Danielian, P.S., and Parker, M.G. (1994). Characterization of ligand-dependent phosphorylation of the estrogen receptor. Mol. Endocrinol. 8, 182-188.

Lammie, G.A., Fantl, V., Smith, R., Schuuring, E., Brookes, S., Michalides, R., Dickson, C., Arnold, A., and Peters, G. (1991). D11S287, a putative oncogene on chromosome 11q13, is amplified and expressed in squamous cell and mammary carcinomas and linked to BCL-1. Oncogene 6, 439-444.

Landel, C.C., Kushner, P.J., and Greene, G.L. (1994). The interaction of human estrogen receptor with DNA is modulated by receptor-associated proteins. Mol. Endocrinol. 8, 1407-1419.

Landgren, E., Blume-Jensen, P., Courtneidge, S.A., and Claesson-Welsh, L. (1995). Fibroblast growth factor receptor-1 regulation of Src family kinases. Oncogene 10, 2027- 2035.

Langdon, S.P., Crew, A.J., Ritchie, A.A., Muir, M., Wakeling, A., Smyth, J.F., and Miller, W.R. (1994). Growth inhibition of oestrogen recpetor-positive human ovarian carcinoma by anti-oestrogens in vitro and in a xenograft model. Eur. J. Cancer 30A, 682- 686.

Lange-Carter, C.A. and Johnson, G.L. (1994). Ras-dependent growth factor regulation of MEK kinase in PC12 cells. Science 265, 1458-1461.

Lax, I., Burgess, W.H., Bellot, F., Ullrich, A., Schlessinger, J., and Givol, D. (1988). Localization of a major receptor-binding comain for epidermal growth factor by affinity labeling. Mol. Cell. Biol. 8, 1831-1834.

Le Goff, P., Montano, M.M., Schodin, D.J., and Katzenellenbogen, B.S. (1994). Phosphorylation of the human estrogen receptor. Identification of hormone-regulated sites and examination of their influence on transcriptional activity. J. Biol. Chem. 269, 4458-4466.

Lee, C.S.L., Hall, R.E., Alexander, I.E., Koga, M., Shine, J., and Sutherland, R.L. (1990). Inverse relationship between estrogen receptor and epidermal growth factor receptor levels in human breast cancer cell lines. Growth Factors 3, 97-103.

Leevers, S.J., Paterson, H.F., and Marshall, C.J. (1994). Requirement for Ras in Raf activation is overcome by targeting raf to the plasma membrane. Nature 369, 411-414.

Lehtola, L., Partanen, J., Sistonen, L., Korhonen, J., Warri, A., Harkönen, P., Clarke, R., and Alitalo, K. (1992). Analysis of tyrosine kinase mRNAs including four FGF receptor mRNAs expressed in MCF-7 breast-cancer cells. Int. J. Cancer 50, 598-603.

LeJeune, S., Leek, R., Horak, E., Plowman, G., Greenall, M., and Harris, A.L. (1993). Amphiregulin, epidermal growth factor receptor, and estrogen receptor expression in human primary breast cancer. Cancer Res. 53, 3579-3602.

Lemoine, N.R., Staddon, S., Dickson, C., Barnes, D.M., and Gullick, W.J. (1990). Absence of activating transmembrane mutations in the c-erbB-2 protooncogene in human breast cancer. Oncogene 5, 237-239.

Leung, D.W., Cachianes, G., Kuang, W., Goeddel, D.V., and Ferrara, N. (1989). Vascular endothelial growth factor is a secreted angiogenic mitogen. Science 246, 1306- 1309.

Levenson, A.S. and Jordan, V.C. (1994). Transfection of human estrogen receptor (ER) cDNA into ER-negative mammalian cell lines. J. Steroid Biochem. Molec. Biol. 51, 229- 239.

Lewis, S., Locker, A., Todd, J.H., Bell, J.A., Nicholson, R., Elston, C.W., Blamey, R.W., and Ellis, I.O. (1990). Expression of epidermal growth factor receptor in breast carcinoma. J. Clin. Pathology 43, 385-389.

Li, N., Batzer, A., Daly, R., Yajnik, V., Skolnik, E., Chardin, P., Bar-Sagi, D., Margolis, B., and Schlessinger, J. (1993). Guanine-nucleotide-releasing factor hSos1 binds to Grb2 and links receptor tyrosine kinases to Ras signalling. Nature 363, 85-88.

Li, P., Wood, K., Mamon, H., Haser, W., and Roberts, T. (1991). Raf-1: A kinase currently without a cause but not lacking in effects. Cell 64, 479-482.

Li, S. and Shipley, G.D. (1991). Expression of multiple species of basic fibroblast growth factor mRNA and protein in normal and tumor-derived mammary epithelial cells in culture. Cell Growth Differ. 2, 195-202.

Lin, A., Minden, A., Martinetto, H., Claret, F.X., Lange-Carter, C., Mercurio, F., Johnson, G.L., and Karin, M. (1995). Identification of a dual specificity kinase that activates the Jun kinases and p38-Mpk2. Science 268, 286-290.

Liotta, L.A. and Stracke, M.L. (1988). Tumor invasion and metastases: Biochemical mechanisms. Cancer Treat. Res. 40, 223-238.

Liotta, L.A., Steeg, P.S., and Stetler-Stevenson, W.G. (1991). Cancer metastasis and angiogenesis: An imbalance of positive and negative regulation. Cell 64, 327-336 .

Lippman, M.E. and Dickson, R.B. (1989). Mechanisms of normal and malignant breast epithelial growth regulation. J. Steroid Biochem. 34, 107-121.

Lippman, M.E., Dickson, R.B., Gelmann, E.P., Rosen, N., Knabbe, C., Bates, S., Bronzert, D., Huff, K.K., and Kasid, A. (1987). Growth regulation of human breast carcinoma occurs through regulated growth factor secretion. J. Cell. Biochem. 35, 1-16.

Lipsett, M.B. (1975). Hormones, nutrition, and cancer. Cancer Res. 35, 3359-3361.

Liscia, D.S., Merlo, G.R., Garrett, C., French, D., Mariani-Costantini, R., and Callahan, R. (1989). Expression of int-2 mRNA in human tumors amplified at the int-2 locus. Oncogene 4, 1219-1224.

Liu, B., Earl, H.M., Baban, D., Shoaibi, M., Fabra, A., Kerr, D.J., and Seymour, L.W. (1995a). Melanoma cell lines express VEGF receptor KDR and respond to exogenously added VEGF. Biochem. Biophys. Res. Commun. 217, 721-727.

Liu, Y., El-Ashry, D., Chen, D., Ding, I.Y.F., and Kern, F.G. (1995b). MCF-7 breast cancer cells overexpressing transfected c-*erb*B-2 have and in vitro growth advantage in estrogen-depleted conditions and reduced estrogen dependence and tamoxifen sensitivity in vivo. Breast Cancer Res. Treat. 34, 97-117.

Livenh, E., Reiss, N., Berent, E., Ullrich, A., and Schlessinger, J. (1987). An insertional mutant of epidermal growth factor receptor allows dissection of diverse receptor functions. EMBO J. 6, 2669-2676.

Livneh, E., Prywes, R., Kashles, O., Reiss, N., Sasson, I., Mory, Y., Ullrich, A., and Schlessinger, J. (1986). Reconstitution of human epidermal growth factor receptors and its deletion mutants in cultured hamster cells. J. Biol. Chem. 261, 12490-12497.

Logan, A. and Logan, S.D. (1991). Studies on the mechanisms of signalling and inhibition by pertussis toxin of fibroblast growth factor-stimulated mitogenesis in Balb/c 3T3 cells. Cell. Signal. 3, 215-223.

Lowenstein, E.J., Daly, R.J., Batzer, A.G., Li, W., Margolis, B., Lammers, R., Ullrich, A., Skolnik, E.Y., Bar-Sagi, D., and Schlessinger, J. (1992). The SH2 and SH3 domain- containing protein Grb2 links receptor tyrosine kinases to Ras signaling. Cell 70, 431-442.

Luqmani, Y.A., Bansal, G.S., Mortimer, C., Buluwela, L., and Coombes, R.C. (1996). Expression of FGFR2 BEK and K-SAM mRNA variants in normal and malignant human breast. Eur. J. Cancer 32A, 518-524.

Luqmani, Y.A., Graham, M., and Coombes, R.C. (1992). Expression of basic fibroblast growth factor, FGFR1 and FGFR2 in normal and malignant human breast, and comparison with other normal tissues. Br. J. Cancer 66, 273-280.

Luqmani, Y.A., Mortimer, C., Yiangou, C., Johnston, C.L., Bansal, G.S., Sinnett, D., Law, M., and Coombes, R.C. (1995). Expression of 2 variant forms of fibroblast growth factor receptor 1 in human breast. Int. J. Cancer 64, 274-279.

Lykkesfeldt, A.E., Larsen, S.S., and Briand, P. (1995). Human breast cancer cell lines resistant to pure anti-estrogens are sensitive to tamoxifen treatment. Int. J. Cancer 61, 529-534.

Macaulay, V.M., Fox, S.B., Zhang, H., Whitehouse, R.M., Leek, R.D., Gatter, K.C., Bicknell, R., and Harris, A.L. (1997). Breast cancer angiogenesis and tamoxifen resistance. Endocrine-Related Cancer 2, 97-103.

Macdonald, S.G., Crews, C.M., Wu, L., Driller, J., Clark, R., Erikson, R.L., and McCormick, F. (1993). Reconstitution of the Raf-1-MEK-ERK signal transduction pathway in vitro. Mol. Cell. Biol. 13, 6615-6620.

Maciag, T., Zhan, X., Garfinkel, S., Friedman, S., Prudovsky, I., Jackson, A., Wessendorf, J., Hu, X., Gamble, S., and Shi, J. (1994). Novel mechanisms of fibroblast growth factor 1 function. Recent Prog. Horm. Res. 49, 105-123.

MacNichol, A.M., Muslin, A.J., and Williams, L.T. (1993). Raf-1 kinase is essential for early *Xenopus* development and mediates the induction of mesoderm by FGF. Cell 73, 571-583.

Mader, S., Kumar, V., de Verneuil, H., and Chambon, P. (1989). Three amino acids of the oestrogen receptor are essential to its ability to distinguish and oestrogen- from a glucocorticoid-responsive element. Nature 338, 271-274.

Maher, P.A. (1996). Nuclear translocation of fibroblast growth factor (FGF) receptors in response to FGF-2. J. Cell Biol. 134, 529-536.

Mahfoudi, A., Roulet, E., Dauvois, S., Parker, M.G., and Wahli, W. (1995). Specific mutations in the estrogen receptor change the properties of anitestrogens to full agonists. Proc. Natl. Acad. Sci. USA 92, 4206-4210.

Maminta, M.L., Molteni, A., and Rosen, S.T. (1991). Stable expression of the human estrogen receptor in HeLa cells by infection: Effect of estrogen on cell proliferation and c-*myc* expression. Mol. Cell. Endo. 78, 61-69.

Mandriota, S.J., Seghezzi, G., Vassalli, J.D., Ferrara, N., Wasi, S., Mazzieri, R., Mignatti, P., and Pepper, M.S. (1995). Vascular endothelial growth factor increases urokinase receptor expression in vascular endothelial cells. J. Biol. Chem. 270, 9709-9716.

Manni, A., Wright, C., Badger, B., Bartholomew, M., Herlyn, M., Mendelsohn, J., Masui, H., and Demers, L. (1990). Role of transforming growth factor-α-related peptides in the autocrine/paracrine control of experimental breast cancer growth in vitro by estradiol, prolactin, and progesterone. Breast Cancer Res. Treat. 15, 73-83.

Mansour, S.J., Matten, W.T., Hermann, A.S., Candia, J.M., Rong, S., Fukasawa, K., Vande Woude, G.F., and Ahn, N.G. (1994). Transformation of mammalian cells by constitutively active MAP kinase kinase. Science 265, 966-970.

Marais, R., Light, Y., Paterson, H.F., and Marshall, C.J. (1995). Ras recruits Raf-1 to the plasma membrane for activation by tyrosine phosphorylation. EMBO J. 14, 3136-3145.

Marchionni, M.A., Goodearl, A.D., Chen, M.S., Bermingham-McDonogh, O., Kirk, C., Hendricks, M., Danehy, F., Misumi, D., Sudhalter, J., and Kobayashi, K. (1993). Glial growth factors are alternatively spliced *erbB2* ligands expressed in the nervous system [see comments]. Nature 362, 312-318.

Marshall, E. (1993). A statistical portrait of breast cancer. Science 259, 619.

Martinez-Lacaci, I., Saceda, M., Plowman, G.D., Johnson, G.R., Normanno, N., Salomon, D.S., and Dickson, R.B. (1995). Estrogen and phorbol esters regulate amphiregulin expression by two separate mechanisms in human breast cancer cell lines. Endocrinology 136, 3983-3992.

Massague, J. (1993). Transforming growth factor α. J. Biol. Chem. 256, 21393-21396.

McCormick, F. (1993). How receptors turn Ras on. Nature 363, 15-16.

McLeskey, S.W., Ding, I.Y., Lippman, M.E., and Kern, F.G. (1994). MDA-MB-134 breast carcinoma cells overexpress fibroblast growth factor (FGF) receptors and are growth-inhibited by FGF ligands. Cancer Res. 54, 523-530.

McLeskey, S.W., Kurebayashi, J., Honig, S.F., Zwiebel, J., Lippman, M.E., Dickson, R.B., and Kern, F.G. (1993). Fibroblast growth factor 4 transfection of MCF-7 cells produces cell lines that are tumorigenic and metastatic in ovariectomized or tamoxifen-treated athymic nude mice. Cancer Res. 53, 2168-2177.

McLeskey, S.W., Zhang, L., El-Ashry, D., Trock, B.J., Lopez, C.A., Kharbana, S., Tobias, C.A., Lorant, L.A., Hannum, R.S., Dickson, R.B., and Kern, F.G. (1998). Tamoxifen-resistant FGF-transfected MCF-7 cells are crossresistant in vivo to the antiestrogen ICI 182,780 and two aromatase inhibitors. Clin. Cancer Res. 4, 697-712.

Melnyk, O., Shuman, M.A., and Kim, K.J. (1996). Vascular endothelial growth factor promotes tumor dissemination by a mechanism distinct from its effect on primary tumor growth. Cancer Res. 56, 921-924.

Menedez-Botet, C.J. and Schwartz, M.K. (1993). Estrogen and progesterone receptor proteins in patients with breast cancer. Adv. Clin. Chem. 30, 185-225.

Migliaccio, A., Castoria, G., de Falco, A., Di Domenico, M., Galdiero, M., Nola, E., Chambon, P., and Auricchio, F. (1991). In vitro phosphorylation and hormone binding activation of the synthetic wild type human estradiol receptor. J. Steroid Biochem. Mol. Biol. 38, 407-413.

Migliaccio, A., Di Domenico, M., Green, S., de Falco, A., Kajtaniak, E.L., Blasi, F., Chambon, P., and Auricchio, F. (1989). Phosphorylation on tyrosine of in vitro synthesized human estrogen receotpr activates its hormone binding. Mol. Endocrinol. 3, 1061-1069.

Migliaccio, A., Pagano, M., De Goeij, C.C., Di Domenico, M., Castoria, G., Sluyser, M., and Auricchio, F. (1992). Phosphorylation and estradiol binding of estrogen receptor in hormone-dependent and hormone-independent GR mouse mammary tumors. Int. J. Cancer 51, 733-739.

Mignatti, P., Morimoto, T., and Rifkin, D.B. (1991). Basic fibroblast growth factor released by single, isolated cells stimulates their migration in an autocrine manner. Proc. Natl. Acad. Sci. USA 88, 11007-11011.

Mignatti, P. and Rifkin, D.B. (1991). Release of basic fibroblast growth factor, an angiogenic factor devoid of secretory signal sequence: A trivial phenomenon or a novel secretion mechanism. J. Cell. Biochem. 47, 201-207.

Miki, T., Bottaro, D.P., Fleming, T.P., Smith, C.L., Burgess, W.H., Chan, A.M., and Aaronson, S.A. (1992). Determination of ligand-binding specificity by alternative splicing: two distinct growth factor receptors encoded by a single gene. Proc. Natl. Acad. Sci. USA 89, 246-250.

Millauer, B., Longhi, M.P., Plate, K.H., Shawver, L.K., Risau, W., Ullrich, A., and Strawn, L.M. (1996). Dominant-negative inhibition of FLk-1 suppresses the growth of many tumor types in vivo. Cancer Res. 56, 1615-1620.

Millauer, B., Shawver, L.K., Plate, K.H., Risau, W., and Ullrich, A. (1994). Glioblastoma growth inhibited in vivo by a dominant-negative Flk-1 mutant. Nature 367, 576-579.

Millauer, B., Wizigmann-Voos, S., Schnurch, H., Martinez, R., M:ller, N.P., Risau, W., and Ullrich, A. (1993). High affinity VEGF binding and developmental expression suggest Flk-1 as a major regulator of vasculogenesis and angiogenesis. Cell 72, 835-846.

Minden, A., Lin, A., Claret, F.X., Abo, A., and Karin, M. (1995). Selective activation of the JNK signaling cascade and c-Jun transcriptional activity by the small GTPases Rac and Cdc42Hs. Cell 81, 1147-1157.

Minden, A., Lin, A., McMahon, M., Lange-Carter, C., Derijard, B., Davis, R.J., Johnson, G.L., and Karin, M. (1994a). Differential activation of ERK and JNK mitogen-activated protein kinases by Raf-1 and MEKK. Science 266, 1719-1723.

Minden, A., Lin, A., Smeal, T., Derijard, B., Cobb, M., Davis, R., and Karin, M. (1994b). c-Jun N-terminal phosphorylation correlates with activation of the JNK subgroup but not the ERK subgroup of mitogen-activated protein kinases. Mol. Cell. Biol. 14, 6683-6688.

Miyamoto, M., Naruo, K., Seko, C., Matsumoto, S., Kondo, T., and Kurokawa, T. (1993). Molecular cloning of a novel cytokine cDNA encoding the ninth member of the fibroblast growth factor family, which has a unique secretion property. Mol. Cell. Biol. 13, 4251-4259.

Mohammadi, M., Dikic, I., Sorokin, A., Burgess, W.H., Jaye, M., and Schlessinger, J. (1996). Identification of six novel autophosphorylation sites on fibroblast growth factor receptor 1 and elucidation of their importance in receptor activation and signal transduction. Mol. Cell. Biol. 16, 977-989.

Mohammadi, M., Dionne, C.A., Li, W., Li, N., Spivak, T., Honegger, A.M., Jaye, M., and Schlessinger, J. (1992). Point mutation in FGF receptor eliminates phosphatidylinositol hydrolysis without affecting mitogenesis. Nature 358, 681-684.

Montgomery, R.B., Moscatello, D.K., Wong, A.J., Cooper, J.A., and Stahl, W.L. (1995). Differential modulation of mitogen-activated protein (MAP) kinase/extracellular signal- related kinase and MAP kinase activities by a mutant epidermal growth factor receptor. J. Biol. Chem. 270, 30562-30566.

Moodie, S.A., Willumsen, B.M., Weber, M.J., and Wolfman, A. (1993). Complexes of Ras-GTP with Raf-1 and mitogen-activated protein kinase kinase. Science 260, 1658- 1661.

Mori, T., Morimoto, T., Komaki, K., and Monden, Y. (1991). Comparison of estrogen receptor and epidermal growth factor receptor content of primary and involved nodes in human breast cancer. Cancer 68, 532-537.

Moriai, T., Kobrin, M.S., Hope, C., Speck, L., and Korc, M. (1994). A variant epidermal growth factor receptor exhibits altered type a transforming growth factor binding and transmembrane signaling. Proc. Natl. Acad. Sci. USA 91, 10217-10221.

Morrison, D.K., Kaplan, D.R., Rapp, U., and Roberts, T.M. (1988). Signal transduction from membrane to cytoplasm: Growth factors and membrane-bound oncogene products increase Raf-1 phosphorylation and associated protein kinase activity. Proc. Natl. Acad. Sci. USA 85, 8855-8859.

Moscatelli, D., Presta, M., and Rifkin, D. (1986). Purification of a factor from human placenta that stimulates capillary endothelial cell protease production, DNA synthesis, and migration. Proc. Natl. Acad. Sci. USA 83, 2091-2095.

Moscatello, D.K., Holgado-Madruga, M., Godwin, A.K., Ramirez, G., Gunn, G., Zoltick, P.W., Biegel, J.A., Hayes, R.L., and Wong, A.J. (1995). Frequent expression of a mutant epidermal growth factor receptor in multiple human tumors. Cancer Res. 55, 5536-5539.

Murgue, B., Tsunekawa, S., Rosenberg, I., deBeaumont, M., and Podolsky, D.K. (1994). Identification of a novel variant form of fibroblast growth factor receptor 3 (FGFR3 IIIb) in human colonic epithelium. Cancer Res. 54, 5206-5211.

Murphy, L., Murphy, L.J., Dubik, D., Bell, G., and Shiu, R. (1988). Epidermal growth factor gene expression in human breast cancer cells: Regulation of expression by progestins. Cancer Res. 48, 4555-4560.

Muthukrishnan, L., Warder, E., and McNeil, P.L. (1991). Basic fibroblast growth factor is efficiently released from a cytolsolic storage site through plasma membrane disruptions of endothelial cells. J. Cell Physiol. 148, 1-16.

Neilson, K.M. and Friesel, R.E. (1995). Constitutive activation of fibroblast growth factor receptor-2 by a point mutation associated with crouzon syndrome. J. Biol. Chem. 270, 26037-26040.

Neufeld, G., Mitchell, R., Ponte, P., and Gospodarowicz, D. (1988). Expression of human basic fibroblast growth factor cDNA in baby hamster kidney-derived cells results in autonomous cell growth. J. Cell Biol. 106, 1385-1394.

Newby, J.C., Johnston, S.R.D., Smith, I.E. and Dowsett, M. (1997). Expression of epidermal growth factor receptor and c-erbB2 during the development of tamoxifen resistance in human breast cancer,. Clin. Cancer Res. 3, 1643-1651.

Nicholson, S., Halcrow, P., Sainsbury, J.R.C., Angus, B., Chambers, P., Farndon, J.R., and Harris, A.L. (1988). Epidermal growth factor receptor (EGFr) status associated with failure of primary endocrine therapy in elderly postmenopausal patients with breast cancer. Br. J. Cancer 58, 810-814.

Nicholson, S., Richard, J., Sainsbury, C., Halcrow, P., Kelly, P., Angus, B., Wright, C., Henry, J., Farndon, J.R., and Harris, A.L. (1991). Epidermal growth factor receptor (EGFr); results of a 6-year follow-up study in operable breast cancer with emphasis on the node negative subgroup. Br. J. Cancer 63, 146-150.

Nicholson, S., Sainsbury, J.R.C., Halcrow, P., Chambers, P., Farndon, J.R., and Harris, A.L. (1989). Expression of epidermal growth factor receptors associated with lack of response to endocrine therapy in recurrent breast cancer. Lancet 1, 182-184.

Nicholson, S., Wright, C., Sainsbury, J.R.C., Halcrow, P., Kelly, P., Angus, B., Farndon, J.R., and Harris, A.L. (1990). Epidermal growth factor receptor (EGFr) as a marker for poor prognosis in node-negative breast cancer patients: *neu* and tamoxifen failure. J. Steroid Biochem. Molec. Biol. 37, 811-814.

Nishikawa, R., Ji, X.-D., Harmon, R.C., Lazar, C.S., Gill, G.N., Cavenee, W.K., and Huang, H.-J.S. (1994). A mutant epidermal growth factor receptor common in human glioma confers enhanced tumorigenicity. Proc. Natl. Acad. Sci. USA 91, 7727-7731.

Normanno, N., Cheng-Fing, Q., Gullick, W.J., Persico, M.G., Yarden, Y., Wen, D., Plowman, G.D., Kenney, N., Johnson, G.R., Kim, N., Brandt, R., Martinez-Lacaci, I., Dickson, R.B., and Salomon, D.S. (1993). Expression of amphiregulin, cripto-1, and heregulin in human breast cancer cells. Int. J. Oncol. 2, 903-911.

O'Connell, P., Pekkel, V., Fuqua, S., Osborne, C.K., and Allred, D.C. (1994). Molecular genetic studies of early breast cancer evolution. Breast Cancer Res. Treat. 32, 5-12.

O'Malley, B.W. (1992). The steroid receptor superfamily: More exictement predicted for the future. Mol. Endocrinol. 4, 363-369.

Oberg, C., Waltenberger, J., Claesson-Welsh, L., and Welsh, M. (1994). Expression of protein tyrosine kinases in islet cells: Possible role of the flk-1 receptor for beta-cell maturation from duct cells. Growth Factors 10, 115-126.

Ogasawara, M. and Sirbasku, D.A. (1988). A new serum-free method for measuring growth factor activities for human breast cancer cells in culture. In Vitro Cell. Dev. Biol. 24, 911-920.

Okada, S., Yamauchi, K., and Pessin, J.E. (1995). Shc isoform-specific tyrosine phosphorylation by the insulin and epidermal growth factor receptors. J. Biol. Chem. 270, 20737-20741.

Olofsson, B., Pajusola, K., Kaipainen, A., von Euler, G., Joukov, V., Saksela, O., Orpana, A., Pettersson, R.F., Alitalo, K., and Eriksson, U. (1996a). Vascular endothelial growth factor B, a novel growth factor for endothelial cells. Proc. Natl. Acad. Sci. USA 93, 2576- 2581.

Olofsson, B., Pajusola, K., von Euler, G., Chilov, D., Alitalo, K., and Eriksson, U. (1996b). Genomic organization of the mouse and human genes for vascular endothelial growth factor B (VEGF-B) and characterization of a second splice isoform. J. Biol. Chem. 271, 19310-19317.

Onate, S.A., Tsai, S.Y., Tsai, M.-J., and O'Malley, B.W. (1995). Sequence and characterization of a coactivator for the steroid hormone receptor superfamily. Science 270, 1354-1371.

Ong, S.H., Goh, K.C., Lim, Y.P., Low, B.C., Klint, P., Claesson-Welsh, L., Cao, X., tan, Y.H., and Guy, G.R. (1996). Suc1-associated neurotrophic factor target (SNT) protein is a major FGF-stimulated tyrosine phosphorylated 90-kDa protein, which binds to the SH2 domain of GRB2. Biochem. Biophys. Res. Commun. 225, 1021-1026.

Ornitz, D.M., Xu, J., Colvin, J.S., McEwen, D.G., MacArthur, C.A., Coulier, F., Gao, G., and Goldfarb, M. (1996). Receptor specificity of the fibroblast growth factor family. J. Biol. Chem. 271, 15292-15297.

Osborne, C.K. and Arteaga, C.L. (1990). Autocrine and paracrine growth regulation of breast cancer: Clinical implications. Breast Cancer Res. Treat. 15, 3-11.

Osborne, C.K., Coronado, E., Allred, D.C., Wiebe, V., and DeGregorio, M. (1991). Acquired tamoxifen resistance: Correlation with reduced breast tumor levels of tamoxifen and isomerizaion of trans-4-hydorxytamoxifen. J. Natl. Cancer Inst. 83, 1477-1482.

Osborne, C.K., Coronado-Heinsohn, E.B., Hilsenbeck, S.G., McCue, B.L., Wakeling, A.E., McClelland, R.A., Manning, D.L., and Nicholson, R.I. (1995). Comparison of the effects of a pure steroidal antiestrogen with those of tamoxifen in a model of human breast cancer. J. Natl. Cancer Inst. 87, 746-750.

Ottaviano, Y.L., Issa, J.-P., Parl, F.F., Smith, H.S., Baylin, S.B., and Davidson, N.E. (1994). Methylation of the estrogen receptor gene CpG island marks loss of estrogen receptor expression in human breast cancer cells. Cancer Res. 54, 2552-2555.

Parker, M.G. (1993). Action of "pure" antiestrogens in inhibiting estrogen receptor action. Breast Cancer Res. Treat. 26, 131-137.

Partanen, J., Makela, T.P., Eerola, E., Korhonen, J., Hirvonen, H., Claesson-Welsh, L., and Alitalo, K. (1991). FGFR-4, a novel acidic fibroblast growth factor receptor with a distinct expression pattern. EMBO J. 10, 1347-1354.

Partanen, J., Vainikka, S., Korhonen, J., Armstrong, E., and Alitalo, K. (1992). Diverse receptors for fibroblast growth factors. Prog. Growth Factor Res. 4, 69-83.

Pawson, T. and Gish, G.D. (1992). SH2 and SH3 domains: from structure to function. Cell 71, 359-362.

Peles, E., Bacus, S.S., Koski, R.A., Lu, H.S., Wen, D., Ogden, S.G., Levy, R.B., and Yarden, Y. (1992). Isolation of the in vivoneu/HER-2 stimulatory ligand: A 44 kda glycoprotein that induces differentiation of mammary tumor cells. Cell 69, 205-216.

Pelicci, G., Lanfrancone, L., Grignan', F., McGlade, J., Cavallo, F., Forni, G., Nicoletti, I., Pawson, T., and Pelicci, P.G. (1992). A novel transforming protein (SHC) with an SH2 domain is implicated in mitogenic signal transduction. Cell 70, 93-104.

Pelley, R.J., Maihle, N.J., Boerkoel, C., Shu, H.K., Carter, T.H., Moscovici, C., and Kung, H.J. (1989). Disease tropism of c-erbB: Effects of carboxyl-terminal tyrosine and internal mutations on tissue-specific transformation. Proc. Natl. Acad. Sci. USA 86, 7164- 7168.

Penault-Llorca, F., Bertucci, F., Adelaide, J., Parc, P., Coulier, F., Jacquemier, J., Birnbaum, D., and deLapeyriere, O. (1995). Expression of FGF and FGF receptor genes in human breast cancer. Int. J Cancer 61, 170-176.

Perren, T.J. (1991). c-*erb*B-2 oncogene as a prognostic marker in breast cancer [editorial]. Br. J. Cancer 63, 328-332.

Peters, G., Brookes, S., Smith, R., Placzek, M., and Dickson, C. (1989). The mouse homolog of the hst/k-FGF gene is adjacent to int-2 and is activated by proviral insertion in some virally induced mammary tumors. Proc. Natl. Acad. Sci. USA 86, 5678-5682.

Peters, K.G., Marie, J., Wilson, E., Ives, H.E., Escobedo, J., Del Rosario, M., Mirda, D., and Williams, L.T. (1992). Point mutation of an FGF receptor abolishes phosphatidylinositol turnover and Ca^{2+} flux but not mitogenesis. Nature 358, 678-681.

Picard, D. and Yamamoto, K.R. (1987). Two signals mediate hormone-dependent nuclear localization of the glucocorticoid receptor. EMBO J. 6, 3333-3340.

Pietras, R.J., Arboleda, J., Reese, D.M., Wongvipat, N., Pegram, M.D., Ramos, L., Gorman, C.M., Parker, M.G., Sliwkowski, M.X., and Slamon, D.J. (1995). HER-2 tyrosine kinase pathway targets estrogen receptor and promotes hormone-independent growth in human breast cancer cells. Oncogene 10, 2435-2446.

Pike, L.J., Marquardt, H., Todaro, G.J., Gallis, B., Gasnellie, J.E., Bornstein, P., and Kerbs, E.G. (1982). Transforming growth factor and epidermal growth factor stimulate the phosphorylation of a synthetic, tyrosine-containing peptide in a similar manner. J. Biol. Chem. 257, 14628-14631.

Piva, R., Rimondi, A.P., Hanau, S., Maestri, I., Alvisi, A., Kumar, V.L., and del Senno, L. (1990). Different methylation of oestrogen receptor DNA in human breast carcinomas with and without oestrogen receptor. Br. J. Cancer 61, 270-275.

Plowman, G.D., Culouscou, J., Whitney, G.S., Green, J.M., Carolton, G.W., Foy, L., Neubauer, M.G., and Shoyab, M. (1993a). Ligand-specific activation of HER4/p180^{erbB4}, a fourth member of the epidermal growth factor receptor family. Proc. Natl. Acad. Sci. USA. 90, 1746-1750.

Plowman, G.D., Green, J.M., Culouscou, J.M., Carlton, G.W., and Rothwell, V.M. (1993b). Heregulin induces tyrosine phosphorylation of HER4/p180 (ErbB4). Nature 366, 473-475.

Plowman, G.D., Whitney, G.S., Neubauer, M.G., Green, J.M., McDonald, V.L., Todaro, G.J., and Shoyab, M. (1990). Molecular cloning and expression of an additional epidermal growth factor receptor-related gene. Proc. Natl. Acad. Sci. USA 87, 4905-4908.

Potgens, A.J., Lubsen, N.H., van Altena, M.C., Vermeulen, R., Bakker, A., Schoenmakers, J.G., Ruiter, D.J., and de Waal, R.M. (1994). Covalent dimerization of vascular permeability factor/vascular endothelial growth factor is essential for its biological activity. evidence from cys to ser mutations. J. Biol. Chem. 269, 32879-32885.

Power, R.F., Mani, S.K., Codina, J., Conneely, O.M., and O'Malley, B.W. (1991). Dopaminergic and ligand-independent activation of steroid hormone receptors. Science 254, 1636-1639.

Prudovsky, I.A., Savion, N., La Vallee, T.M., and Maciag, T. (1996). The nuclear trafficking of extracellular fibroblast growth factor (FGF)-1 correlates with the perinuclear association of the FGF receptor-1 α isoforms but not the FGF receptor-1 β isoforms. J. Biol. Chem. 271, 14198-14205.

Qian, X., Dougall, W.C., Hellman, M.E., and Greene, M.I. (1994). Kinase-deficient in vivoneu proteins suppress epidermal growth factor receptor function and abolish cell transformation. Oncogene 9, 1507-1514.

Quarto, N., Finger, F.P., and Rifkin, D.B. (1991a). The NH2-terminal extension of high molecular weight bFGF is a nuclear targeting signal. J. Cell. Physiol. 147, 311-318.

Quarto, N., Talarico, D., Florkiewicz, R., and Rifkin, D.B. (1991b). Selective expression of high molecular weight basic fibroblast growth factor confers a unique phenotype to NIH 3T3 cells. Cell Regul. 2, 699-708.

Qiu, R.G., Chen, J., Kim, D., McCormick, F., and Symons, M. (1995). An essential role for Rac in Ras transformation. Nature 374, 457-459.

Quinn, T.P., Peters, K.G., de Vries, C., Ferrara, N., and Williams, L.T. (1993). Fetal liver kinase 1 is a receptor for vascular endothelial growth factor and is selectively expressed in vascular endothelium. Proc. Natl. Acad. Sci. USA 90, 7533-7537.

Ramakrishnan, S., Olson , T.A., Bautch, V.L., and Mohanraj, D. (1996). Vascular endothelial growth factor-toxin conjugate specifically inhibits KDR/Flk-1 positive entothelial cell proliferation in vitro and angiogenesis in vivo. Cancer Res. 56, 1324-1330.

Rapp, U.R. (1991). Role of Raf-1 serine/threonine protein kinase in growth factor signal transduction. Oncogene 6, 495-500.

Rastinejad, F., Polverini, P.J., and Bouck, N.P. (1989). Regulation of the activity of a new inhibitor of angiogenesis by a cancer suppressor gene. Cell 56, 345-355.

Richards, J., Guzman, R., Konrad, M., Yang, J., and Nandi, S. (1982). Growth of mouse mammary gland end buds cultured in a collagen gel matirx. Exp. Cell. Res. 141, 433-443.

Riedel, H., Massoglia, S., Schlessinger, J., and Ullrich, A. (1988). Ligand activation of overexpressed epidermal growth factor receptors transforms NIH 3T3 mouse fibroblasts. Proc. Natl. Acad. Sci. USA 85, 1477-1481.

Riese, D.J., Bermingham, Y., van Raaij, T.M., Buckley, S., Plowman, G.D., and Stern, D.F. (1996). β cellulin activates the epidermal growth factor receptor and *erb*B- 4, and induces cellular response patterns distinct from those stimulated by epidermal growth factor or neuregulin Oncogene, 12, 345-353.

Riese, D.J., van Raaij, T.M., Plowman, G.D., Andrews, G.C., and Stern, D.F. (1995). The cellular response to neuregulins is governed by complex interactions of the *erb*B-4 receptor family. Mol. Cell. Biol. 15, 5770-5776.

Rochefort, H., Cavailles, V., Augereau, P., Capony, F., Maudelonde, T., Touitou, I., and Garcia, M. (1989). Overexpression and hormonal regulation of procathepsin D in mammary and endometrial cancer. J. Steroid Biochem. 34, 177-182.

Rogelj, S., Weinberg, A., Fanning, P., and Klagsbrun, M. (1988). Basic fibroblast growth factor fused to a signal peptide transforms cells. Nature 31, 173-175.

Sainsbury, J.R., Farndon, J.R., Needham, G.K., Malcom, A.J., and Harris, A.L. (1987). Epidermal-growth-factor receptor status as predictor of early reccurrence of and death from breast cancer. Lancet 1, 1398-1402.

Sainsbury, J.R.C., Farndon, J.R., Sherbet, G.V., and Harris, A.L. (1985). Epidermal growth factor receptors and oestrogen receptors in human breast cancer. Lancet 1, 364- 366.

Saleh, M., Stacker, S.A., and Wilks, A.F. (1996). Inhibition of growth of C6 glioma cells in vivo by expression of antisense vascular endothelial growth factor sequence. Cancer Res. 56, 393-401.

Salomon, D.S., Normanno, N., Ciardiello, F., Brandt, R., Shoyab, M., Todaro, G.J. (1995). The role of amphiregulin in breast cancer. Breast Cancer Res. Treat. 33, 103-114.

Salomonsson, M., Haggblad, J., O'Malley, B.W., and Sitbon, G.M. (1994). The human estrogen receptor hormone binding domain dimerizes independently of ligand activation. J. Steroid Biochem. Mol. Biol. 48, 447-452.

Samanta, A. and Greene, M.I. (1995). A kinase associated with chromatin that can be activated by ligand-p185c-*neu* or epidermal growth factor-receptor interactions. Proc. Natl. Acad. Sci. USA 92, 6582-6586.

Samanta, A., LeVea, C.M., Dougall, W.C., Qian, X.L., and Greene, M.I. (1994). Ligand and p185 (C-*neu*) Density govern receptor interactions and tyrosine kinase activation. Proc. Natl. Acad. Sci. USA 91, 1711-1715.

Sasada, R., Kurokawa, T., Iwane, M., and Igarashi, K. (1988). Transformation of mouse BALB/c3T3 cells with human basic fibroblast growth factor cDNA. Mol. Cell. Biol. 8, 588-594.

Sasaoka, T., Langlois, W.J., Leitner, J.W., Draznin, B., and Olefsky, J.M. (1994). The signaling pathway coupling epidermal growth factor receptors to activation of $p21^{ras}$. J. Biol. Chem. 269, 32621-32625.

Savagner, P., Valles, A.M., Jouanneau, J., Yamada, K.M., and Thiery, J.P. (1994). Alternative splicing in fibroblast growth factor receptor 2 is associated with induced epithelial-mesenchymal transition in rat bladder carcinoma cells. Mol. Biol. Cell 5, 851- 862.

Schreiber, A.B., Winkler, M.E., and Derynck, R. (1986). Transforming growth factor-α: A more potent angiogenic mediator than epidermal growth factor. Science 232, 1250-1253.

Schwabe, J.W., Chapman, L., Finch, J.T., and Rhodes, D. (1993). The crystal structure of the estrogen receptor DNA-binding domain bound to DNA: How receptors discriminate between their response elements. Cell 75, 567-578.

Seetharam, L., Gotoh, N., Maru, Y., Neufeld, G., Yamaguchi, S., and Shibuya, M. (1995). A unique signal transduction from flt tyrosine kinase, a receptor for vascular endothelial growth factor VEGF. Oncogene 10, 135-147.

Seger, R. and Kerbs, E.G. (1995). The MAPK signaling cascade. FASEB J. 9, 726-735.

Seibert, K., Shafie, S.M., Triche, T.J., Whang-Peng, J.J., O'Brien, S.J., Toney, J.H., Huff, K.K., and Lippman, M.E. (1983). Clonal variation of MCF-7 breast cancer cells in vitro and in athymic nude mice. Cancer Res. 43, 2223-2239.

Selva, E., Raden, D.L., Davis, R.J. (1993). Mitogen-activatd protein kinase stimulation by a tyrosine kinase-negative epidermal growth factor receptor. J. Biol. Chem. 268, 2250-2254.

Shalaby, F., Rossant, J., Yamaguchi, T.P., Gertsenstein, M., Wu, X.F., Breitman, M.L., and Schuh, A.C. (1995). Failure of blood-island formation and vasculogenesis in flk-1- deficient mice. Nature 376, 62-66.

Shaoul, E., Reich-Slotky, R., Berman, B., and Ron, D. (1995). Fibroblast growth factor receptors display both common and distinct signaling pathways. Oncogene 10, 1553-1561.

Sheng, Z., Wu, K., Carraway, K.L., and Fregien, N. (1992). Molecular cloning of the transmembrane component of the 13762 mammary adenocarcinoma sialomucin complex. A new member of the epidermal growth factor superfamily. J. Biol. Chem. 267, 16341- 16346.

Sherblom, A.P. and Carraway, K.L. (1980). A complex of two cell surface glycoproteins from ascites mammary adenocarcinoma cells. J. Biol. Chem. 255, 12051-12059.

Shi, E., Kan, M., Xu, J., Wang, F., Hou, J., and McKeehan, W.L. (1993). Control of fibroblast growth factor receptor kinase signal transduction by heterodimerization of combinatorial splice variants. Mol. Cell. Biol. 13, 3907-3918.

Shimizu, N. and Kondo, I. (1982). Hyperproduction of EGF receptors in human A431 cells is regulated by a translocaiton chromosome t(7:11)(q22;q23). Cytogenic. Cell. Genet. 32, 316.

Shing, Y., Christofori, G., Hanahan, D., Ono, Y., Sasada, R., Igarashi, K., and Folkman, J. (1993). β cellulin: A mitogen from pancreatic β cell tumors. Science 259, 1604- 1607.

Shoyab, M., Plowman, G.D., McDonald, V.L., Bradley, J.G., and Todaro, G.J. (1989). Structure and function of human amphiregulin: A member of the epidermal growth factor family. Science 243, 1074-1076.

Shu, H.K., Pelley, R.J., and Kung, H.J. (1990). Tissue-specific transformation by epidermal growth factor receptor: A single point mutation within the ATP-binding pocket of the erB product increases its intrinsic kinase activity and activates its sarcomagenic potential. Proc. Natl. Acad. Sci. USA 87, 9103-9107.

Silfversward, C., Gustafsson, J.A., Gustafsson, S.A., Humla, S., Nordenskjold, B., Wallgren, A., and Wrange, O. (1980). Estrogen receptor concentrations in 269 cases of histologically classified human breast cancer. Cancer 45, 2001-2005.

Skolnik, E.Y., Margolis, B., Mohammadi, M., Lowenstein, E., Fischer, R., Drepps, A., Ullrich, A., and Schlessinger, J. (1991). Cloning of PI3 kinase-associated p85 using a novel method for expression/cloning of target proteins for receptor tyrosine kinases. Cell 65, 83-90.

Slamon, D.J., Clark, G.M., Wong, S.G., Levin, W.J., Ullrich, A., and McGuire, W.L. (1987). Human breast cancer: Correlation of relapse and survival with amplification of the HER-2/*neu* oncogene. Science 235, 177-182.

Slamon, D.J., Godolphin, W., Jones, L.A., Holt, J.A., Wong, S.G., Keith, D.E., Levin, W.J., Stuart, S.G., Udove, J., Ullrich, A., and Press, M.F. (1989). Studies of the HER- 2/*neu* proto-oncogene in human breast and ovarian cancer. Science 244, 707-712.

Sliwkowski, M.X., Schaefer, G., Akita, R.W., Lofgren, J.A., Fitzpatrick, V.D., Nuijens, A., Fendly, B.M., Cerione, R.A., Vandlen, R.L., and Carraway, K.L.I. (1994). Coexpression of *erb*B2 and *erb*B3 proteins reconstitutes a high affinity receptor for heregulin. J. Biol. Chem. 269, 14661-14665.

Smallwood, P.M., Munoz-Sanjuan, I., Tong, P., Macke, J.P., Hendry, S.H., Gilbert D.J., Copeland, N.G., Jenkins, N.A., and Nathans, J. (1996). Fibroblast growth factor (FGF) homologous factors:

New members of the FGF family implicated in nervous system development. Proc. Nat. Acad. Sci. USA. 93, 9850-9857.

Smith, H.S., Lu, Y., Deng, G., Martinez, O., Krams, S., Ljung, B.M., Thor, A., and Lagios, M. (1993). Molecular aspects of early stages of breast cancer progression. J. Cell. Biochem. (suppl.) 17G, 144-152.

Smith, J., Yelland, A., Baillie, R., and Coombes, R.C. (1994). Acidic and basic fibroblast growth factors in human breast tissue. Eur. J. Cancer 30A, 496-503.

Soker, S., Fidder, H., Neufeld, G., and Klagsbrun, M. (1996). Characterization of novel vascular endothelial growth factor (VEGF) receptors on tumor cells that bind $VEGF_{165}$ via its exon 7-encoded domain. J. Biol. Chem. 271, 5761-5767.

Soker, S., Svahn, C.M., and Neufeld, G. (1993). Vascular endothelial growth factor is inactivated by binding to α 2-macroglobulin and the binding is inhibited by heparin. J. Biol. Chem. 268, 7685-7691.

Sorokin, A., Lemmon, M.A., Ullrich, A. and Schlessinger, J. (1994). Stabilization of an active dimeric form of the epidermal growth factor receptor by introduction of an inter- receptor disulfide bond. J. Biol. Chem. 269, 9752-9759.

Souttou, B., Gamby, C., Crepin, M., and Hamelin, R. (1996). Tumoral progression of human breast epithelial cells secreting FGF2 and FGF4. Int. J. Cancer 68, 1-7.

Souttou, B., Hamelin, R., and Crepin, M. (1994). FGF2 as an autocrine growth factor for immortal human breast epithelial cells. Cell Growth Differ. 5, 615-623.

Spivak-Kroizman, T., Lemmon, M.A., Dikic, I., Ladbury, J.E., Pinchasi, D., Huang, J., Jaye, M., Crumley, G., Schlessinger, J., and Lax, I. (1994). Heparin-induced oligomerization of FGF molecules is responsible for FGF receptor dimerization, activation, and cell proliferation. Cell 79, 1015-1024.

Steck, P.A. and Nicolson, G.L. (1983). Cell surface glycoproteins of 13762NF mammary adenocarcinoma clones of differing metastatic potentials. Exp. Cell Res. 147, 255-267.

Stern, D.F. and Kamps, M.P. (1988). EGF-stimulated tyrosine phosphorylation of p185*neu*: A potential model for receptor interactions. EMBO J. 7, 995-1001.

Stern, D.F., Kamps, M.P., and Cao, H. (1988). Oncogenic activation of p185*neu* stimulates tyrosine phosphorylation in vivo. Mol. Cell. Biol. 8, 3969-3973.

Stewart, A.J., Westley, B.R., and May, F.E.B. (1992). Modulation of the proliferative response of breast cancer cells to growth factors by oestrogen. Br. J. Cancer 66, 640-648.

Strawn, L.M., McMahon, G., App, H., Schreck, R., Kuchler, W.R., Longhi, M.P., Hui, T.H., Tang, C., Levitzki, A., Gazit, A., Chen, I., Keri, G., Orfi, L., Risau, W., Flamme, I., Ullrich, A., Hirth, K.P., and Shawver, L.K. (1996). Flk-1 as a target for tumor growth inhibition. Cancer Res. 56, 3540-3545.

Takahashi, K., Suzuki, K., Kawahara, S., and Ono, T. (1989). Growth stimulation of human breast epithelial cells by basic fibroblast growth factor in serum-free medium. Int. J. Cancer 43, 870-874.

Tanaka, A., Miyamoto, K., Minamino, N., Takeda, M., Sato, B., Matsuo, H., and Matsumoto, K. (1992). Cloning and characterization of an androgen-induced growth factor essential for the androgen-dependent growth of mouse mammary carcinoma cells. Proc. Natl. Acad. Sci. USA 89, 8928-8932.

Tang, C.K., Perez, C., Grunt, T., Waibel, C., Cho, C., and Lupu, R. (1996). Involvement of heregulin-β-2 in the acquisition of the hormone-independent phenotype of breast cancer cells. Cancer Res. 56, 3350-3358.

Tavormina, P.L., Shiang, R., Thompson, L.M., Zhu, Y.Z., Wilkin, D.J., Lachman, R.S., Wilcox, W.R., Rimoin, D.L., Cohn, D.H., and Wasmuth, J.J. (1995). Thanatophoric dysplasia (types I and II) caused by distinct mutations in fibroblast growth factor receptor 3. Nat. Genet. 9, 321-328.

Terenius, L. (1970). Two modes of interaction between oestrogen and anti-oestrogen. Acta Endocrinol. 64, 47-58.

Terman, B.I., Dougher-Vermazen, M., Carrion, M.E., Dimitrov, D., Armellino, D.C., Gospodarowicz,

D., and Böhlen, P. (1992). Identification of the KDR tyrosine kinase as a receptor for vascular endothelial cell growth factor. Biochem. Biophys. Res Commun. 187, 1579-1586.

Theillet, C., Le Roy, X., De Lapeyriere, O., Grosgeorges, J., Adnane, J., Raynaud, S.D., Simony-Lafontaine, J., Goldfarb, M., Escot, C., and Birnbaum, D. (1989). Amplification of FGF-related genes in human tumors: Possible involvement of HST in breast carcinomas. Oncogene 4, 915-922.

Thomas, K.A., Rios-Candelore, M., Gimenez-Gallego, G., DiSalvo, S., Bennett, C., Rodkey, J., and Fitzpatrick, S. (1985). Pure brain-derived acidic fibroblast growth factor is a potent angiogenic vascular endothelial cell mitogen with sequence homology to interleukin 1. Proc Natl Acad Sci USA 82, 6409-6413.

Tischer, E., Mitchell, R., Hartman, T., Silva, M., Gospodarowicz, D., Fiddes, J.C., and Abraham, J.A. (1991). The human gene for vascular endothelial growth factor. Multiple protein forms are encoded through alternative exon splicing. J. Biol. Chem. 266, 11947- 11954.

Todaro, G.J., Fryling, C., and DeLarco, J.E. (1980). Transforming growth factor produced by certain human tumor cells: Polypeptides that interact with epidermal growth factor receptors. Proc. Natl. Acad. Sci. USA 77, 5258-5262.

Toi, M., Hoshina, S., Takayanagi, T., and Tominaga, T. (1994). Association of vascular endothelial growth factor expression with tumor angiogenesis and with early relapse in primary breast cancer. Jpn. J. Cancer Res. 85, 1045-1049.

Toi, M., Inada, K., Suzuki, H., and Tominaga, T. (1995). Tumor angiogenesis in breast cancer: Its importance as a prognostic indicator and the association with vascular endothelial growth factor expression. Breast Cancer Res. Treat. 36, 193-204 .

Toi, M., Kondo, S., Suzuki, H., Yamamoto, Y., Inada, K., Imazawa, T., Taniguchi, T., and Tominaga, T. (1996). Quantitative analysis of vascular endothelial growth factor in primary breast cancer. Cancer 77, 1101-1106.

Toi, M., Osaki, A., Yamada, H., and Toge, T. (1991). Epidermal growth factor receptor expression as a prognostic indicator in breast cancer. Eur. J. Cancer 27, 977-980.

Tonelli, Q.J. and Sorof, S. (1980). Epidermal growth factor requirement for development of cultured mammary gland. Nature 285, 250-252.

Tora, L., White, J., Brou, C., Tasset, D., Webster, N., Scheer, E., and Chambon, P. (1989). The human estrogen receptor has two independent nonacidic transcriptional activation functions. Cell 59, 477-487.

Tzukerman, M.T., Esty, A., Santiso-Mere, D., Danielian, P., Parker, M.G., Stein, R.B., Pike, J.W., and McDonnell, D.P. (1994). Human estrogen receptor transactivational capacity is determined by both cellular and promoter context and mediated by two functionally distinct intramolecular regions. Mol. Endocrinol. 8, 21-30.

Ullrich, A., Coussens, L., and Hayflick, J. (1984). Human epidermal growth factor receptor cDNA and abberant expression of the amplified gene in A431 epidermoid carcinoma cells. Nature 309, 418-425.

Ullrich, A. and Schlessinger, J. (1990). Signal transduction by receptors with tyrosine kinase activity. Cell 61, 203-212.

Unemori, E.N., Ferrara, N., Bauer, E.A., and Amento, E.P. (1992). Vascular endothelial growth factor induces interstitial collagenase expression in human endothelial cells. J. Cell Physiol. 153, 557-562.

Vainikka, S., Joukov, V. Wennstrom, S., Bergman, M., Pelicci, P.G., and Alitalo, K. (1994). Signal transduction by fibroblast growth factor receptor-4 (FGFR-4). Comparison with FGFR-1. J. Biol. Chem., 269, 18320-18326.

Vainikka, S., Joukov, V., Klint, P., and Alitalo, K. (1996). Association of a 85-kDa serine kinase with activated fibroblast growth factor receptor-4. J. Biol. Chem. 271, 1270-1273.

Valverius, E.M., Ciardiello, F., Heldin, N.E., Blondel, B., Merlo, G., Smith, G., Stampfer, M.R., Lippman, M.E., Dickson, R.B., and Salomon, D.S. (1990). Stromal influences on transformation of human mammary epithelial cells overexpressing c-myc and SV40 T. J. Cell. Physiol. 145, 207-216.

Van Aelst, L., Barr, M., Marcus, S., Polverino, A., and Wegler, M. (1993). Complex formation between RAS and RAF and other protein kinases. Proc. Natl. Acad. Sci. USA 90, 6213-6217.

van Agthoven, T., Timmermans, M., Foekens, J.A., Dorssers, L.C., and Henzen- Logmans, S.C. (1994a). Differential expression of estrogen, progesterone, and epidermal growth factor receptors in normal, benign, and malignant human breast tissues using dual staining immunohistochemistry. Am. J. Pathol. 144, 1238-1246.

van Agthoven, T., van Agthoven, T.L.A., Dekker, A., Foekens, J.A., and Dorssers, L.C.J. (1994b). Induction of estrogen independence of ZR-75-1 human breast cancer cells by epigenetic alterations. Mol. Endocrinol. 8, 1474-1483.

Velu, T.J., Beguinot, L., Vass, W.C., Willingham, M.C., Merlino, G.T., Pastan, I., and Lowy, D. (1987). Epidermal growth factor-dependent transformation by a human EGF receptor proto-oncogene. Science 238, 1408-1410.

Vickers, P.J., Dickson, R.B., Shoemaker, R., and Cowan, K.H. (1988). A multidrug- resistant MCF-7 human breast cancer cell line which exhibits cross-resistance to antiestrogens and hormone-independent tumor growth in vivo. Mol. Endocrinol. 2, 886- 892.

Vignon, F., Capony, F., Chambon, M., Freiss, G., Garcia, M., and Rochefort, H. (1986). Autocrine growth stimulation of the MCF 7 breast cancer cells by the estrogen-regulated 52 k protein. Endocrinology 118, 1537-1545.

Visscher, D.W., De Mattia, F., Ottosen, S., Sarkar, F.H., and Crissman, J.D. (1995). Biologic and clinical significance of basic fibroblast growth factor immunostaining in breast carcinoma. Mod. Pathol. 8, 665-670.

Vojtek, A.B., and Cooper, J.A. (1995). Rho family members: Activators of MAP kinase cascades. Cell 82, 527-529.

Wada, T., Qian, X., and Greene, M.I. (1990). Intermolecular association of the $p185^{neu}$ protein and EGF receptor modulates EGF receptor function. Cell 61, 1339-1347.

Wakeling, A.E., Dukes, M., and Bowler, J. (1991). A potent specific pure antiestrogen with clinical potential. Cancer Res. 51, 3867-3873.

Waltenberger, J., Claesson-Welsh, L., Siegbahn, A., Shibuya, M., and Heldin, C.H. (1994). Different signal transduction properties of KDR and Flt1, two receptors for vascular endothelial growth factor. J. Biol. Chem. 269, 26988-26995.

Wang, F., Kan, M., Yan, G., Xu, J., and McKeehan, W.L. (1995a). Alternately spliced NH2-terminal immunoglobulinlike loop I in the ectodomain of the fibroblast growth factor (FGF) receptor 1 lowers affinity for both heparin and FGF-1. J. Biol. Chem. 270, 10231-10235.

Wang, H., Peters, G.A., Zeng, X., Tang, M., Ip, W., and Khan, S.A. (1995b). Yeast two- hybrid system demonstrates that estrogen receptor dimerization is ligand-dependent in vivo. J. Biol. Chem. 270, 23322-23329.

Wang, J.-K., Gao, G., and Goldfarb, M. (1994). Fibroblast growth factor receptors have different signaling and mitogenic potentials. Mol. Cell Biol. 14, 181-188.

Wang, J.-K., Xu, H., Li, H.-C., and Goldfarb, M. (1996). Broadly expressed SNT-like proteins link FGF receptor stimulation to activators of Ras. Oncogene 13, 721-729.

Warren, R.S., Yuan, H., Matli, M.R., Gillett, N.A., and Ferrara, N. (1995). Regulation by vascular endothelial growth factor of human colon cancer tumorigenesis in a mouse model of experimental liver metastasis. J. Clin. Invest. 95, 1789-1797.

Washburn, T., Hocutt, A., Brautigan, D.L., and Korach, K.S. (1991). Uterine estrogen receptor in vivo: Phosphorylation of nuclear specific forms on serine residues. Mol. Endocrinol. 5, 235-242.

Watanabe, S., Lazar, E., and Sporn, M.B. (1987). Transformation of normal rat kidney (NRK) cells by an infectious retrovirus carrying a synthetic rat type α transforming growth factor gene. Proc. Natl. Acad. Sci. USA 84, 1258-1262.

Watanabe, T., Shintani, A., Nakata, M., Shing, Y., Folkman, J., Igarashi, K., and Sasada, R. (1994). Recombinant human β cellulin: Molecular structure, biological activities, and receptor interaction. J. Biol. Chem. 269, 9966-9973.

Waters, S.B., Holt, K.H., Ross, S.E., Syu, L.-J., Guan, K.-L., Saltiel, A.R., Koretzky, G.A., and Pessin, J.E. (1995). Desensitization of Ras activation by a feedback disassociation of the Sos-Grb2 complex. J. Biol. Chem. 270, 20883-20886.

Webster, M.K., D'Avis, P.Y., Robertson, S.C., and Donoghue, D.J. (1996). Profound ligand-independent kinase activation of fibroblast growth factor receptor 3 by the activation loop mutation responsible for a lethal skeletal dysplasia, thanatophoric dysplasia type II. Mol. Cell. Biol. 16, 4081-4087.

Webster, M.K. and Donoghue, D.J. (1996). Constitutive activation of fibroblast growth factor receptor 3 by the transmembrane domain point mutation found in achondroplasia. EMBO J. 15, 520-527.

Webster, N.J.G., Green, S., Jin, J.-R., and Chambon, P. (1988). The hormone-binding domains of the estrogen and glucocorticoid receptors contain an inducible transcription activation function. Cell 54, 199-207.

Weidner, N., Folkman, J., Pozza, F., Bevilacqua, P., Allred, E.N., Moore, D.H., Meli, S., and Gasparini, G. (1992). Tumor angiogenesis: A new significant and independent prognostic indicator in early-stage breast carcinoma [see comments]. J. Natl. Cancer Inst. 84, 1875-1887.

Weidner, N., Semple, J.P., Welch, W.R., and Folkman, J.T. (1991). Tumor angiogenesis and metastasis—correlation in invasive breast carcinoma. N. Engl. J. Med. 324, 1-8.

Weigel, R.J. and deConinck, E.C. (1993). Transcriptional control of estrogen receptor in estrogen receptor-negative breast carcimona. Cancer Res. 53, 3472-3474.

Weiner, D.B., Liu, J., Cohen, J.A., Williams, W.V., and Greene, M.I. (1989). A point mutation in the *neu* oncogene mimics ligand induction of receptor aggregation. Nature 339, 230-231.

Wellstein, A., Lupu, R., Zugmaier, G., Flamm, S.L., Cheville, A.L., Delli Bovi, P., Basilico, C., Lippman, M.E., and Kern, F.G. (1990). Autocrine growth stimulation by secreted Kaposi fibroblast growth factor but not by endogenous basic fibroblast growth factor. Cell Growth Differ. 1, 63-71.

Welshons, W.V., Cormier, E.M., Wolf, M.F., Williams, P.O.J., and Jordan, V.C. (1988). Estrogen receptor distribution in enucleated breast cancer cell lines. Endocrinology 122, 2379-2386.

Welshons, W.V. and Jordan, V.C. (1987). Adaption of estrogen-dependent MCF-7 cells to low estrogen (phenol red-free) culture. Eur. J. Cancer Clin. Oncol. 23, 1935-1939.

Wen, D., Peles, E., Cupples, R., Suggs, S.V., Bacus, S.S., Luo, Y., Trail, G., Hu, S., Silbiger, S.M., and Levy, R.B. (1992). *Neu* differentiation factor: A transmembrane glycoprotein contianing an EGF domain and an immunoglobulin homology unit. Cell 69, 559-572.

Wen, D., Suggs, S.V., Karunagaran, D., Liu, N., Cupples, R.L., Luo, Y., Janssen, A.M., Ben-Baruch, N., Trollinger, D.B., Jacobsen, V.L., Meng, S.-Y., Lu, H.S., Hu, S., Chang, D., Yang, W., Yanghara, D., Koski, R.A., and Yarden, Y. (1994). Structural and functional aspects of the multiplicity of *neu* differentiation factors. Mol. Cell. Biol. 14, 1909-1919.

Whitmarsh, A.J., Shore, P., Sharrocks, A.D., and Davis, R.J. (1995). Integration of MAP kinase signal transduction pathways at the serum response element. Science 269, 403-407.

Wiebe, V.J., Osborne, C.K., McGuire, W.L., and DeGregorio, M.W. (1992). Identification of estrogenic tamoxifen metabolite(s) in tamoxifen-resistant human breast tumors. J. Clin. Oncol. 10, 990-994.

Wikstrand, C.J., Hale, L.P., Batra, S.K., Hill, L., Humphrey, P.A., Kurpad, S.N., McLendon, R.E., Moscatello, D., Pegram, C.N., Reist, C.J., Traweek, T., Wong, A.J., Zalutsky, M.R., and Bigner, D.D. (1996). Monoclonal antibodies against EGFRvIII are tumor specific and react with breast and lung carcinomas and malignant gliomas. Cancer Res. 55, 3140-3148.

Winkler, M.E., O'Connor, L.O., Winget, M., and Fendly, B. (1989). Epidermal growth factor and transforming growth factor bind differently to the epidermal growth factor receptor. Biochemistry 28, 6373-6378.

Wittliff, J.L. (1984). Steroid-hormone receptors in breast cancer. Cancer 53, 630-643.

Wolf, D.M. and Jordan, V.C. (1994). The estrogen receptor from a tamoxifen stimulated MCF-7 tumor variant contains a point mutation in the ligand binding domain. Breast Cancer Res. Treat. 31, 129-138.

Wood, K.W., Sarnecki, C., Roberts, T.M., and Blenis, J. (1992). Ras mediates nerve growth factor receptor modulation of three signal-transducing protein kinases: MAP kinase, Raf-1 and RSK. Cell 1041-1050.

Wu, D., Wang, L., Chi, Y., Sato, G.H., and Sato, J.D. (1990). Human epidermal growth factor receptor residue covalently cross-linked to epidermal growth factor. Proc. Natl. Acad. Sci. USA 87, 3151-3155.

Wu, J., Dent, P., Jelinek, T., Wolfman, A., Weber, M.J., and Sturgill, T.W. (1993). Inhibition of EGF-activated MAP kinase signaling by adenosine 3', 5'-monophosphate. Science 262, 1065-1069.

Wu, K., Salas, P.J., Yee, L., Fregien, N., and Carraway, K.L. (1994). Tissue and tumor expression of a cell surface glycoprotein complex containing an integral membrane glycoprotein activator of p185*neu*. Oncogene 9, 3139-3147.

Yaich, L., Dupont, W.D., Cavener, D.R., and Paul, F.F. (1992). Analysis of the *Pvu* II restriction fragment-length polymorphism and exon structure of the estrogen receptor gene in breast cancer and peripheral blood. Cancer Res. 52, 77-83.

Yamaguchi, F., Saya, S., Bruner, J.M., and Morrison, R.S. (1994). Differential expression of two fibroblast growth factor-receptor genes is associated with malignant progression in human astrocytomas. Proc. Natl. Acad. Sci. USA 91, 484-488.

Yamazaki, H., Ohba, Y., Tamaoki, N., and Shibuya, M. (1990). A deletion mutation within the ligand binding domain is responsible for activation of epidermal growth factor receptor gene in human brain tumors. Jpn. J. Cancer Res. 81, 773-779.

Yan, G., Fukabori, Y., McBride, G., Nikolaropolous, S., and McKeehan, W.L. (1993). Exon switching and activation of stromal and embryonic fibroblast growth factor (FGF)- receptor genes in prostate epithelial cells accompany stromal independence and malignancy. Mol. Cell. Biol. 13, 4513-4522.

Yan, G., Wang, F., Fukabori, Y., Sussman, D., Hou, J., and McKeehan, W.L. (1992). Expression and transforming activity of a variant of the heparin-binding fibroblast growth factor receptor (flg) gene resulting from splicing of the α exon at an alternate 3'- acceptor site. Biochem. Biophys. Res. Commun. 183, 423-430.

Yan, M., Dai, T., Deak, J.C., Kyriakis, J.M., Zon, L.I., Woodgett, J.R., and Templeton, D.J. (1994). Activation of stress-activated protein kinase by MEKK1 phosphorylation of its activator SEK1. Nature 372, 798-800.

Yang, K.-P. and Samaan, N.A. (1983). Reduction of estrogen receptor concentration in MCF-7 human breast carcinoma cells following exposure to chemotheraputic drugs. Cancer Res. 43, 3534-3538.

Yarden, R.I., Lauber, A.H., El-Ashry, D., and Chrysogelos, S.A. (1994). Estrogen regulation of EGF receptor mRNA in human breast cancer cells [Abstract]. J. Cell. Biochem. 18B, 363.

Yayon, A., Klagsbrun, M., Esko, J.D., Leder, P., and Ornitz, D.M. (1991). Cell surface, heparinlike molecules are required for binding of basic fibroblast growth factor to its high affinity receptor. Cell 64, 841-848.

Yee, D., Cullen, K.J., Paik, S., Perdue, J.F., Hampton, B., Schwartz, A., Lippman, M.E., and Rosen, N. (1988). Insulinlike growth factor II mRNA expression in human breast cancer. Cancer Res. 48, 6691-6696.

Yiangou, C., Gomm, J.J., Coope, R.C., Law, M., Luqmani, Y.A., Shousa, S., Coombes, R.C., and Johnston, C.L. (1997). Fibroblast growth factor 2 in breast cancer: Occurrence and prognostic significance. Br. J. Cancer 75, 28-33.

Yoshiji, H., Gomez, D.E., Shibuya, M., and Thorgeirsson, U.P. (1996). Expression of vascular endothelial growth factor, its receptor, and other angiogenic factors in human breast cancer. Cancer Res. 56, 2013-2016.

Zajchowski, D.A., Sager, R., and Webster, L. (1993). Estrogen inhibits the growth of estrogen receptor-negative, but not estrogen receptor-positive, human mammary epithelial cells expressing a recombinant estrogen receptor. Cancer Res. 53, 5004-5011.

Zhan, X., Hu, X., Friesel, R., Maciag, T. (1993). Long term growth factor exposure and differential tyrosine phosphorylation are required for DNA synthesis in BALB/c 3T3 cells. J. Biol. Chem. 268, 9611-9620.

Zhan, X., Plourde, C., Hu, X., Friesel, R., and Maciag, T. (1994). Association of fibroblast growth factor receptor-1 with c-Src correlates with association between c-Src and cortactin. J. Biol. Chem. 269, 20221-20224.

Zhang, H., Craft, P., Scott, P.A.E., Ziche, M., Weich, H.A., Harris, A.L., and Bicknell, R. (1995). Enhancement of tumor growth and vascular density by transfection of vascular endothelial cell growth factor into MCF-7 human breast carcinoma cells. J. Natl. Cancer Inst. 87, 213-219.

Zhang, L., Kharbanda, S., Chen, D., Miller, D.L., Ding, I.Y.F., Hanfelt, J., McLeskey, S.W., and Kern, F.G. (1997). MCF-7 breast carcinoma cells overexpressing FGF-1 form vascularized, metastatic tumors in ovariectomized or tamoxifen-treated nude mice. Oncogene 15, 2093-2108.

Zhang, Q.-X., Borg, A., and Fuqua, S.A.W. (1991). An exon 5 deletion variant of the estrogen receptor frequently coexpressed with wild-type estrogen receptor in human breast cancer. Cancer Res. 53, 5882-5884.

Zimmer, Y., Givol, D., and Yayon, A. (1993). Multiple structural elements determine ligand binding of fibroblast growth factor receptors. J. Biol. Chem. 268, 7899-7903.

Zinck, R., Cahill, M.A., Kracht, M., Sachsenmaier, C., Hipskind, R.A., and Nordheim, A. (1995). Protein synthesis inhibitors reveal differential regulation of mitogen-activated protein kinase and stress-activated protein kinase pathways that converge on Elk-1. Mol. Cell. Biol. 15, 4930-4938.

BIOLOGY OF HIGH RISK BENIGN BREAST LESIONS

Fred Raymond Miller

Advances in Oncobiology
Volume 2, pages 81-116.

ISBN: 0-7623-0388-3

INTRODUCTION

The natural history of breast cancer indicates that it develops over years—even decades—and may progress through recognizable stages of proliferative breast disease. Although breast cancers are slow growing, they metastasize early. The ability to recognize and treat high-risk, precursor breast lesions is therefore desirable.

Many human carcinomas evolve via a sequence of changes from benign hyperplasia through atypical hyperplasia to carcinoma in situ and eventually to fully malignant invasive tumors with the potential to metastasize. In the case of colorectal neoplasia, the well recognized polyp/carcinoma sequence is associated with a series of specific genetic alterations (Fearon and Vogelstein, 1990). Although the colon cancer model has become a paradigm for solid tumor development, definition of critical genetic events in breast cancer has been hampered by its often lengthy natural history and architectural complexity. In the human breast, a spectrum of microscopic changes has been termed proliferative breast disease (PBD). The progression of histopathological features of PBD has been correlated with increased risk for the development of invasive carcinoma (Page and Dupont, 1990), but the focal and microscopic lesions of PBD provide scant tissue for genetic or other biological analyses. Although hyperplastic lesions are observed in human breast (Wellings et al., 1975; Dupont and Page, 1985; Dupont et al., 1993), their role in disease progression is not understood.

Patients with the most severe form of PBD, atypical hyperplasia, do have a four- to fivefold increased relative risk of developing breast cancer (Page and Dupont, 1990; Palli et al., 1991; London et al., 1992; Dupont et al., 1993), but it has been unclear whether these lesions are precursors of cancer or simply markers of breasts likely to give rise to independent neoplastic lesions. Carcinoma in situ (CIS), with an associated tenfold increased risk for breast cancer, is accepted as a precursor lesion because subsequent development of invasive cancer is frequently in the same breast in which CIS was detected earlier. In addition, risk factors for CIS and invasive cancer are nearly identical (Kerlikowske et al., 1997). On the other hand, the risk for cancer in patients with biopsies diagnosed as atypical hyperplasia is nearly equal in the contralateral as in the ipsilateral breast. Thus, it has been suggested that atypical hyperplasia may reflect a field effect or condition rather than being a precursor lesion. However, occurrence of cancer in the ipsilateral breast is slightly higher, and the mean time to occurrence is lower in the breast in which a proliferative lesion is detected compared to the contralateral breast. For example, a study of 116 benign breast disease biopsy patients who subsequently developed invasive breast cancer during a 25-year follow-up indicated that 56% of breast cancers occurred in the ipsilateral breast and 44% occurred in the contralateral breast and the mean time till occurrence of cancer following biopsy was 11 years for the ipsilateral breast and 14 years for the contralateral breast cancers (Krieger and Hiatt, 1992). These data suggest that some proliferative breast disease (PBD) lesions are precursors because if all hyperplastic lesions were simply identifying a field condition, incidence and time till occurrence should be equal in contralateral and ipsilateral breasts.

Other benign lesions, not part of the proliferative, hyperplastic sequence, have also been found to be associated with increased risk for later development of breast cancer. Fibroadenomas, particularly ones associated with cysts, sclerosing adenosis, papillary apocrine changes, or epithelial calcification are associated with elevated risk for developing breast cancer (Dupont et al., 1994). Remarkably, the increased risk for cancer in fibroadenoma patients remains raised for decades, whereas the risk of developing invasive cancer declines with time following diagnosis of atypical hyperplasia such that after 10 years there is no increased risk (Dupont and Page, 1989). The apparent transient increased risk may also be due to the existence of multiple types of hyperplasia, only some of which are precursor lesions and progress within the 10-year period.

It is possible that the true precursor hyperplastic lesions can be identified. For example, 17% of patients having atypical hyperplasia with sclerosing adenosis subsequently develop cancer versus the 4% of cases without sclerosing adenosis that do so (Tavassoli and Norris, 1990); moreover, invasive cancer that develops in women previously having atypical hyperplasia within papillomas is nearly always ipsilateral to the biopsy site (Page et al., 1996).

ORTHOTOPIC TISSUE INTERACTIONS

Most of what is known about the role of mammary tissue interactions in progression is based on studies with preneoplastic mouse mammary tissues. These lesions require normal mammary stroma for growth and progression to carcinoma but are inhibited by normal mammary epithelium (DeOme et al., 1959, 1978; Faulkin and DeOme, 1960; Daniel et al., 1968; Medina, 1975, 1976, 1996). Mouse mammary tumor cells still respond to stroma but are no longer stroma-dependent and are stimulated, rather than inhibited, by normal mammary epithelium (Miller et al., 1981; Miller and McInerney, 1988). Furthermore, mammary tumors metastasize more readily from mammary gland than from other sites (Miller, 1981; Unemori et al., 1984; Vaage, 1988). The growth and metastasis of xenografted human breast cancer cells are also enhanced by implantation into mammary fatpads in nude mice (Price, 1990; Price et al., 1990).

These observations suggest that the cleared, that is, epithelium-free, mammary fatpads of immune-deficient mice might be ideal sites for xenograft transplants of normal and preneoplastic human breast epithelium. An early study reported successful transplantation of pieces of breast fibrocystic disease into cleared fatpads (Outzen and Custer, 1975). Three cases were described and serial transplantation through four transplant generations of a lesion, described as cystic hyperplasia with ductal proliferation, indicated that 16 volume doublings were obtained. Another study found that normal-appearing lobules persisted nearly as well at subcutaneous sites (54 of 78 implanted lesions; 69%) as in mammary fatpads (51/64; 80%) of nude mice (Jensen and Wellings, 1976). None of the explants increased in size, but intraductal proliferation occurred more readily in fatpads (25/51; 49%) than at subcutaneous sites (4/54; 7%). Histologically normal lobules from cancer patients were no more likely to proliferate than lobules from noncancerous patients when age was taken into consideration (nearly all noncancerous patients were under 50). However, lobules from cancer patients over age 50 were significantly more likely to proliferate than lobules from cancer patients under age 50 (15/28 vs. 5/33; 54% vs. 15%).

Another point of interest is that 20/22 (91%) cancer-associated lobules that proliferated were vascularized as compared with only 19/48(40%) of lobules that failed to do so. For lobules from noncancerous breasts, vascularization was observed in 3/7 (43%)proliferating lobules and in 9/28 (32%) lobules that did not proliferate. Thus, angiogenic activity may be indicative of precursor proliferative breast lesions.

ANGIOGENESIS OF EARLY BREAST LESIONS

Angiogenesis is known to be important in metastasis, and microvasculature density, defined as vessels per unit area, has been reported to be a reliable prognostic indicator in early stage breast cancer (Weidner et al., 1992). It has been suspected that angiogenesis might be the necessary switch for progression from CIS to invasive carcinoma. However, increased angiogenic activity appears to be an even earlier event in progression. Preneoplastic mammary lesions induce an angiogenic response when transplanted into the anterior chamber of the eye (Brem et al., 1977, 1978). In one study (Brem et al., 1978), fragments of human invasive breast cancers and CIS induced angiogenesis equally well (two thirds of the fragments were positive). Fragments from normal breast tissue, nonproliferative fibrocystic disease, fibroadenomas, and other benign lesions were not angiogenic. However, nearly one third of fragments from hyperplastic lesions were angiogenic.

Immunohistochemical studies have identified apparent angiogenic activity in ductular carcinoma in situ (DCIS) at an incidence similar to that found in hyperplastic lesions. Pure DCIS was associated with vascular cuffing detected by staining for factor VIII in 38% of cases (Guidi et al., 1994) and DCIS mixed with invasive carcinoma was associated with vascular cuffing in 23% of cases (Weidner et al., 1991). Vascularity was found to increase sequentially with progression from normal to proliferative disease without atypia (PDWA) to atypical hyperplasia to DCIS (Heffelfinger et al., 1996). Furthermore, comedo DCIS was more highly vascularized than other types. These findings are somewhat at odds with an earlier study in which size of blood vessels increased around areas of hyperplasia and DCIS but new vessels were not apparent (Ottinetti and Sapino, 1988). The latter authors suggested that the expansion of the epithelial mass impeded blood flow, causing stasis and vessel dilation, and was inconsequential because the phenomenon was detected in both DCIS and benign hyperplasias.

Both vascular concentration (number of vessels per μm^2 lesion field) and vascular density (vascular area per area of lesion field) were determined in a case control study of 24 cancer patients and 24 controls for whom benign biopsy specimens were available (Guinebretiere et al., 1994). All had been diagnosed as fibrocystic disease, but there were only four cases of atypical hyperplasia in each group. Factor VIII staining of the most proliferative slide from each patient and quantitative image analysis revealed that both vascular density and vascular concentration correlated with subsequent development of invasive cancer. Thus, in this study, vascular parameters identified high risk fibrocystic disease without atypia.

INVASION AND PROTEASES

An increase in expression of enzymes capable of degrading basement membranes and extracellular matrix would be expected to occur in the progression to invasive cancer. This has been observed for cathepsin D (Zhao et al., 1993; Schultz et al., 1994), type IV collagenases (Davies et al., 1993; Polette et al., 1993), and stromelysin 3 (Basset et al., 1990; Hahnel et al., 1993, 1994; Kawami et al., 1993; Wolf et al., 1993). However, altered expression can also be detected in benign breast lesions.

Three forms of cathepsin D with molecular weights of 27 kDa, 31 kDa, and 52 kDa were described in patient sera, and the proportion of the 31-kDa form was found to increase from 1.8% in normal controls to 13% in patients with proliferative breast disease and to 24% in cancer patients (Schultz et al., 1994). Quantitative in situ hybridization indicated a significant increase in cathepsin D mRNA in breast cancer cells compared to epithelial cells of benign mastopathies, but cathepsin D expression was higher in hyperplastic ductal epithelial cells than in lobules (Zhao et al., 1993). However, lobular carcinomas are more frequently positive for cathepsin D expression than are invasive ductal breast carcinomas (Domagala et al., 1993), so the significance of the differential expression in hyperplastic ductular and lobular epithelium is questionable.

Type IV collagenases may also be expressed in benign lesions. One study reported that 2/6 benign lesions and 13/17 invasive breast cancers expressed the 72 kDa, type IV collagenase (Polette et al., 1993), and another reported the detection by in situ hybridization of both 72-kDa and 92-kDa forms in breast cancers, fibroadenomas, and benign breast ducts and acini (Soini et al., 1994). The epithelium predominantly expressed the 92-kDa form whereas stromal cells and endothelial cells more strongly expressed the 72-kDa form. The benign lesions were not well described in these studies and interpretation is difficult. Production of a 62-kDa activated form of the 72-kDa gelatinase was reported to increase in benign breast lesions (9 fibroadenomas and 2 fibrocystic disease cases), and production continued to increase with progression to DCIS and invasive carcinoma (Lee et al., 1996).

Although stromelysin 3 is expressed by the stromal cells in close proximity to epithelial breast lesions, altered expression is probably a response to factors provided by the epithelium (Basset et al., 1993; Patel and Schrey, 1995). Few studies have examined expression of stromelysin in benign lesions, although fibroadenomas were reported to be negative (Hahnel et al., 1994). Uninvolved breast epithelium from cancer patients was also negative (Hahnel et al., 1993). Expression of stromelysin 3 by stromal cells occurs in most invasive cancers and in some DCIS (Hahnel et al., 1993, 1994; Engel et al., 1994). Levels of stromelysin are significantly higher in invasive cancer than CIS and higher in invasive ductular carcinoma than in invasive lobular carcinoma (Engel et al., 1994). Perhaps stromelysin 3 expression will identify a subset of CIS lesions that progress to invasive carcinoma more rapidly.

ESTROGEN RECEPTOR

To many, progression of breast cancer and loss of estrogen receptor (ER) are synonymous. The perception is that normal breast tissue is ER^+, is growth stimulated by estrogen, and is dependent upon estrogen for growth. Patients with ER^+ invasive carcinoma have better prognosis than those with ER-negative cancer. Furthermore, loss of receptor makes the cancer more difficult to treat because hormonal manipulation is not useful. However, in the normal breast very few epithelial cells are ER^+. The number varies with the menstrual cycle but by immunohistochemical methods they are fewer than 10% of the normal epithelial cell population (Petersen et al., 1987; Jacquemier et al., 1990). Expression of ER increases in benign breast lesions. Immunohistochemical analysis of ER expression in biopsy specimens from premenopausal patients found 6% of normal cells positive, 12% of epithelial cells in fibroadenomas, 13% in atypical lobular hyperplasia, and 31% in CIS (Jacquemier et al., 1990). However, in postmenopausal cases, 26% of normal cells, 25% of fibroadenoma cells, 73% of epithelial cells from proliferative lesions, and 30% from DCIS were positive.

In a case control study in which ER^+ status of benign lesions was defined as the presence of any nuclear staining, 84% of cases versus 57% of controls were ER^+ (Khan et al., 1994). Progesterone receptor status was similar—86% positive for both. The benign lesions were not further defined but the adjusted odds ratio for ER^+ benign lesions in this study was 6.5 (Khan et al., 1994). In a study of 140 benign breast lesions of various types, including fibroadenomas, papillomas, and ductal hyperplasias, 70% showed some nuclear staining (Giri et al., 1989).

Many immunohistochemical studies use the H score system which multiplies the % of cells positive by the intensity of staining (1^+, 2^+, or 3^+) for each positive cell. Thus, the maximum score is 300 (all cells 3^+). Using an H score of 20 as the criterion for classification as ER^+, Schmitt (1995) found that the proportion of positive cases was similar for PBD and breast cancer, but proliferation indices, determined by

MIB-1 antibody staining, were higher in ER^+ than in ER-negative hyperplastic lesions. On the other hand, MIB-1 staining of DCIS and invasive cancer cases was greater in ER-negative than in ER^+ cases, suggesting alteration in mitogenic response to estrogen with progression rather than alteration of receptor expression (Schmitt, 1995). However, due to the low H score used in this study, 80% of normal biopsies were also judged to be positive.

It has been reported that at least 25% of the epithelial nuclei stain strongly in 30% of benign breast lesions (Giri et al., 1989). This level of staining corresponds to biochemical assay values in excess of 10 fmol/mg protein (Giri et al., 1987), a level frequently used clinically as the cut off for classifying cancer as ER positive or negative. This study (Giri et al., 1989) indicates that 30% of benign lesions would have H scores of 50 or more (at least 25% of cells at 2^+ or more). Normal tissues, including areas co-existing with DCIS, were negative by this criterion whereas 12/39 fibroadenomas, 13/32 papillomas, 7/20 sclerosing adenoses, 12/49 ductal hyperplasias, and 21/48 DCIS were positive (Giri et al., 1989). Since an H score of 50 corresponds to 10 fmol/mg protein receptor protein, a level expressed by approximately half of breast cancers at diagnosis, it is clear that progression often involves increased ER in PBD and that loss of ER is a later event in progression.

The early expression of receptor most likely enables cells to respond to the mitogenic signal of estrogen thus promoting expansion of initiated epithelial cells. Loss of ER at later stages in breast cancer may reflect a loss of dependency on estrogen as a mitogenic signal. Amplification of erbB-2 is strongly associated with loss of ER expression in DCIS as well as in invasive carcinoma (Pavelic et al., 1992; Poller et al., 1993) and may reflect a role for an alternative mitogenic signal. Overexpression of other growth factors may enhance hormone-independent growth in the absence of estrogen. Overexpression of heregulin-β2 in MCF-7 cells leads to estrogen independence and resistance to antiestrogens (Tang, C.K. et al., 1996). Some tumor suppressor genes may function by maintaining hormonal regulation, the deregulation of which has been observed to correlate with loss of heterozygosity(LOH). LOH at 17q21 has been significantly correlated with absence of both estrogen and progesterone receptors in breast cancer, and LOH at 17p13.3 with loss of progesterone receptor (Ito et al., 1995). Hormonal regulation may be a central feature for many of the molecular alterations that occur in breast cancer.

MOLECULAR BIOLOGY OF PROLIFERATIVE BREAST DISEASE

A number of genes have been identified as putative oncogenes in breast cancer based on expression in breast cancer and correlation with prognosis or as tumor suppressor genes by LOH. A genetic sequence of events has been deduced for colon carcinoma (Fearon and Vogelstein, 1990). Although it is generally believed that breast cancer, too, is the culmination of numerous genetic alterations, a genetic sequence involved in progression for breast has not been deciphered. The road to cancer in the breast may take many routes. Perhaps complementation groups such as those described for immortalization (Pereira-Smith and Smith, 1988) exist such that one gene increases the probability of cancer only in the presence of second, specific genetic alteration(s). Undoubtedly genes yet to be identified, as well as the genes to be discussed, are involved.

ErbB-1

ErbB-1 encodes for the 170-kDa protein epidermal growth factor receptor (EGFR). More than 5% of cells overexpress EGFR in 56% of early invasive breast cancers, and expression may be associated with early relapse (Gasparini et al., 1992). In particular, cases that are EGFR positive with a growth fraction exceeding 7.5% as determined by antibody against Ki-67 are much more likely to relapse. EGFR may be overexpressed in about 25% of DCIS cases (Moller et al., 1989), but a possible role in the development of PBD is suggested by the report that EGFR was detected more frequently in breast smears of complex breast cyst fluids (12/17) than in smears of simple cyst fluids (5/23; Athanassiadou et al., 1992). Simple cysts are those with flattened epithelium and complex cysts have apocrine or hyperplastic epithelium. The prevalence of overexpressed EGFR in fine-needle aspirates obtained from high risk patients (35%) was significantly greater than from women of low risk (4%; Fabian et al., 1994). High risk patients were defined as those with a first-degree relative with breast cancer, prior breast cancer, or precancerous mastopathy. There is also evidence that overexpression of EGFR is common in fibroadenomas (Zelada-Hedman et al., 1994).

An association of EGFR expression and p53 expression has been described leading to the suggestion that the two interact in the pathogenesis of cancer (Horak et al., 1991). In that study, p53-positive tumors (staining-detectable with Pab240) were found to have a mean concentration of 31 fmol EGFR per mg of membrane protein compared to 14 fmol/mg for p53-negative tumors. However, nearly half of the p53-positive cases were EGFR negative. A number of studies have found a negative correlation between expression of EGFR and ER (Sainsbury et al., 1987; Horak et al., 1991; Koenders et al., 1991) but many cases express both receptors (Koenders et al., 1991). The latter study found that, although erbB-1 levels decreased with increasing ER levels, 173/531 cases (33%) expressed both using at least 10 fmol/mg cytosolic protein as the criterion for ER status and 0.5 fmol/mg of membrane protein as the criterion for EGFR expression (median values were 40 fmol/mg). Using different criteria (5 fmol/mg for ER and 3 fmol/mg for EGFR), 25% of normal breast samples and 33% of benign/fibrocystic samples were found to be positive for both ER and EGFR (Barker et al., 1989). However, individual cells within breast cancers apparently are never positive for both (Sharma et al., 1994). Thus, EGFR and ER expression appear to be mutually exclusive at the cellular level.

Normal breast epithelium, fibrocystic disease, and fibroadenomas all stained with weak to moderate intensity with antibodies to both erbB-2 and EGFR (Tsutsumi et al., 1990). Invasive carcinomas overexpressed one or the other, but not both, in 13 of 36 cases. In a single case, both erbB-2 and EGFR were overexpressed but in a reciprocal manner in which erbB-2 was overexpressed in the DCIS component and the EGFR was overexpressed in the invasive component (Tsutsumi et al., 1990). Based on this case, the authors suggest that erbB-2 loss at the stage of invasion might explain the decrease in incidence of erbB-2 expression in invasive cancer from DCIS. However, this case seems to be the rare exception (see below). Both erbB-2 and EGFR were detected in only 6/126 breast cancers by immunohistochemistry, but these patients had significantly worse prognoses than did patients with invasive cancers expressing one or the other or neither (Toi et al., 1994). This study did not address whether EGFR and erbB-2 were detected in the same or separate cells within the six $EGFR^+$/$erbB\text{-}2^+$ cancers.

ErbB-2

The *erbB-2* gene encodes for a 185-kDa receptor tyrosine kinase with homology to the epidermal growth factor receptor erbB-1, erbB-3, and erbB-4. The natural ligand for erbB-2 has not been identified, but erbB-2 phosphorylation may occur by transmodulation when other members of the erbB kinase family react with their ligands (e.g., epidermal growth factor; Akiyama et al., 1988; Stern and Kamps, 1988; Johnson et al., 1993) and neu differentiation factor (Culouscou et al., 1993; Plowman et al., 1993; Carraway et al., 1994; Kita et al., 1994; Sliwkowski et al., 1994; Tzahar et al., 1994). This transmodulation occurs through heterodimerization of erbB-2 with the other kinases.

Although overexpression of *erbB-2* is associated with poor prognosis in human breast cancer, erbB-2 is rarely detected in benign lesions (De Potter et al., 1989; Lizard-Nacol et al., 1995; Millikan et al., 1995). *ErbB-2* is not overexpressed in either lobular CIS (Gusterson et al., 1988b; Ramachandra et al., 1990; Porter et al., 1991) or invasive lobular carcinoma (Porter et al., 1991; Barbareschi et al., 1992). However, overexpression of *erbB-2* is frequent in DCIS and invasive ductal carcinoma. The precipitous rise in expression at the stage of DCIS suggests that overexpression of *erbB-2* may be sufficient to produce this step in progression.

Although erbB2 is detected in DCIS more frequently than in invasive cancer (Gusterson et al., 1988a, b; De Potter et al., 1989; Kobayashi et al., 1992), studies of mixed lesions containing both in situ and invasive components invariably report that the two components express *erbB-2* coordinately (Ramachandra et al., 1990; Liu et al., 1992; Iglehart et al., 1995). Thus, DCIS in which *erbB-2* is not amplified gives rise to invasive carcinoma in which *erbB-2* is not amplified and DCIS in which *erbB-2* is amplified gives rise to invasive carcinoma in which *erbB-2* is amplified. Since *erbB-2* is not amplified in the majority of invasive breast cancers, these data suggest that DCIS in which *erbB-2* is not amplified is more likely to progress to invasive carcinoma than is *erbB-2*–positive DCIS. The latter conclusion is somewhat counterintuitive because *erbB-2* expression by invasive breast cancer is associated with poor prognosis (Slamon et al., 1987, 1989; Paik et al., 1990; Gullick et al., 1991; Gusterson et al., 1992; Sauer et al., 1992). If *erbB-2*–expressing DCIS had a poor prognosis as indicated by more rapid/frequent progression to invasive cancer, the incidence of *erbB-2* expression might be expected to be higher in invasive cancer than in DCIS.

Both the incidence of *erbB-2* amplification and the propensity of DCIS lesions to progress to invasive cancer vary with the histotype of DCIS. Comedo DCIS is more frequently *erbB-2*–positive than is cribriforming DCIS (Bartkova et al., 1990; Ramachandra et al., 1990). Ramachandra and colleagues found that 19% of pure cribriforming DCIS and 55% of mixed lesions containing both cribriforming in situ and invasive components expressed *erbB-2*. On the other hand, 62% of pure comedo DCIS lesions were *erbB-2*–expressing compared to only 49% of mixed lesions (Ramachandra et al., 1990). These results indicate that *erbB-2*–positive cribriforming DCIS are more likely to progress to invasive cancers than are *erbB-2*–negative lesions, whereas *erbB-2*–negative comedo DCIS are more likely to progress than are *erbB-2*–positive comedo DCIS. Once again, this suggests complementation groups, a requirement for coexpression of particular combinations of genotypic and phenotypic characteristics in order for cancer to develop and progress.

p53

Several studies using antibodies to detect p53 protein indicate that p53 mutations are rare in benign lesions (Bartek et al., 1990; Heyderman and Dagg, 1991; Barbareschi et al., 1992; Eriksson et al., 1994; Stephenson et al., 1994; Umekita et al., 1994; Schmitt et al., 1995; Younes et al., 1995; Chitemerere et al., 1996; Siziopikou et al., 1996). The total incidence of staining in these combined studies was 47/646 for benign lesions but most studies included fibroadenomas, which are frequently positive (Younes et al., 1995), as well as hyperplasias. In studies of hyperplastic lesions, only two of 80 specimens (2.5%) of PDWA and one of 108 atypical hyperplastic lesions were positive (Heyderman and Dagg, 1991; Umekita et al., 1994; Schmitt et al., 1995; Siziopikou et al., 1996). DCIS is more frequently positive, with a total of 58 of 300 lesions staining for p53 (Bartek et al., 1990; Eriksson et al., 1994; Umekita et al., 1994; Leal et al., 1995; Schmitt et al., 1995; Chitemerere et al., 1996; Albonico et al., 1996; Siziopikou et al., 1996), and the proportion of invasive cancer-positive for p53 is higher yet; a total of 180 of 734 in total from a number of studies (Bartek et al., 1990; Heyderman and Dagg, 1991; Eriksson et al., 1994; Seth et al., 1994; Stephenson et al., 1994; Umekita et al., 1994; Younes et al., 1995; Nakopoulou et al., 1996). Seth et al. (1994) revealed that benign lesions, previously removed from two patients with p53-positive invasive cancers, were negative for p53.

The fact that CIS and carcinoma are similar in *p53* expression but that hyperplastic lesions are rarely positive suggest that *p53* mutation, like *erbB-2*, is not an early event in breast disease.

myc

C-myc is amplified in about 20% of breast cancers (Garcia et al., 1989; Berns et al., 1992). Interestingly, only 1% of cases are positive for both *erbB-2* and *c-myc* (Garcia et al., 1989; Berns et al., 1992), again indicating at least two distict molecular pathways to cancer. Although *erbB-2* amplification is a useful prognostic factor for ER-negative cases, *c-myc* amplification is a useful prognostic factor for ER^+ cases (Berns et al., 1992). Unlike *erbB-2*, *c-myc* amplification occurs early in proliferative breast disease (Agnantis et al., 1992; Escot et al., 1993; Hehir et al., 1993; Pechoux et al., 1994; Brem et al., 1977).

Myc expression is induced by estrogen and appears to be hyper-responsive in atypical hyperplasia in that expression is very high during the follicular phase, but not the luteal phase, of the menstrual cycle in atypical hyperplasia relative to other benign lesions (Escot et al., 1993). This study also found that *myc* expression was significantly higher in benign lesions of patients with a first-degree family history of breast cancer.

C-myc amplification alters the reponse of breast epithelial cells to growth factors. EGF and FGF, which are normally mitogenic, transform cells that overexpress *c-myc* (reviewed in Dickson et al., 1992). Overexpression of *c-myc* in benign breast lesions may identify a high-risk subset as suggested by a finding of Hehir and colleagues: 62% of benign lesions from patients subsequently developing breast cancer were *c-myc*–positive, compared to 13% from patients who remained cancer-free. (1993).

Cyclin D1/PRAD1

Cyclin Dl/PRADl is a cell cycle regulator necessary for G1 progression. The protein is overexpressed in one-third to one-half of invasive breast cancers as detected by immunohistochemistry (Bartkova et al., 1994, 1995; Zhang et al., 1994; McIntosh et al., 1995; Zukerberg et al., 1995). Overexpression appears to first occur at DCIS (Bartkova et al., 1995; Weinstat-Saslow et al., 1995) and, in 37 cases of mixed DCIS and invasive carcinoma, was concordant in both invasive and noninvasive components (Bartkova et al., 1995). Elevated levels of PRADl mRNA were detected by in situ hybridization in proliferative disease, but PDWA and atypical hyperplasia were similar (18% in both; Weinstat-Saslow et al., 1995). The latter study revealed elevated PRADl mRNA in 76% of noncomedo DCIS, 87% of comedo DCIS, and 83% of invasive cancer. Thus, overexpression of PRADl, like p53 and erbB2, is associated with rapid progression to the DCIS stage.

Ras

Ras genes, *Ha-ras, N-ras,* and *Ki-ras*, encode for 21-kDa proteins important in signal transduction. Following binding of growth factors to cell surface receptor tyrosine kinases, p21 is transiently activated by phosphorylation and activates downstream serine/threonine kinases such as mitogen-activated protein (MAP) kinases and Raf-1, which in turn translocate to the nucleus where nuclear transcription factors such as Fos and Jun are phosphorylated/regulated. Activating mutations of *ras* result in a p21 protein that remains phosphorylated, maintaining active signal transduction pathways , in the absence of external stimuli.

Ras mutations are common in human cancer, occurring in approximately 30% of all cancers combined. However, mutated *ras* is much more prevalent in some types than in others. Whereas 90% of pancreatic cancers have a mutated *ras* gene (Almoguera et al., 1988), fewer than 5% of breast cancers do so (Kraus et al., 1984; Kozma et al., 1987; Rochlitz et al., 1989; Prosperi et al., 1990; Clark and Der, 1995). Nevertheless, *ras* may be very important in breast cancer. The p21 ras protein is frequently overexpressed in breast cancer (Horan Hand et al., 1984; DeBortoli et al., 1985) and overexpression is significantly correlated with poor clinical prognosis (Clair et al., 1987; Querzoli et al., 1988; Dati et al., 1991; Bland et al., 1995). Overexpression of p21 ras protein occurs early in the development of breast disease (Ohuchi et al., 1986; Spandidos et al., 1987; Agnantis et al., 1992; Going et al., 1992; Pechoux et al., 1994). Normal breast epithelium (Spandidos et al., 1987) and lactating breast epithelium (Thor et al., 1986) do not express sufficient p21 ras to be identified, and only 1% of cells in fibrocystic disease without hyperplasia stain positive for ras (Ohuchi et al., 1986). Cystic disease associated with apocrine metaplasia is more often positive (Agnantis et al., 1992). As the stage of proliferative disease advances, the proportion of cells positive for ras increases from 18% for PDWA to 34% for atypical hyperplasia, 54% for CIS, and 62% for invasive carcinoma (Ohuchi et al., 1986). Differences in the Ohuchi study were statistically significant at each sequential stage: fibrocystic disease versus PDWA, $P<0.01$; PDWA versus atypical hyperplasia, $P<0.05$; and atypical hyperplasia versus carcinoma (Pooled invasive and CIS), $P<0.01$; there was no difference in ras expression between CIS and invasive carcinoma. A more recent study confirmed the stepwise increase in ras expression based both on numbers of positive cells and intensity of staining (Going et al., 1992). As in the earlier study, significant overexpression of *ras* was detected in PDWA.

Since numerous studies implicate ras function in PBD and breast cancer, it is curious that mutated ras is so infrequent. Perhaps breast is very susceptible to even minor disruptions in the *ras* pathway that could result from a number of genetic events, whereas pancreatic cells, for example, are made malignant only by full *ras* activation. If we assume that the rate of mutation of ras in breast cells is the same as in pancreatic cells, the number of breast cancers with mutated ras should be similar to the number of pancreatic cancers with mutated ras, suggesting that the low percentage of mutant ras breast cancers reflects the greater number of alternative complementation groups that can lead to this disease. In the United States, the annual incidence of pancreatic cancers in white women is 7.5 per 100,000 compared to 113 breast cancers per 100,000 women (age adjusted) (Devesa et al., 1995). Thus, the incidence of mutant ras in breast cancer (5.65 cases per 100,000, assuming 5% have mutant ras) is similar to the incidence of mutant ras in pancreatic cancers (6.75 cases per 100,000, assuming 90% have mutant *ras*). Furthermore, mutations in a number of the components of the ras signal transduction pathway, including upstream receptor tyrosine kinases (eg., PDGFR, EGFR) and downstream serine/threonine kinases or transcription factors, are functionally equivalent to ras mutation (Clark and Der, 1995). Perhaps breast epithelium has a number of genes (e.g., receptor tyrosine kinases) active that pancreatic cells do not so that more genes are at risk in breast cancer cells which, if mutated, are surrogates for activated *ras*.

BRCA1

BRCA1 mutations are present in approximately three-fourths of families with ovarian and breast cancer and in about half of families with only breast cancer. The molecule has a zinc-binding ring finger domain suggesting a function in regulation of transcription, but the protein has also been reported to be a secreted product that is able to inhibit tumor cell growth (Ezzell, 1996; Holt et al., 1996; Jensen, R.A. et al., 1996). Athough a single mutation has been found in 1% of the Ashkenase Jewish population, over 100 mutations have been identified, many of which occur in only one or two families (Collins, 1996). Not all BRCA1 mutations are equal. Mutations in either the amino terminal or carboxy terminal conserved domains result in tumors with higher proliferation rates than mutations at other sites (Sobol et al., 1996). Furthermore, cancer grade segregates by family, some giving rise to more proliferative, aggressive invasive carcinoma than others (Eisinger et al., 1996).

In a study of 202 cases from 36 BRCA1 breast cancer families, only four pure CIS lesions were found, two of which were sporadic cancers, that is, their hosts were not carriers of the mutant alleles (Sun et al., 1996). This finding suggests that, if present in BRCA1 patients, DCIS is a very transient stage that rapidly progresses to invasive carcinoma.

BRCA1 is seldom mutated in sporadic breast cancer (Xu and Solomon, 1996). However, allelic imbalance at the *BRCA1* locus was seen in 74% of DCIS cases (Munn et al., 1996). Thus, the role of BRCA1 in sporadic breast cancer is still undetermined.

BRCA2

BRCA2 is a second hereditary breast cancer gene and is involved in site-specific breast cancer families. BRCA2 may account for most of the male breast cancer families but was reported to be mutated in only eight of 49 families with breast cancer only (Phelan et al., 1996). Like BRCA1, the role of BRCA2 in sporadic breast cancer is unclear. Although loss of heterozygosity at the *BRCA2* locus is common in sporadic breast cancer, mutations in BRCA2 are rare (Lancaster et al., 1996; Miki et al., 1996).

Other Suppressor Genes

A number of other suppressor genes have been implicated in breast cancer. *Nm23*, originally characterized as a metastasis suppressor gene, has been reported to be uniformly expressed in all epithelial cells of benign breast lesions (35 cases that included fibrocystic disease and fibroadenomas but were not further described) and noncomedo DCIS (18 cases including solid, cribriform, and micropapillary types), but some cells were negative for *Nm23* in all 7 cases of comedo DCIS (Royds et al., 1993). For invasive cancers, the percent of negative cases increased with grade (Royds et al., 1993), supporting observations that comedo DCIS is associated with more aggressive cancers.

An immunohistochemical study of retinoblastoma susceptibility gene (RB1) expression reported that 16 of 56 invasive cancers contained some negative cells, but all epithelial cells in benign breast lesions were uniformly positive (Varley et al., 1989). The benign lesions were not described but fibrocystic disease and fibroadenomas were included and no DCIS cases were examined.

If LOH occurs randomly, most events that result in LOH probably are irrelevant to the development of breast disease and cancer. The baseline incidence for LOH appears to be less than 5% based on a study using randomly selected probes, that is, probes for chromosomal regions not known to be associated with cancer (Chen, L.C. et al., 1992). Significantly increased incidences of LOH have been described for more than half of the chromosome arms (Table 1), and more than one locus has been implicated on several arms in invasive breast cancer (Callahan et al., 1993; Cleton-Jansen et al., 1994; Kirchweger et al., 1994; Merlo et al., 1994; Dorion-Bonnet et al., 1995; Gudmundsson et al., 1995; Nagai et al., 1995; Radford et al., 1995; Yaremko et al., 1995; Munn et al., 1996; Sheng et al., 1996; Spirin et al., 1996). Thus, a great number of suppressor genes may be involved in breast cancer. To determine which are important in early stages of breast cancer development, studies have compared LOH in low grade invasive carcinomas to LOH in high grade cancers or examined DCIS, sometimes comparing comedo to noncomedo histologic types. Only a few studies have utilized benign breast lesions.

LOH is frequently observed in DCIS. Low nuclear grade DCIS averaged 1.2 chromosomal arms lost per case whereas intermediate and high grade DCIS averaged a loss of 5.6 chromosomal arms (Fujii et al., 1996b). Losses of 16q and 17p were common in all grades of DCIS suggesting a role in initiation of the DCIS stage (Fujii et al., 1996b). Microdissection and microsatellite length polymorphism analysis revealed loss of 16q in 89% of DCIS cases with the most frequent loss in the region q23.3–24.1 (20 of 26 informative cases; Chen, T. et al., 1996). In a number of studies of cases of mixed DCIS and invasive cancer, LOH was concordant in the two lesions as one would expect if DCIS is a precursor for the invasive component (O'Connell et al., 1994; Koreth et al., 1995; Zhuang et al., 1995). Additional chomosomal arms are lost and heterogeneity develops with progression (Fujii et al., 1996a, b). Other analyses of DCIS indicate frequent LOH at 8p, 13q, and

Table 1. Loss of Heterozygosity in Breast Disease.

LOH	*Familial Invasive Cancer*	*Sporadic Invasive Cancer*	*DCIS*	*PBD*
1p		Genuardi et al., 1989; Devilee et al.,1991a; Callahan et al., 1992; O'Connell et al., 1994	Munn et al., 1995; Fujii et al., 1996b	
1q		Chen, L.C. et al., 1989, 1992; Callahan et al., 1992; Bieche et al., 1993	Fujii et al., 1996b)	Dietrich et al., 1995
2p		O'Connell et al., 1994	O'Connell et al., 1994	O'Connell et al., 1994
2q		O'Connell et al., 1994		
3p		Ali et al., 1989; Sato et al., 1990; Devilee et al., 1991b; Callahan et al., 1992; Chen, L.C. et al., 1992; O'Connell et al., 1994		Dietrich et al., 1995; Panagopoulos et al., 1996
4q		O'Connell et al., 1994	O'Connell et al., 1994	O'Connell et al., 1994
5q		Thompson et al., 1993		
6q		Devilee et al., 1991b; O'Connell et al., 1994; Hankins et al., 1996; Sheng et al., 1996	Fujii et al., 1996b; Hankins et al., 1996	
7q		Bieche et al., 1992; Callahan et al., 1992; Champeme et al., 1995		
8p	Lindblom et al., 1993b	O'Connell et al., 1994; Yaremko et al., 1995	Radford et al., 1995	
8q		O'Connell et al., 1994		
9p		Eiriksdottir et al., 1995	Fujii et al., 1996b	
11p		Ali et al., 1987; Mackay et al., 1988a; Callahan et al., 1992; Gudmundsson et al., 1995	Fujii et al., 1996b	

continued

Table 1. Continued

LOH	*Familial Invasive Cancer*	*Sporadic Invasive Cancer*	*DCIS*	*PBD*
11q		Carter et al., 1994; Hampton et al., 1994; O'Connell et al., 1994; Gudmundsson et al., 1995; Koreth et al., 1995	Koreth et al., 1995; Zhuang et al., 1995; Fujii et al., 1996b	Dietrich et al., 1995
12p		Spirin et al., 1996		
13q	Wooster et al., 1994	Lundberg et al., 1987; T'Ang, A. et al., 1988; Callahan et al., 1992; Chen, L.C. et al., 1992; Tamura et al., 1994	Radford et al., 1995; Fujii et al., 1996b	
14q		O'Connell et al., 1994		
15q		Wick et al., 1996		
16q	Lindblom et al., 1993a, b	Cleton-Jansen et al., 1994; O'Connell et al., 1994; Dorion-Bonnet et al., 1995; Skirnisdottir et al., 1995	Radford et al., 1995; Tsuda et al., 1995; Fujii et al., 1996b	O'Connell et al., 1994; Lakhani et al., 1995
17p	Lindblom et al., 1993b	Mackay et al., 1988b; Cropp et al., 1990; Devilee et al., 1990; Chen, L.C. et al., 1991, 1992; Callahan et al., 1992; Merlo et al., 1994; O'Connell et al., 1994; Ito et al., 1995	Radford et al., 1995; Tsuda et al., 1995; Fujii et al., 1996b	Lakhani et al., 1995
17q	Smith et al., 1992; Lindblom et al., 1993b	Devilee et al., 1990, 1991b; Callahan et al., 1992; O'Connell et al., 1994; Ito et al., 1995	O'Connell et al., 1994; Radford et al., 1995	O'Connell et al., 1994
18p		Chen, L.C. et al., 1992		
18q		Devilee et al., 1991c; Callahan et al., 1992; Thompson et al., 1993; Kashiwaba et al., 1995; Schenk et al., 1996		
19p	Lindblom et al., 1993b			
22q		Chen, L.C. et al., 1992		

17p as well as 16q (Radford et al., 1995; Stratton et al., 1995). The latter two studies consisted primarily of comedo DCIS. A recent study restricted to noncomedo DCIS found LOH in 40% of cases at 3p, 11q, 13q, and 16q (Man et al., 1996). There were two areas involved on 11q and the 3p loss appeared to be loss of the FHIT gene (fragile histidine triad), which is of interest because FHIT as well as another gene located in the chromosmal band 3p14.2, PTPRG (human receptor tyrosine phosphatase λ gene), were deleted in three samples of atypical hyperplasia but not in fibroadenoma as assessed by RT-PCR (Panagopoulos et al., 1996).

PCR was used to study LOH in 60 breast cancer cases for which PDWA and/or atypical hyperplasia coexisted within the surgical specimen (O'Connell et al., 1994). This study found frequent LOH at 17q21, 16q21, and 2pter in hyperplastic lesions (16–25%). The most valuable information from this study was the finding that LOH patterns were frequently shared among different levels of progression within the same lesion: 50% of hyperplastic lesions shared LOH with coexistent DCIS and 80% of DCIS shared LOH with coexistent invasive cancer. Thus, early proliferative lesions are very likely to be precursors in the progression to breast cancer.

Correlations of different sets of genetic alterations have been described in a number of studies: LOH on chromosomes 11p, 17p, and 18q frequently occurred in the same tumor in one study (Callahan et al., 1992); LOH at 11p and 11q were independent of each other, but 90% of tumors with LOH at 11q also showed LOH at 17p in another (Carter et al., 1994); and concordant loss of 17p and 16q and of 13q and 17p as well as loss of 17p with erbB-2 amplification were seen in a third study (Sato et al., 1991). These findings lend credence to the existence of complementation groups in breast cancer development.

Understanding of the genetic basis of the development of proliferative breast disease and progression obviously requires much more work. The use of LOH patterns to determine complementation groups could be very informative in deciphering the role of various gene deletions, amplifications, and deletions and altered transcriptional/translational regulation observed in various studies.

EXPERIMENTAL MODEL FOR HUMAN PROLIFERATIVE BREAST DISEASE

Recently, we established a new and unique model of early human breast cancer progression. This model, which is called "MCF10AT", consists of preneoplastic human breast epithelial cells that are able to grow in immune-deficient mice where they undergo a sequence of progressive histological changes culminating in cases of frank neoplasia in about 25% of the animals (Miller et al., 1993; Dawson et al., 1996). MCF10AneoT is a cell line derived from a normal human breast epithelial line, which spontaneously immortalized (Soule et al., 1990) and was then transfected with mutated H-ras (Basolo et al., 1991). MCF10AneoT is a heterogeneous mixture containing a subpopulation that is a multipotent human breast epithelial stem cell that can be selected in vivo by its ability to form persistent lesions in nude/beige mice (Miller et al., 1993). Cell lines established from lesions after serial transplantation into nude/beige mice (MCF10AT# where # represents transplant generation) show more severe atypia at earlier times post-transplantation than does MCF10AneoT (Dawson et al., 1996). Although ER can not be detected (by Abbot antibody) on cells in xenograft lesions, Dr. P.V.M. Shekhar has shown that the *ras*-transfected MCF10AT cells (but not MCF10A) do express functional ER (Manuscript submitted).

The reproducible establishment of representative stages in early breast cancer progression from the MCF10AT model offers an unprecedented opportunity to analyze critical events in human breast carcinogenesis. Normal ducts form initially but become proliferative and sporadically progress to carcinoma in some cases. The ducts formed in xenografts are bilayered, being composed of both myoepithelial and luminal epithelial layers. These normal structures may persist for months in immune-deficient mice. The MCF10AT model clearly is a precursor cell line that forms all intermediate stages of proliferative breast disease. Results with our model strengthen the perception of progression from hyperplasia to atypical hyperplasia to invasive breast carcinoma.

We confirmed the invasive phenotype in the MCF10AT lesions classified as invasive carcinoma by demonstrating the loss of basement membrane utilizing a silver stain method (Tait et al., 1996a). Unlike the MCF10AneoT (i.e., *ras*-transfected cells prior to in vivo selection), all variants derived from xenografts formed intermediate lesions that resemble the lesions seen in human PBD including (PDWA), atypical hyperplasia, and DCIS (Dawson et al., 1996).

It is a widely held point of view that mutations in differentiated epithelial cells are of little importance due to limited growth potential of such cells and that cancer results only from mutations in stem cells. Indeed, we have found that MCF10AT lines contain stem cells able to differentiate in xenografts into myoepithelial as well as luminal epithelial cells (Tait et al., 1996b). Although ducts formed by the stem cell appear to be normal in most respects, cytokeratin differentiation is abnormal. MCF10AT cells express both cytokeratin 14 and cytokeratin 18 (Pauley et al., 1996), a pattern consistent with a stem cell able to generate both cytokeratin 14–expressing myoepithelial cells and cytokeratin 18–expressing luminal cells (Taylor-Papadimitriou et al., 1989). However, neither myoepithelial nor luminal epithelial cells in the xenograft lesions consistently express either cytokeratin 14 or 18 (not shown). MCF10AT stem cells are unique, differing from previously described human breast and rat mammary stem cells in their capacity to produce organized ductular structures with myoepithelium properly oriented between a basement membrane and the luminal epithelium in situ in xenografts. Although our failure to consistently detect cytokeratins 14 and 18 in situ may be due to technical difficulties with the specific antibodies used, the antibodies did stain the appropriate cell layers in control, formalin-fixed, paraffin-embedded human breast tissue. The failure of

MCF10AT cells to differentiate fully in xenografts may be critical to their preneoplastic nature.

As discussed previously, although *ras* mutations are rare in breast cancer, overexpression of *ras* p21 is frequently detected. Mutation of *ras* is just one of a number of ways that ras signal transduction pathways are activated. The relative infrequency with which *ras* mutations are seen in human breast cancer may reflect a very large number of active growth factor receptors in breast epithelium which, whether mutant or amplified, result in overexpression of *ras*. The mechanism(s) by which *ras* is overexpressed in human breast cancer is not known. Normal c-Ha-ras–transfected cells, MCF10AneoN (Basolo et al., 1991), have an extra copy of *c-Ha-ras* but the protein is not overexpressed (unpublished) and the cells do not form xenograft lesions (Miller et al., 1993). Insertion of the mutant *ras* resulted in increased p21 protein and transformation of MCF10A cells. Although the preneoplastic phenotype in xenografts was observed following ras transfection, we have found that not all clones of MCF10AneoT are able to form lesions in vivo (Miller et al., 1996). Restriction size fragment analysis demonstrated, however, that clones unable to form preneoplastic lesions retained the activated *c-Ha-ras* and confirmed that the insertion site of the activated *c-Ha-ras* was the same for these clones as for the variants selected for ability to form lesions in vivo. Furthermore, Western blotting with antibodies specific for the codon 12 valine *c-Ha-ras* demonstrated that mutant p21 protein was comparable in nonlesion-forming clones and lesion-forming variants as well (Miller et al., 1996). Thus, activated *c-Ha-ras* is not sufficient to produce the preneoplastic phenotype of the MCF10AT human breast stem cells. This was an important finding because it offers the opportunity to identify the genetic alteration(s) involved in this critical early stage in proliferative breast disease by comparing the nonlesion-forming clones with the preneoplastic variants selected by their ability to form preneoplastic lesions.

The karyotype of the MCF10AT cells in vivo is remarkably stable despite long-term passage in xenografts. The only alterations have occurred during extended *in vitro* maintenance. An additional marker chromosome, t(3;17)(p13;p12), appeared in MCF10A cells in vitro between the time of immortalization and the time of transfection with T24 *ras* (Table 2). Likewise, MCF10AneoT cells were initially identical to the untransfected MCF10A cells but acquired an additional, apparently normal copy of chromosome 9 prior to the first xenograft experiment. Subsequent MCF10AT# variants isolated from xenograft lesions have been unchanged regardless of the histology of the lesion from which the cells were derived (Table 2).

However, we do *not* suggest that progression to carcinoma occurs without karyotypic alterations in the MCF10AT model. Breast cancer cells are notoriously difficult to establish in culture from solid tumors; therefore we hypothesize that the preneoplastic stem cell with the depicted karyotype has a growth advantage in vitro. Thus, even in xenografts that appear to be pure carcinoma, the precursor stem cell exists covertly and grows when primary cultures are established in vitro. The fact that these variants established from xenografts have been consistently preneoplas-

Table 2. Karyotypes of MCF10A and MCF10AT Series

MCF-10A (initial): 46,XX,der(3)t(3;9)(p13;p22),der(6)t(6;19)(p25;q12),der(9)t(3;9)(p13;p22)t(3;5)(p26;q31)
MCF-10A (later stage) and MCF10AneoT (initial): 46,XX,der(3)t(3;9)(p13;p22),t(3;17)(p13;p12),der(6)t(6;19)(p25;q12),der(9)t(3;9)(p13;p22)t(3;5)(p26;q31)
MCF10AneoT (prior to xenografting) and MC10AT# variants: 47,XX,der(3)t(3;9)(p13;p22),t(3;17)(p13;p12),der(6)t(6;19)(p25;q12),+9,der(9)t(3;9)(p13;p22)t(3;5)(p26;q31)

tic stem cells rather than carcinomas supports this concept. When clones are established directly from xenografts by modifying the tissue disaggregation method, other karyotypes have been isolated. Thus, we believe that cells with other karyotypes do develop within the lesions in vivo but are overgrown by the dominant stem cell in vitro. We have recently isolated tumorigenic variants (i.e., carcinomas form directly with no precursor lesion) by serially transplanting lesions by trocar before establishing cells in culture.

Although the MCF10AT xenografts progress to carcinoma with approximately a 25% incidence, rapidly growing xenografts are rare. Carcinoma has generally been diagnosed only after removal of the lesion and histologic examination. Indeed, the largest lesion to date has weighed 200 mg and most weigh less than 50 mg. When one considers the natural history of human breast cancer, which takes years to develop, the slow growth of these relatively early carcinomas is not surprising. However, results with the MIB-1 antibody against the Ki-67 nuclear proliferation antigen clearly demonstrate that the proliferative rate increases with progression in the MCF10AT xenograft model from 5% in simple hyperplastic lesions to 10% in CIS and 20% in invasive cancers. These rates resemble those reported for lesions from human patients (Barbareschi et al., 1992; Pavelic et al., 1992; Jensen, V. et al., 1995).

The MCF10AT model should be valuable to analyze genetic alterations in the development of breast cancer. Specific genes can be experimentally altered in the cells to determine the effect on the incidence of cancer in xenografts, and new genes may be identified by comparing cells before and after spontaneous progression in xenografts.

A number of genes implicated in breast cancer are being analyzed in the MCF10AT system. Immunohistochemical studies with DO7 antibody to *p53* (detects both wild type and mutant forms) found that staining was not a feature of hyperplastic lesions, CIS, or adenocarcinomas developing in MCF10AT xenografts; significant increased expression of *p53* was detected only in squamous carcinomas (Iravani et al., 1996). A number of investigators have found that *p53* is not mutated in MCF10A (Diella et al., 1993; Gudas et al., 1995; Merlo et al., 1995) nor MCF10AT cells that form preneoplastic lesions in xenografts (Chen, Y.Q. et al.,

1995; Shekhar et al., 1995). Taken together, these data suggest that mutated *p53* does not play a significant role in producing the preneoplastic phenotype or in progression to adenocarcinoma but may be important in generation of squamous carcinoma. However, the expression of a conformationally altered, but not mutated, *p53* does increase with successive transplant generations (MCF10AT1/ MCF10AT2B/ MCF10AT3B) and is accompanied by diminished normal *p53* function (Shekhar et al., 1995).

Although *erbB-2* is rarely amplified in hyperplastic human biopsy specimens, *erbB-2* overexpression can be detected beginning at the stage of atypical hyperplasia in the MCF10AT xenograft model (Iravani et al., 1996). As in human cancer, the proportion of *erbB-2*–overexpressing MCF10AT lesions increases at the stage of CIS and decreases somewhat in invasive adenocarcinomas. Although *erbB-2*–expression increases two- to threefold with in vivo passage (Wang and Miller, unpublished), MCF10AneoT cells transfected with *erbB-2* (MCF10HE) are not tumorigenic (Ciardiello et al., 1992), nor do they form persistent lesions in immune-deficient mice (Miller, unpublished). As discussed previously, because clones of MCF10AneoT may not form persistent preneoplastic xenograft lesions, we hypothesize that a defect in addition to the T24 *ras* is necessary to produce the preneoplastic phenotype observed in MCF10AneoT and MCF10AT variants. We suggest that the MCF10HE cells, which were cloned from MCF10AneoT after transfection with the erbB-2 expression construct (Ciardiello et al., 1992), lack that hypothetical genetic defect. Growth in vivo selects for the second genetic defect and clones of MCF10AT variants do form lesions.

CONCLUSION

The number of genetic alterations seen in breast cancer makes the task of constructing a model of genetic progression daunting. In general, the clinician pays little attention to any given genetic factor and bases patient management primarily upon tumor size, stage, and hormone receptor status. It is imperative that we decipher the accumulating genetic data in breast cancer in order to apply this information in optimizing patient care.

Patterns of LOH and gene expression in cancer as well as in experimental cell models support the notion that breast cancer can develop from a number of pathways of interdependent genetic changes. Thus, it is possible that future studies will allow the construction of several complementation groups that lead to breast cancer. The groups are likely to be degenerative such that one gene defect may be involved in a number of pathways. Other defects may be specific to one group and thus occur relatively rarely. Depending upon the complementation group, prognoses may vary. A number of prognostic studies have sought to use multiple markers and multivariate analysis to identify such genetic groups in breast cancer. In the same way, we may be able to identify early precursor lesions, based on the particular patterns of genetic alterations that are very likely to progress to aggressive cancers. It may be possible to stratify patients with atypical hyperplasia into groups with much higher risk than four- to fivefold and groups with very little increased risk. Once these genes that are associated with early stages of breast disease are identified, we may begin formulation of complementation groups combining these with genetic alterations seen in later stages of breast cancer.

Patterns of concurrent genetic alterations may identify complementation groups, but other clues to the role of different genetic aberrations in proliferative breast disease and cancer are provided by the incidence of individual genetic lesions at different stages of progression. For example, if a genetic alteration predisposes a cell to progression following the additional genetic event(s), that first alteration should appear in a greater proportion of the precursor lesions than in the subsequent stage of progression, whereas a genetic event that precipitates progression to a more advanced stage should appear more frequently in the advanced lesion than in the precursor. This suggests that *p53*, *erbB-2*, and PRAD1, all of which are frequently detectable in CIS, may precipitate progression to CIS because expression at precursor stages are rare.

For those genetic events that may participate in multiple complementation groups, these arguments have limitations and one can not be dogmatic in application of these principles. The case of *erbB-2* overexpression suggests that *erbB-2* precipitates progression to DCIS but predisposes DCIS to progression to invasive cancer because *erbB-2* is rarely overexpressed in PBD, frequently expressed in DCIS, and less frequently overexpressed in invasive cancer. However, an argument was made earlier in this review that *erbB-2*–negative cribriforming DCIS are more likely to progress to invasive cancer than are *erbB-2*–positive cribriform DCIS, whereas *erbB-2*–positive comedo DCIS are more likely to progress than erbB-2–negative comedo DCIS. If comedo and cribriforming DCIS result from two different complementation groups and DCIS that overexpress erbB-2 are a different group than DCIS that do not, at least four different complementation groups are implicated. It is also interesting that *erbB-2* and *c-myc* are rarely overexpressed simultaneously, a fact that suggests separate complementation groups. Perhaps *c-myc*–positive DCIS progression to invasive cancer explains the higher likelihood of *erbB-2*–negative cribriform DCIS than of *erbB-2*–positive DCIS to progress to invasive cancer.

A third pattern of genetic alteration with sequential stages of progression is exhibited by *ras*. Overexpression of *ras* increases with each successive stage. One possible interpretation is that *ras* overexpression is an essential element of many complementation groups. An alternative possibility is that *ras* overexpression is a consequence of several other genetic changes and that many complementation groups include an event that acts through the ras signal transduction pathway.

Finally, a gene may seemingly increase and then decrease at subsequent stages of progression. This appears to be true for ER expression. As suggested for the *ras* gene, this probably indicates that gene expression is an indicator of other genetic alterations rather than a true member of a complementation group that, in concert, produces invasive breast cancer.

Another level of complexity may involve the differentiation state of the normal breast epithelial cell in which the initial genetic event in a complementation group occurs. It has been suggested that ductules, type 1 lobules, type 2 lobules, type 3 lobules, and type 4 lobules represent successive stages of differentiation with loss of stem cells (Russo and Russo, 1996). The Russos suggest that only type 1 lobules give rise to atypical ductular hyperplasia, DCIS, and invasive ductular carcinoma, that type 2 lobules give rise to atypical lobular hyperplasia and lobular carcinomas, and that type 3 lobules give rise to fibroadenomas, sclerosing adenoses, and apocrine cysts (Russo and Russo, 1996).

Although the task to unravel the complexity of breast cancer genetics is a formidable one, the availability of the MCF10AT preneoplastic human breast stem cells may provide the tool necessary to test suspected complementation groups directly. Although technically difficult, it may now be possible not only to overexpress genes following transfection/transduction with appropriate constructs, but also to induce point mutations in endogenous genes via homologous recombination using chimeric nucleotides (Kotani and Kmiec, 1994; Kmiec, 1995; Cole-Strauss et al., 1996; Yoon et al., 1996). It would be foolish to presume that the MCF10AT will allow detection of all complementation groups. MCF10AT is already a preneoplastic precursor and may be restricted in the number of routes it can follow. However, identification and verification of even a few such complementation groups would be a significant achievement.

ACKNOWLEDGMENTS

The author gratefully acknowledges the advice and suggestions of Dr. Gloria Heppner. The authors studies have been supported by USPHS grant CA28366 from the National Cancer Institute.

REFERENCES

Agnantis, N.J., Mahera, H., Maounis, N., and Spandidos, D.A. (1992). Immunohistochemical study of ras and myc oncoproteins in apocrine breast lesions with and without papillomatosis. Eur. J. Gynaecol. Oncol. 13, 309-315.

Akiyama, T., Saito, T., Ogawara, H., Toyoshima, K., and Yamamoto, T. (1988). Tumor promoter and epidermal growth factor stimulate phosphorylation of the *c-erbB-2* gene product in MKN-7 human adenocarcinoma cells. Mol. Cell. Biol. 8, 1019-1026.

Albonico, G., Querzoli, P., Ferretti, S., Magri, E., and Nenci, I. (1996). Biophenotypes of breast carcinoma in situ defined by image analysis of biological parameters. Pathol. Res. Pract. 192, 117-123.

Ali, I.U., Lidereau, R., Theillet, C., and Callahan, R. (1987). Reduction to homozygosity of genes on chromosome 11 in human breast neoplasia. Science 238, 185-188.

Ali, I.U., Lidereau, R., and Callahan, R. (1989). Presence of two members of *c-erbA* receptor gene family (*c-erbA*β and *cerbA2*) in smallest region of somatic homozygosity on chromosome 3p21-p25 in human breast carcinoma. J. Natl. Cancer Inst. 81, 1815-1820.

Almoguera, C., Shibata, D., Forrester, K., Martin, J., Arnheim, N., and Perucho, M. (1988). Most human carcinomas of the exocrine pancreas contain mutant *c-K-ras* genes. Cell 53, 549-554.

Athanassiadou, P.P., Veneti, S.Z., Kyrkou, K.A., Intzes, K.S., Mouzaka, L.I., and Papadimitriou, G.K. (1992). Presence of epidermal growth factor receptor in breast smears of cyst fluids: Relationship to electrolyte ratios and pH concentrations. Cancer Detect. Prev. 16, 113-118.

Barbareschi, M., Leonardi, E., Mauri, F.A., Serio, G., and Palma, P.D. (1992). p53 and c-erbB-2 protein expression in breast carcinomas. An immunohistochemical study including correlations with receptor status, proliferation markers, and clinical stage in human breast cancer. Am. J. Clin. Pathol. 98, 408-418.

Barker, S., Panahy, C., Puddefoot, J.R., Goode, A.W., and Vinson, G.P. (1989). Epidermal growth factor receptor and oestrogen receptors in the nonmalignant part of the cancerous breast. Br. J. Cancer 60, 673-677.

Bartek, J., Bartkova, J., Vojtesek, B., Staskova, Z., Rejthar, A., Kovarik, J., and Lane, D.P. (1990). Patterns of expression of the p53 tumour suppressor in human breast tissues and tumours in situ and in vitro. Int. J. Cancer 46, 839-844.

Bartkova, J., Barnes, D.M., Millis, R.R., and Gullick, W.J. (1990). Immunohistochemical demonstration of c-erbB-2 protein in mammary ductal carcinoma in situ. Hum. Pathol. 21, 1164-1167.

Bartkova, J., Lukas, J., Muller, H., Lutzhoft, D., Strauss, M., and Bartek, J. (1994). Cyclin D1 protein expression and function in human breast cancer. Int. J. Cancer 57, 353-361.

Bartkova, J., Lukas, J., Strauss, M., and Bartek, J. (1995). Cyclin D1 oncoprotein aberrantly accumulates in malignancies of diverse histogenesis. Oncogene 10, 775-778.

Basolo, F., Elliott, J., Tait, L., Chen, X.Q., Maloney, T., Russo,I.H., Pauley, R., Momiki, S., Caamano, J., Klein-Szanto, A.J.P., Koszalka, M., and Russo, J. (1991). Transformation of human breast epithelial cells by c-Ha-ras oncogene. Mol.Carcinogen. 4, 25-35.

Basset, P., Bellocq, J.P., Wolf, C., Stoll, I., Hutin, P., Limacher, J.M., Podhajcer, O.L., Chenard, M.P., Rio, M.C., and Chambon, P. (1990). A novel metalloproteinase gene specifically expressed in stromal cells of breast carcinomas. Nature 348, 699-704.

Basset, P., Wolf, C., and Chambon, P. (1993). Expression of the *stromelysin-3* gene in fibroblastic cells of invasive carcinomas of the breast and other human tissues: A review. Breast Cancer Res. Treat. 24, 185-193.

Berns, E.M.J.J., Klijn, J.G.M., van Putten, W.L.J., van Staveren, I.L., Portengen, H., and Foekens, J.A. (1992). *c-myc* amplification is a better prognostic factor than *HER2/neu* amplification in primary breast cancer. Cancer Res. 52, 1107-1113.

Bieche, I., Champeme, M.H., Matifas, F., Hacene, K., Callahan, R., and Lidereau,R. (1992). Loss of heterozygosity on chromosome 7q and aggressive primary breast cancer. Lancet 339, 139-143.

Bieche, I., Champeme, M.-H., Matifas, F., Cropp, C.S., Callahan, R., and Lidereau,R. (1993). Two distinct regions involved in 1p deletion in human primary breast cancer. Cancer Res. 53, 1990-1994.

Bland, K.I., Konstadoulakis, M.M., Vezeridis, M.P., and Wanebo, H.J. (1995). Oncogene protein co-expression. Value of Ha-ras, c-myc, c-fos, and p53 as prognostic discriminants for breast carcinoma. Ann. Surg. 221, 706-718.

Brem, S.S., Gullino, P.M., and Medina, D. (1977). Angiogenesis: A marker for neoplastic transformation of mammary papillary hyperplasia. Science 195, 880-882.

Brem, S.S., Jensen, H.M., and Gullino, P.M. (1978). Angiogenesis as a marker of preneoplastic lesions of the human breast. Cancer 41, 239-244.

Callahan, R., Cropp, C.S., Merlo, G.R., Liscia, D.S., Cappa, A.P., and Lidereau, R. (1992). Somatic mutations and human breast cancer. A status report. Cancer 69, 1582-1588.

Callahan, R., Cropp, C., Merlo, G.R., Diella, F., Venesio, T., Lidereau, R., Cappa, A.P., and Lisicia, D.S. (1993). Genetic and molecular heterogeneity of breast cancer cells. Clin. Chim. Acta 217, 63-73.

Carraway, 3rd, K.L., Sliwkowski, M.X., Akita, R., Platko, J.V., Guy, P.M., Nuijens, A., Diamonti, A.J., Vandlen, R.L., Cantley, L.C., and Cerione, R.A. (1994). The *erbB3* gene is a receptor for heregulin. J. Biol. Chem. 269, 14303-14306.

Carter, S.L., Negrini, M., Baffa, R., Gillum, D.R., Rosenberg, A.L., Schwartz, G.F., and Croce, C.M. (1994). Loss of heterozygosity at 11q22-q23 in breast cancer. Cancer Res. 54, 6270-6274.

Champeme, M.H., Bieche, I., Beuzelin, M., and Lidereau, R. (1995). Loss of heterozygosity on 7q31 occurs early during breast tumorigenesis. Genes, Chromosomes, and Cancer 12, 304-306.

Chen, L.C., Dollbaum, C., and Smith, H.S. (1989). Loss of heterozygosity on chromosome 1q in human breast cancer. Proc. Natl. Acad. Sci. (USA) 86, 7204-7207.

Chen, L.C., Neubauer, A., Kurisu, W., Waldman, F.M., Ljung, B.-M., Goodson, W.I.I.I., Goldman, E.S., Moore, D., Balazs, M., Liu, E., Mayall, B.H., and Smith, H.S. (1991). Loss of heterozygosity on the short arm of chromosome 17 is associated with high proliferative capacity and DNA aneupoidy in primary human breast cancer. Proc. Natl. Acad. Sci. (USA) 88, 3847-3851.

Chen, L.C., Kurisu, W., Ljung, B.M., Goldman, E.S., Moore, Jr., D., and Smith, H.S. (1992). Heterogeneity for allelic loss in human breast cancer. J. Natl. Cancer Inst. 84, 506-510.

Chen, T., Sahin, A., and Aldaz, C.M. (1996). Deletion map of chromosome 16q in ductal carcinoma in situ of the breast: Refining a putative tumor suppressor gene region. Cancer Res. 56, 5605-5609.

Chen, Y.Q., Cipriano, S.C., Sarkar, F.H., Ware, J.L., and Arenkiel, J.M. (1995). p53-independent induction of p21(WAF1) pathway is preserved during tumor progression. Int. J. Oncol. 7, 889-893.

Chitemerere, M., Andersen, T.I., Holm, R., Karlsen, F., Borresen, A.-L., and Nesland, J.M. (1996). TP53 alterations in atypical ductal hyperplasia and ductal carcinoma in situ of the breast. Breast Cancer Res. Treat. 41, 103-109.

Ciardiello, F., Gottardis, M., Basolo, F., Pepe, S., Normanno, N., Dickson, R.B., Bianco, A.R., and Salomon, D.S. (1992). Additive effects of *c-erbB-2*, *c-Ha-ras*, and *transforming growth factor-á* genes on in vitro transformation of human mammary epithelial cells. Mol. Carcinogen. 6, 43-52.

Clair, T., Miller, W.R., and Cho-Chung, Y.S. (1987). Prognostic significance of the expression of a ras protein with a molecular weight of 21,000 by human breast cancer. Cancer Res. 47, 5290-5293.

Clark, G.J., and Der, C.J. (1995). Aberrant function of the ras signal transduction pathway in human breast cancer. Breast Cancer Res. Treat. 35, 133-144.

Cleton-Jansen, A.M., Moerland, E.W., Kuipers-Dijkshoorn, N.J., Callen, D.F., Sutherland, G.R., Hansen, B., Devilee, P.,and Cornelisse, C.J. (1994). At least two different regions are involved in allelic imbalance on chromosome arm 16q in breast cancer. Genes, Chromosomes, and Cancer 9, 101-107.

Cole-Strauss, A., Yoon, K., Xiang, Y., Byrne, B.C., Rice, M.C., Gryn, J., Holloman, W.K., and Kmiec, E.B. (1996). Correction of the mutation responsible for sickle cell anemia by an RNA-DNA oligonucleotide. Science 273, 1386-1389.

Collins,F.S. (1996). *BRCA1* - Lots of mutations, lots of dilemmas. N. Engl. J. Med. 334, 186-188.

Cropp, C.S., Lidereau, R., Campbell, G., Champeme, M.H., and Callahan, R. (1990). Loss of heterozygosity on chromosomes 17 and 18 in breast carcinoma: Two additional regions identified. Pnas. 87, 7737-7741.

Culouscou, J.M., Plowman, G.D., Carlton, G.W., Green, J.M., and Shoyab, M. (1993). Characterization of a breast cancer cell differentiation factor that specifically activates the HER4/p180erbB4 receptor. J. Biol. Chem. 268, 18407-18410.

Daniel, C.W., DeOme, K.B., Young, J.T., Blair, P.B., and Faulkin, L.J.J. (1968). The in vivo life span of normal and preneoplastic mouse mammary glands: A serial transplantation study. Proc. Natl. Acad. Sci. (USA) 61, 53-60.

Dati, C., Muraca, R., Tazartes, O., Antoniotti, S., Perroteau, I., Giai, M., Cortese, P., Sismondi, P., Saglio, G., and De Bortoli, M. (1991). c-erbB-2 and ras expression levels in breast cancer are correlated and show a co-operative association with unfavorable clinical outcome. Int. J. Cancer 47, 833-838.

Davies, B., Miles, D.W., Happerfield, L.C., Naylor, M.S., Bobrow, L.G., Rubens, R.D., and Balkwill, F.R. (1993). Activity of type IV collagenases in benign and malignant breast disease. Br. J. Cancer 67, 1126-1131.

Dawson, P.J., Wolman, S.R., Tait, L., Heppner, G.H., and Miller, F.R. (1996). MCF10AT: A model for the evolution of cancer from proliferative breast disease. Am. J. Path. 148, 313-319.

DeBortoli, M.E., Abou-Issa, H., Haley, B.E., and Cho-Chung, Y.S. (1985). Amplified expression of p21 ras protein in hormone-dependent mammary carcinomas of humans and rodents. Biochem. Biophys. Res. Commun. 127, 699-706.

DeOme, K.B., Faulkin, Jr., L.J., Bern, H.A., and Blair, P.B. (1959). Development of mammary tumors from hyperplastic alveolar nodules transplanted into gland-free mammary fatpads of female C3H mice. Cancer Res. 19, 515-520.

DeOme, K.B., Miyamoto, M.J., Osborn, R.C., Guzman, R.C., and Lum, K. (1978). Detection of inapparent nodule-transformed cells in the mammary gland tissues of virgin female BALB/cfC3H mice. Cancer Res. 38, 2103-2111.

De Potter, C.R., Van Daele, S., Van De Vijver, M.J., Pauwels, C., Maertens, G., De Boever, J., Vanderkerckhove, D., and Roels, H. (1989). The expression of the neu oncogene product in breast lesions and in normal fetal and adult human tissues. Histopathology 15, 351-362.

Devesa, S.S., Blot, W.J., Stone, B.J., Miller, B.A., Tarone, R.E., and Fraumeni, J.F. Jr. (1995). Recent cancer trends in the United States. J. Natl. Cancer Inst. 87, 175-182.

Devilee, P., Cornelisse, C.J., Kuipers-Dijkshoorn, N., Jonker, C., and Pearson, P.L. (1990). Loss of heterozygosity on 17p in human breast carcinomas: Defining the smallest common region of deletion. Cytogenetics and Cell Genetics 53, 52-54.

Devilee, P., van Vliet, M., Bardoel, A., Kievits, T., Kuipers-Dijkshoorn, N., Pearson, P.L., and Cornelisse, C.J. (1991a). Frequent somatic imbalance of marker alleles for chromosome 1 in human primary breast carcinoma. Cancer Res. 51, 1020-1025.

Devilee, P., van Vliet, M., van Sloun, P., Dijkshoorn, N.K., Hermans, J., Pearson, P.L., and Cornelisse, C.J. (1991b). Allelotype of human breast carcinoma: A second major site for loss of heterozygosity is on chromosome 6q. Oncogene 6, 1705-1711.

Devilee, P., Vliet, M., van Kuipers-Dijkshoorn, N., Pearson, P.L., and Cornelisse, C.J. (1991c). Somatic genetic changes on chromosome 18 in breast carcinoma: Is the *DCC* gene involved? Oncogene 6, 311-315.

Dickson, R.B., Salomon, D.S., and Lippman, M.E. (1992). Tyrosine kinase receptor-nuclear protooncogene interactions in breast cancer. Cancer Treat. Res. 61, 249-273.

Diella, F., Normanno, N., Merlo, D.S., and Callahan, R. (1993). Absence of p53 point mutations in nontransformed human mammary epithelial cell lines. Life Sci. Adv. Biochem. 12, 47-51.

Dietrich, C.U., Pandis, N., Teixeira, M.R., Bardi, G., Gerdes, A.-M., Andersen, J.A., and Heim, S. (1995). Chromosome abnormalities in benign hyperproliferative disorders of epithelial and stromal breast tissue. Int. J. Cancer 60, 49-53.

Domagala, W., Markiewski, M., Kubiak, R., Bartkowiak, J., and Osborn, M. (1993). Immunohistochemical profile of invasive lobular carcinoma of the breast: Predominantly vimentin and p53 protein negative, cathepsin D, and oestrogen receptor positive. Virchows. Arch. A. Pathol. Anat. Histopathol. 423, 497-502.

Dorion-Bonnet, F., Mautalen, S., Hostein, I., and Longy, M. (1995). Allelic imbalance study of 16q in human primary breast carcinomas using microsatellite markers. Genes, Chromosomes, and Cancer 14, 171-181.

Dupont, W.D., and Page, D.L. (1985). Risk factors for breast cancer in women with proliferative breast disease. N. Engl. J. Med. 312, 146-151.

Dupont, W.D., and Page, D.L. (1989). Relative risk of breast cancer varies with time since diagnosis of atypical hyperplasia. Hum. Pathol. 20, 723-725.

Dupont, W.D., Parl, F.F., Hartmann, W.H., Brinton, L.A., Winfield, A.C., Worrell, J.A., Scuyler, P.A., and Plummer, W.D. (1993). Breast cancer risk associated with proliferative breast disease and atypical hyperplasia. Cancer 71, 1258-1265.

Dupont, W.D., Page, D.L., Parl, F.F., Vnencak-Jones, C.L., Plummer, W.D.J., Rados, M.S., and Schuyler, P.A. (1994). Long-term risk of breast cancer in women with fibroadenoma. N. Engl. J. Med. 331, 10-15.

Eiriksdottir, G., Sigurdsson, A., Jonasson, J.G., Agnarsson, B.A., Sigurdsson, H., Gudmundsson, J., Bergthorsson, J.T., Barkardottir, R.B., Egilsson, V., and Ingvarsson, S. (1995). Loss of heterozygosity on chromosome 9 in human breast cancer: Association with clinical variables and genetic changes at other chromosome regions. Int. J. Cancer 64, 378-382.

Eisinger, F., Stoppa-Lyonnet, D., Longy, M., Kerangueven, F., Noguchi, T., Bailly, C., Vincent-Salomon, A., Jacquemier, J., Birnbaum, D., and Sobol, H. (1996). Germ line mutation at *BRCA1* affects the histoprognostic grade in hereditary breast cancer. Cancer Res. 56, 471-474.

Engel, G., Heselmeyer, K., Auer, G., Backdahl, M., Eriksson, E., and Linder, S. (1994). Correlation between stromelysin-3 mRNA level and outcome of human breast cancer. Int. J. Cancer 58, 830-835.

Eriksson, E.T., Schimmelpenning, H., Aspenblad, U., Zetterberg, A., and Auer, G.U. (1994). Immunohistochemical expression of the mutant p53 protein and nuclear DNA content during the transition from benign to malignant breast disease. Hum. Pathol. 25, 1228-1233.

Escot, C., Simony-Lafontaine, J., Maudelonde, T., Puech, C., Pujol, H., and Rochefort, H. (1993). Potential value of increased myc but not erbB2 RNA levels as a marker of high-risk mastopathies. Oncogene 8, 969-974.

Ezzell, C. (1996). BRCA1 shock: Breast-cancer gene encodes a secreted protein. J. NIH Res. 8, 21-22.

Fabian, C.J., Zalles, C., Kamel, S., Kimler, B.F., McKittrick, R., Tranin, A.S., Zeiger, S., Moore, W.P., Hassanein, R.S., Simon, C. et al. (1994). Prevalence of aneuploidy, overexpressed ER, and overexpressed EGFR in random breast aspirates of women at high and low risk for breast cancer. Breast Cancer Res. Treat. 30, 263-274.

Faulkin, Jr., L.J., and DeOme, K.B. (1960). Regulation of growth and spacing of the gland elements in the mammary fatpad of the C3H mouse. J. Natl. Cancer Inst. 24, 953-969.

Fearon, E.R., and Vogelstein, B. (1990). A genetic model for colorectal tumorigenesis. Cell. 61, 759-767.

Fujii, H., Marsh, C., Cairns, P., Sidransky, D., and Gabrielson, E. (1996a). Genetic divergence in the clonal evolution of breast cancer. Cancer Res. 56, 1493-1497.

Fujii, H., Szumel, R., Marsh, C., Zhou, W., and Gabrielson, E. (1996b). Genetic progression, histological grade, and allelic loss in ductal carcinoma in situ. Cancer Res. 56, 5260-5265.

Garcia, I., Dietrich, P.-Y., Aapro, M., Vauthier, G., Vadas, L., and Engel, E. (1989). Genetic alterations of c-myc, c-erbB-2, and c-Ha-ras protooncogenes and clinical associations in human breast carcinomas. Cancer Res. 49, 6675-6679.

Gasparini, G., Bevilacqua, P., Pozza, F., Meli, S., Boracchi, P., Marubini, E., and Sainsbury, J.R. (1992). Value of epidermal growth factor receptor status compared with growth fraction and other factors for prognosis in early breast cancer. Br. J. Cancer 66, 970-976.

Genuardi, M., Tsihira, H., Anderson, D.E., and Saunders, G.F. (1989). Distal deletion of chromosome 1p in ductal carcinoma of the breast. Am. J. Hum. Genet. 45, 73-82.

Giri, D.D., Dangerfield, V.J.M., Lonsdale, R., Rogers, K., and Underwood, J.C.E. (1987). Immunohistology of oestrogen receptor and D5 antigen: Correlations with radioligand binding assays and enzyme immunoassays. J. Clin. Pathol. 40, 734-740.

Giri, D.D., Dundas, S.A.C., Nottingham, J.F., and Underwood, J.C.E. (1989). Oestrogen receptors in benign epithelial lesions and intraduct carcinomas of the breast: An immunohistological study. Histopathology 15, 575-584.

Going, J.J., Anderson, T.J., and Wyllie, A.H. (1992). Ras p21 in breast tissue: Associations with pathology and cellular localisation. Br. J. Cancer 65, 45-50.

Gudas, J., Nguyen, H., Li, T., Hill, D., and Cowan, K.H. (1995). Effects of cell cycle, wild-type p53 and DNA damage on p21CIP1/Waf1 expression in human breast epithelial cells. Oncogene 11, 253-261.

Gudmundsson, J., Barkardottir, R.B., Eiriksdottir, G., Baldursson, T., Arason, A., Egilsson, V., and Ingvarsson, S. (1995). Loss of heterozygosity at chromosome 11 in breast cancer: Association of prognostic factors with genetic alterations. Br. J. Cancer 72, 696-701.

Guidi, A.J., Fischer, L., Harris, J.R., and Schnitt, S.J. (1994). Microvessel density and distribution in ductal carcinoma in situ of the breast. J. Natl. Cancer Inst. 86, 614-619.

Guinebretiere, J., Le Monique, G., Gaviolle, A., Bahi, J., and Contesso, G. (1994). Angiogenesis and risk of breast cancer in women with fibrocystic disease. J. Natl. Cancer Inst. 86, 635-636.

Gullick, W.J., Love, S.B., Wright, C., Barnes, D.M., Gusterson, B., Harris, A.L., and Altman, D.G. (1991). c-erbB-2 protein overexpression in breast cancer is a risk factor in patients with involved and uninvolved lymph nodes. Br. J. Cancer 63, 434-438.

Gusterson, B.A., Machin, L.G., Gullick, W.J., Gibbs, N.M., Powles, T.J., Elliott, C., Ashley, S., Monaghan, P., and Harrison, S. (1988a). c-erbB-2 expression in benign and malignant breast disease. Br. J. Cancer 58, 453-457.

Gusterson, B.A., Machin, L.G., Gullick, W.J., Gibbs, N.M., Powles, T.J., Price, P., McKinna, A., and Harrison, S. (1988b). Immunohistochemical distribution of c-erbB-2 in infiltrating and in situ breast cancer. Int. J. Cancer 42, 842-845.

Gusterson, B.A., Gelber, R.D., Goldhirsch, A., Price, K.N., Save-Soderborgh, J., Anbazhagan, R., Styles, J., Rudenstam, C.M., Golough, R., Reed, R., Martinez-Tello, F., Tiltman, A., Torhorst, J., Grigolato, P., Bettelheim, R., Neville, A.M., Burki, K., Castiglione, M., Collins, J., Lindtner, J., and Senn, H.J. (1992). Prognostic importance of c-erbB2 expression in breast cancer. J. Clin. Oncol. 10, 1049-1056.

Hahnel, E., Harvey, J.M., Joyce, R., Robbins, P.D., Sterrett, G.F., and Hahnel, R. (1993). Stromelysin-3 expression in breast cancer biopsies: Clinico-pathological correlations. Int. J. Cancer 55, 771-774.

Hahnel, E., Dawkins, H., Robbins, P., and Hahnel, R. (1994). Expression of stromelysin-3 and nm23 in breast carcinoma and related tissues. Int. J. Cancer 58, 157-160.

Hampton, G.M., Mannermaa, A., Winquist, R., Alavaikko, M., Blanco, G., Taskinen, P.J., Kiviniemi, H., Newsham, I., Cavenee, W.K., and Evans, G.A. (1994). Loss of heterozygosity in sporadic human breast carcinoma: A common region between 11q22 and 11q23.3. Cancer Res. 54, 4586-4589.

Hankins, G.R., De Souza, A.T., Bentley, R.C., Patel, M.R., Marks, J.R., Iglehart, J.D., and Jirtle, R.L. (1996). M6P/IGF2 receptor: A candidate breast tumor suppressor gene. Oncogene 12, 2003-2009.

Heffelfinger, S.C., Yassin, R., Miller, M.A., and Lower, E. (1996). Vascularity of proliferative breast disease and carcinoma in situ correlates with histological features. Clinical Cancer Research 2, 1873-1878.

Hehir, D.J., McGreal, G., Kirwan, W.O., Kealy, W., and Brady, M.P. (1993). c-myc oncogene expression: A marker for females at risk of breast carcinoma. J. Surg. Oncol. 54, 207-209.

Heyderman, E., and Dagg, B. (1991). p53 immunostaining in benign breast disease. Lancet 338, 1532.

Holt, J.T., Thompson, M.E., Szabo, C., Robinson-Benion, C., Arteaga, C.L., King, M.-C., and Jensen, R.A. (1996). Growth retardation and tumour inhibition by BRCA1. Nat. Genet. 12, .

Horak, E., Smith, K., Bromley, L., LeJune, S., Greenall, M., Lane, D., and Harris, A.L. (1991). Mutant p53, EGF receptor and c-erbB-2 expression in human breast cancer. Oncogene 6, 2277-2284.

Horan Hand, P., Thor, A., Wunderlich, D., Muraro, R., Caruso, A., and Schlom, J. (1984). Monoclonal antibodies of predefined specificity detect activated *ras* gene expression in human mammary and colon carcinomas. Proc. Natl. Acad. Sci. (USA) 81, 5227-5231.

Iglehart, J.D., Kerns, B.-J., Huper, G., and Marks, J.R. (1995). Maintenance of DNA content and erbB-2 alterations in intraductal and invasive phases of mammary cancer. Breast Cancer Res. Treat. 34, 253-263.

Iravani, S., Dawson, P.J., and Miller, F.R. (1996). Tumor histotype and gene expression in the MCF10AT model. Faseb J. 10, A1416.

Ito, I., Yoshimoto, M., Iwase, T., Watanabe, S., Katagiri, T., Harada, Y., Kasumi, F., Yasuda, S., Mitomi, T., Emi, M., and Nakamura, Y. (1995). Association of genetic alterations on chromosome 17 and loss of hormone receptors in breast cancer. Br. J. Cancer 71, 438-441.

Jacquemier, J.D., Hassoun, J., Torrente, M., and Martin, P.-M. (1990). Distribution of estrogen and progesterone receptors in healthy tissue adjacent to breast lesions at various stages-immunohistochemical study of 107 cases. Breast Cancer Res. Treat. 15, 109-117.

Jensen, H.M., and Wellings, S.R. (1976). Preneoplastic lesions of the human mammary gland transplanted into the nude athymic mouse. Cancer Res. 36, 2605-2610.

Jensen, R.A., Thompson, M.E., Jetton, T.L., Szabo, C.I., van der Meer, R., Helou, B., Tronick, S.R., Page, D.L., King, M.-C., and Holt, J.T. (1996). BRCA1 is secreted and exhibits properties of a granin. Nat. Genet. 12, .

Jensen, V., Ladekarl, M., Holm-Nielsen, P., Melsen, F., and Soerensen, F.B. (1995). The prognostic significance of oncogenic antigen 519 (OA-519) expression and proliferative activity detected by antibody MIB-1 in node-negative breast cancer. J. Pathol. 176, 343-352.

Johnson, G., Kanna, B., Shoyab, M., and Stromberg, K. (1993). Amphiregulin induces tyrosine phosphorylation of the epidermal growth factor and p185erbB2. J. Biol. Chem. 268, 2924-2931.

Kashiwaba, M., Tamura, G., and Ishida, M. (1995). Frequent loss of heterozygosity at the deleted in colorectal carcinoma gene locus and its association with histologic phenotypes in breast carcinoma. Virchows Arch. 426, 441-446.

Kawami, H., Yoshida, K., Ohsaki, A., Kuroi, K., Nishiyama, M., and Toge, T. (1993). Stromelysin-3 mRNA expression and malignancy: Comparison with clinicopathological features and type IV collagenase mRNA expression in breast tumors. AntiCancer Res. 13, 2319-2323.

Kerlikowske, K., Barclay, J., Grady, D., Sickles, E.A., and Ernster, V. (1997). Comparison of risk factors for ductal carcinoma in situ and invasive breast cancer. J. Natl. Cancer Inst. 89, 76-82.

Khan, S.A., Rogers, M.A.M., Obando, J.A., and Tamsen, A. (1994). Estrogen receptor expression of benign breast epithelium and its association with breast cancer. Cancer Res. 54, 993-997.

Kirchweger, R., Zeillinger, R., Schneeberger, C., Speiser, P., Louason, G., and Theillet, C. (1994). Patterns of allele losses suggest the existence of five distinct regions of LOH on chromosome 17 in breast cancer. Int. J. Cancer 56, 193-199.

Kita, Y.A., Barff, J., Luo, Y., Wen, D., Brankow, D., Hu, S., Liu, N., Prigent, S.A., Gullick, W.J., and Nicolson, M. (1994). NDF/heregulin stimulates the phosphorylation of Her3/erbB3. FEBS Letters 349, 139-143.

Kmiec, E.B. (1995). Genetic manipulations in mammalian cells using an RNA/DNA chimeric oligonucleotide. Adv. Drug Deliv. Rev. 17, 333-340.

Kobayashi, S., Iwase, H., Itoh, Y., Fukuoka, H., Yamashita, H., Kuzushima, T., Iwata, H., Masaoka, A., and Kimura, N. (1992). Estrogen receptor, c-erbB-2 and nm23/NDP kinase expression in the intraductal and invasive components of human breast cancers. Jpn. J. Cancer Res. 83, 859-865.

Koenders, P.G., Beex, L.V.A.M., Geurts-Moespot, A., Heuvel, J.J.T.M., Kienhuis, C.B.M., and Benraad, T.J. (1991). Epidermal growth factor receptor-negative tumors are predominantly confined to the subgroup of estradiol receptor-positive human primary breast cancers. Cancer Res. 51, 4544-4548.

Koreth, J., Bethwaite, P.B., and McGee, J.O. (1995). Mutation at chromosome 11q23 in human nonfamilial breast cancer: A microdissection microsatellite analysis. J. Pathol. 176, 11-18.

Kotani, H., and Kmiec, E.B. (1994). A role for RNA synthesis in homologous pairing events. Mol. Cell. Biol. 14, 6097-6106.

Kozma, S.C., Bogaard, M.E., Buse, K., Saurer, S.M., Bos, J.L., Groner, B., and Hynes, N.E. (1987). The human *c-Kirsten ras* gene is activated by a novel mutation in codon 13 in the breast carcinoma cell line MDA-MB-231. Nucleic Acids Res. 15, 5963-5971.

Kraus, M.H., Yuasa, Y., and Aaronson, S.A. (1984). A position 12-activated H-ras oncogene in all HS578T mammary carcinosarcoma cells but not in normal mammary cells of the same patient. Proc. Natl. Acad. Sci. (USA) 81, 5384-5388.

Krieger, N., and Hiatt, R.A. (1992). Risk of breast cancer after benign breast diseases. Variations by histologic type, degree of atypia, age at biopsy, and length of follow-up. Am. J. Epidemiol. 135, 619-631.

Lakhani, S.R., Collins, N., Stratton, M.R., and Sloane, J.P. (1995). Atypical ductal hyperplasia of the breast: Clonal proliferation with loss of heterozygosity on chromosomes 16q and 17p. J. Clin. Pathol. 48, 611-615.

Lancaster, J.M., Wooster, R., Mangion, J., Phelan, C.M., Cochran, C., Gumbs, C., Seal, S., Barfoot, R., Collins, N., Bignell, G., Patel, S., Hamoudi, R., Larsson, C., Wiseman, R.W., Berchuck, A., Iglehart, J.D., Marks, J.R., Ashworth, A., Stratton, M.R., and Futreal, P.A. (1996). *BRCA2* mutations in primary breast and ovarian cancers. Nat. Genet. 13, 238-240.

Leal, C.B., Schmitt, F.C., Bento, M.J., Maia, N.C., and Lopes, C.S. (1995). Ductal carcinoma in situ of the breast. Histologic categorization and its relationship to ploidy and immunohistochemical expression of hormone receptors, p53, and c-erbB-2 protein. Cancer 75, 2123-2131.

Lee, K.S., Rha, S.Y., Kim, S.J., Kim, J.H., Roh, J.K., Kim, B.S., and Chung, H.C. (1996). Sequential activation and production of matrix metalloproteinase-2 during breast cancer progression. Clin. Expl. Metastasis 14, 512-519.

Lindblom, A., Rotstein, S., Skoog, L., Nordenskjold, M., and Larsson, C. (1993a). Deletions on chromosome 16 in primary familial breast carcinomas are associated with development of distant metastases. Cancer Res. 53, 3707-3711.

Lindblom, A., Skoog, L., Rotstein, S., Werelius, B., Larsson, C., and Nordenskjold, M. (1993b). Loss of heterozygosity in familial breast carcinomas. Cancer Res. 53, 4356-4361.

Liu, E., Thor, A., He, M., Barcos, M., Ljung, B.M., and Benz, C. (1992). The HER2 (c-erbB-2) oncogene is frequently amplified in in situ carcinomas of the breast. Oncogene 7, 1027-1032.

Lizard-Nacol, S., Lidereau, R., Collin, F., Arnal, L., Roignot, P., Cuisenier, J., and Guerrin, J. (1995). Benign breast disease: Absence of genetic alterations at several loci implicated in breast cancer malignancy. Cancer Res. 55, 4416-4419.

London, S.J., Connolly, J.L., Schnitt, S.J., and Colditz, G.A. (1992). A prospective study of benign breast disease and the risk of breast cancer. Journal of the American Medical Association 267, 941-944.

Lundberg, C., Skoog, L., Cavenee, W.K., and Nordenskjold, M. (1987). Loss of heterozygosity in human ductal breast tumors indicates a recessive mutation on chromosome 13. Pnas. 84, 2372-2376.

Mackay, J., Elder, P.A., Porteus, D.J., Steel, C.M., Hawkins, R.A., Going, J.J., and Chetty, U. (1988a). Partial deletion of chromosome 11p in breast cancer correlates with size of primary tumour and oestrogen receptor level. Br. J. Cancer 58, 710-714.

Mackay, J., Steel, C.M., Elder, P.A., Forrest, A.P.M., and Evans,H.J. (1988b). Allele loss on short arm of chromosome 17 in breast cancers. Lancet 2, 1384-1385.

Man, S., Ellis, I.O., Sibbering, M., Blamey, R.W., and Brook, J.D. (1996). High levels of allele loss at the FHIT and ATM genes in non-comedo ductal carcinoma in situ and grade I tubular invasive breast cancers. Cancer Res. 56, 5484-5489.

McIntosh, G.G., Anderson, J.J., Milton, I., Steward, M., Parr, A.H., Thomas, M.D., Henry, J.A., Angus, B., Lennard, T.W., and Horne, C.H. (1995). Determination of the prognostic value of cyclin D1 overexpression in breast cancer. Oncogene 11, 885-891.

Medina, D. (1975). Tumor progression. IN: *Cancer*. (Becker, F.F., Ed.), pp. 99-119, Plenum Publishing Corp, New York.

Medina, D. (1976). Preneoplastic lesions in murine mammary cancer. Cancer Res. 36, 2589-2595.

Medina, D. (1996). The mammary gland: A unique organ for the study of development and tumorigenesis. J. Mammary Gland Biol. Neoplasia 1, 5-19.

Merlo, G.R., Venesio, T., Bernardi, A., Cropp, C.S., Diella, F., Cappa, A.P., Callahan, R., and Liscia, D.S. (1994). Evidence for a second tumor suppressor gene on 17p linked to high S-phase index in primary breast carcinomas. Cancer Genet. Cytogenet. 76, 106-111.

Merlo, G.R., Basolo, F., Fiore, L., Duboc, L., and Hynes, N.E. (1995). p53-dependent and -independent activation of apoptosis in mammary epithelial cells reveals a survival function of EGF and insulin. J. Cell Biol. 128, 1185-1196.

Miki, Y., Katagari, T., Kusami, F., Yoshimoto, T., and Nakamura, Y. (1996). Mutation analysis in the *BRCA2* gene in primary breast cancers. Nat. Genet. 13, 245-247.

Miller, F.R. (1981). Comparison of metastasis of mammary tumors growing in the mammary fatpad versus the subcutis. Invasion Metastasis 1, 220-226.

Miller, F.R., and McInerney, D. (1988). Epithelial component of host-tumor interactions in the orthotopic site preference of a mouse mammary tumor. Cancer Res. 48, 3698-3701.

Miller, F.R., Medina, D., and Heppner, G.H. (1981). Preferential growth of mammary tumors in intact mammary fatpads. Cancer Res. 41, 3863-3867.

Miller, F.R., Soule, H.D., Tait, L., Pauley, R.J., Wolman, S.R., Dawson, P.J., and Heppner, G.H. (1993). Xenograft model of human proliferative breast disease. J. Natl. Cancer Inst. 85, 1725-1732.

Miller, F.R., Pauley, R.J., and Wang, B. (1996). Activated *c-Ha-ras* is not sufficient to produce the preneoplastic phenotype of human breast cell line MCF10AT. AntiCancer Res. 16, 1765-1770.

Millikan, R., Hulka, B., Thor, A., Zhang, Y., Edgerton, S., Zhang, X., Pei, H., He, M., Wold, L., Melton, L.J., Ballard, D., Conway, K., and Liu, E.T. (1995). p53 mutations in benign breast tissue. J. Surg. Oncol. 13, 2293-2300.

Moller, P., Mechtersheimer, G., Kaufmann, M., Moldenhauer, G., Momburg, F., Mattfeldt, T., and Otto, H.F. (1989). Expression of epidermal growth factor receptor in benign and malignant primary tumors of the breast. Virch. Arch. A Pathol. Anat. Histopathol. 414, 157-164.

Munn, K.E., Walker, R.A., and Varley, J.M. (1995). Frequent alterations on chromosome 1 in ductal carcinoma in situ of the breast. Oncogene 10, 1653-1657.

Munn, K.E., Walker, R.A., Menasce, L., and Varley, J.M. (1996). Allelic imbalance in the region of the *BRCA1* gene in ductal carcinoma in situ of the breast. Br. J. Cancer 73, 636-639.

Nagai, M.A., Medeiros, A.C., Brentani, M.M., Brentani, R.R., Marques, L.A., Mazoyer, S., and Mulligan, L.M. (1995). Five distinct deleted regions on chromosome 17 defining different subsets of human primary breast tumors. Oncology 52, 448-453.

Nakopoulou, L.L., Alexiadou, A., Theodoropoulos, G.E., Lazaris, A.C., Tzonou, A., and Keramopoulos, A. (1996). Prognostic significance of the co-expression of p53 and c-erbB-2 proteins in breast cancer. J. Pathol. 179, 31-38.

O'Connell, P.O., Pekkel, V., Fuqua,S., Osborne, C.K., and Allred, D.C. (1994). Molecular genetic studies of early breast cancer evolution. Breast Cancer Res. Treat. 32, 5-12.

Ohuchi, N., Thor, A., Page, D.L., Horan Hand, P., Halter, S.A., and Schlom, J. (1986). Expression of the 21,000 molecular weight ras protein in a spectrum of benign and malignant human mammary tissues. Cancer Res. 46, 2511-2519.

Ottinetti, A., and Sapino, A. (1988). Morphometric evaluation of microvessels surrounding hyperplastic and neoplastic mammary lesions. Breast Cancer Res. Treat. 11, 241-248.

Outzen, H.C., and Custer, R.P. (1975). Growth of human normal and neoplastic mammary tissues in the cleared fat pad of the nude mouse. J. Natl. Cancer Inst. 55, 1461-1466.

Page, D.L., and Dupont, W.D. (1990). Anatomic markers of human premalignancy and risk of breast cancer. Cancer 66, 1326-1335.

Page, D.L., Salhany, K.E., Jensen, R.A., and Dupont, W.D. (1996). Subsequent breast carcinoma risk after biopsy with atypia in a breast papilloma. Cancer 78, 258-266.

Paik, S., Hazan, R., Fisher, E.R., Sass, R.E., Fisher, B., Redmond, C., Schlessinger, J., Lippman, M.E., and King, C.R. (1990). Pathologic findings from the National Surgical Adjuvant Breast and Bowel Project: Prognostic significance of erbB-2 protein overexpression in primary breast cancer. J. Clin. Oncol. 8, 103-112.

Palli, D., del Turco, M.R., Simoncini, R., and Bianchi, S. (1991). Benign breast disease and breast cancer: A case-control study in a cohort in Italy. Int. J. Cancer 47, 703-706.

Panagopoulos, I., Pandis, N., Thelin, S., Patersson, C., Mertens, F., Borg, A., Kristoffersson, U., Mitelman, F., and Aman, P. (1996). The *FHIT* and *PTPRG* genes are deleted in benign proliferative breast disease associated with familial breast cancer and cytogenetic rearrangements of chromosome band 3p14. Cancer Res. 56, 4871-4875.

Patel, K.V., and Schrey, M.P. (1995). Human breast cancer cells contain a phosphoramidon-sensitive metalloproteinase which can process exogenous big endothelin-1 to endothelin-1: A proposed mitogen for human breast fibroblasts. Br. J. Cancer 71, 442-447.

Pauley, R.J., Gimotty, P.A., Paine, T.J., Dawson, P.J., and Wolman, S.R. (1996). *INT2* and *ERBB2* amplification and ERBB2 expression in breast tumors from patients with different outcomes. Breast Cancer Res. Treat. 37, 65-76.

Pavelic, Z.P., Pavelic, L., Lower, E.E., Gapany, M., Gapany, S., Barker, E.A., and Preisler, H.D. (1992). c-myc, c-erbB-2, and Ki-67 expression in normal breast tissue and in invasive and noninvasive breast carcinoma. Cancer Res. 52, 2597-2602.

Pechoux, C., Chardonnet, Y., Chignol, M.C., and Noel, P. (1994). Heterogeneous immunoreactivity of frozen human benign and malignant breast lesions to c-myc and c-Ha-ras cellular oncogenes. Histol. Histopathol. 9, 35-44.

Pereira-Smith, O.M., and Smith, J.R. (1988). Genetic analysis of indefinite division in human cells: Identification of four complementation groups. Proc. Natl. Acad. Sci. (USA) 85, 6042-6046.

Petersen, O.W., Hoyer, P.E., and van Deurs, B. (1987). Frequency and distribution of estrogen receptor-positive cells in normal, nonlactating human breast tissue. Cancer Res. 47, 5748-5751.

Phelan, C.M., Lancaster, J.M., Tonin, P., Gumbs, C., Cochran, C., Carter, R., Ghadirian, P., Perret, C., Moslehi, R., Dion, F., Faucher, M.C., Dole, K., Karimi, S., Foulkes, W., Lounis, H., Warner, E., Goss, P., Anderson, D., Larsson, C., Narod, S.A., and Futreal, P.A. (1996). Mutation analysis of the *BRCA2* gene in 49 site-specific breast cancer families. Nat. Genet. 13, 120-122.

Plowman, G.D., Green, J.M., Culouscou, J.M., Carlton, G.W., Rothwell, V.M., and Buckley, S. (1993). Heregulin induces tyrosine phosphorylation of HER4/p180erbB4. Nature 366, 473-475.

Polette, M., Clavel, C., Cockett, M., Girod-de-Bentzmann, S., Murphy, G., and Birembaut, P. (1993). Detection and localization of mRNAs encoding matrix metalloproteinases and their tissue inhibitor in human breast pathology. Invasion Metastasis 13, 31-37.

Poller, D.N., Snead, D.R.J., Roberts, E.C., Galea, M., Bell, J.A., Gilmour, A., Elston, C.W., Blamey, R.W., and Ellis, I.O. (1993). Oestrogen receptor expression in ductal carcinoma in situ of the breast: Relationship to flow cytometric analysis of DNA and expression of the c-erbB-2 oncoprotein. Br. J. Cancer 68, 156-161.

Porter, P.L., Garcia, R., Moe, R., Corwin, D.J., and Gown, A.M. (1991). C-erbB-2 oncogene protein in in situ and invasive lobular breast neoplasia. Cancer 68, 331-334.

Price, J.E. (1990). The biology of metastatic breast cancer. Cancer 66, 1313-1320.

Price, J.E., Polyzos, A., Zhang, R.D., and Daniels, L.M. (1990). Tumorigenicity and metastasis of human breast carcinoma cell lines in nude mice. Cancer Res. 50, 717-721.

Prosperi, M., Dupre, G., Lidereau, R., and Goubin, G. (1990). Point mutation at codon 12 of the *K-ras* gene in a primary breast carcinoma and the MDA-MB-134 human mammary carcinoma cell line. Cancer Lett. 51, 169-174.

Querzoli, P., Marchetti, E., Bagni, A., Marzola, A., Fabris, G., and Nenci, I. (1988). Expression of *p21 ras* gene product in breast cancer relates to histological type and to receptor and nodal status. Breast Cancer Res. Treat. 12, 23-30.

Radford, D.M., Fair, K.L., Phillips, N.J., Ritter, J.H., Steinbrueck,T., Holt, M.S., and Donis-Keller, H. (1995). Allelotyping of ductal carcinoma in situ of the breast: Deletion of loci on 8p, 13q, 16q, 17p, and 17q. Cancer Res. 55, 3399-3405.

Ramachandra, S., Machin, L., Ashley, S., Monaghan, P., and Gusterson, B.A. (1990). Immunohistochemical distribution of c-erbB-2 in in situ breast carcinoma—A detailed morphological analysis. J. Pathol. 161, 7-14.

Rochlitz, C.F., Scott, G.K., Dodson, J.M., Liu, E., Dollbaum, C., Smith, H.S., and Benz, C.C. (1989). Incidence of activating ras oncogene mutations associated with primary and metastatic human breast cancer. Cancer Res. 49, 357-360.

Royds, J.A., Stephenson, T.J., Rees, R.C., Shorthouse, A.J., and Silcocks, P.B. (1993). Nm23 protein expression in ductal in situ and invasive breast carcinoma. J. Natl. Cancer Inst. 85, 727-731.

Russo, I.H., and Russo, J. (1996). Mammary gland neoplasia in long-term rodent studies. Environmental Health Perspectives 104, 938-967.

Sainsbury, J.R.C., Farndon, J.R., Needham, G.K., Malcolm, A.J., and Harris, A.L. (1987). Epidermal growth factor receptor status as predictor of early recurrence of and death from breast cancer. Lancet 1, 1398-1402.

Sato, T., Tanigami, A., Yamakawa, K., Akiyama, F., Kasumi, F., Sakamoto, G., and Nakamura, Y. (1990). Allelotype of breast cancer: Cumulative allele losses promote progression in primary breast cancer. Cancer Res. 50, 7184-7189.

Sato, T., Akiyama, F., Sakamoto, G., Kasumi, F., and Nakamura, Y. (1991). Accumulation of genetic alterations and progression of primary breast cancer. Cancer Res. 51, 5794-5799.

Sauer, R., Schauer, A., Rauschecker, H.F., Schumacher, M., Gatzemeier, W., Schmoor, C., Dunst, J., Seegenschmeidt, M.H., and Marx, D. (1992). Therapy of small breast cancer: A prognostic study on 1036 patients with special emphasis on prognostic factors. Int. J. Radiat. Oncol. Biol. Physics 23, 907-914.

Schenk, M., Leib-Mosch, C., Schenck, I.U., Jaenicke, M., Indraccolo, S., Saeger, H.D., Dallenbach-Hellwig, G., and Hehlmann, R. (1996). Lower frequency of allele loss on chromosome 18q in human breast cancer than in colorectal tumors. J. Mol. Med. 74, 155-159.

Schmitt, F.C. (1995). Multistep progression from an oestrogen-dependent growth towards an autonomous growth in breast carcinogenesis. Eur. J. Cancer 31, 2049-2052.

Schmitt, F.C., Leal, C., and Lopes, C. (1995). p53 protein expression and nuclear DNA content in breast intraductal proliferations. J. Pathol. 176, 233-241.

Schultz, D.C., Bazel, S., Wright, L.M., Tucker, S., Lange, M.K., Tachovsky, T., Longo, S., Niedbala, S., and Alhadeff, J.A. (1994). Western blotting and enzymatic activity analysis of cathepsin D in breast tissue and sera of patients with breast cancer and benign breast disease and of normal controls. Cancer Res. 54, 48-54.

Seth, A., Palli, D., Mariano, J.M., Metcalf, R., Venanzoni, M.C., Bianchi, S., Kottardis, S.D., and Papas, T.S. (1994). p53 gene mutations in women with breast cancer and a previous history of benign breast disease. Eur. J. Cancer 30A, 808-812.

Sharma, A.K., Horgan, K., Douglas-Jones, A., McClelland, R., Gee, J., and Nicholson, R. (1994). Dual immunocytochemical analysis of oestrogen and epidermal growth factor receptors in human breast cancer. Br. J. Cancer 69, 1032-1037.

Shekhar, P.V.M., Welte, R.A., Sarkar, F., Jones, R., Pauley, R.J., and Christman, J.K. (1995). Role of p53 in neoplastic progression of MCF10AneoT in a xenograft model for human breast cancer. Proc. AACR 36, 573.

Sheng, Z.M., Marchetti, A., Buttitta, F., Champeme, M.H., Campani, D., Bistocchi, M., Lidereau, R., and Callahan, R. (1996). Multiple regions of chromosome 6q affected by loss of heterozygosity in primary human breast carcinomas. Br. J. Cancer 73, 144-147.

Siziopikou, K.P., Prioleau, J.E., Harris, J.R., and Schnitt, S.J. (1996). bcl-2 expression in the spectrum of preinvasive breast lesions. Cancer 77, 499-506.

Skirnisdottir, S., Eiriksdottir, G., Baldursson, T., Barkerdottir, R.B., Egilsson, V., and Ingvarrson, S. (1995). High frequency of allelic imbalance at chromosome 16q22-23 in human breast cancer: Correlation with high PgR and low S phase. Int. J. Cancer 64, 112-116.

Slamon, D.J., Clark, G.M., Wong, S.G., Levin, W.J., Ullrich, A., and McGuire, W.L. (1987). Human breast cancer: Correlation of relapse and survival with amplification of the Her-2/neu oncogene. Science 235, 177-182.

Slamon, D.J., Godolphin, W., Jones, L.A., Holt, J.A., Wong, S.G., Keith, D.E., Levin, W.J., Stuart, S.G., Udove, J., Ullrich, A., and Press, M.F. (1989). Studies of the HER-2/neu proto-oncogene in human breast and ovarian cancer. Science 244, 707-712.

Sliwkowski, M.X., Schaefer, G., Akita, R.W., Lofgren, J.A., Fitzpatrick, V.D., Nuijens, A., Fendly, B.M., Cerione, R.A., Vandlen, R.L., and Carraway, 3rd, K.L. (1994). Coexpression of erbB2 and erbB3 proteins reconstitutes a high affinity receptor for heregulin. J. Biol. Chem. 269, 14661-14665.

Smith, S.A., Easton, D.F., Evans, D.G.R., and Ponder, B.A.J. (1992). Allele losses in the region 17q12-21 in familial breast and ovarian cancer involve the wild type chromosome. Nat. Genet. 2, 128-131.

Sobol, H., Stoppa-Lyonnet, D., Bressac-de-Paillerets, B., Peyrat, J.P., Kerangueven, F., Janin,N., Noguchi,T., Eisinger,F., Guinebretiere,J.M., Jacquemier,J.,and Birnbaum,D. (1996). Truncation at conserved terminal regions of BRCA1 protein is associated with highly proliferating hereditary breast cancers. Cancer Res. 56, 3216-3219.

Soini, Y., Hurskainen, T., Hoyhtya, M., Oikarinen, A.,and Autio-Harmainen, H. (1994). 72 kDa and 92 kDa type IV collagenase, type IV collagen, and laminin mRNAs in breast cancer: A study by in situ hybridization. J.Histochem. Cytochem. 42, 945-951.

Soule, H.D., Maloney, T.M., Wolman, S.R., Peterson, W.D.J., Brenz, R., McGrath, C.M., Russo, J., Pauley, R.J., Jones, R.F.,and Brooks, S.C. (1990). Isolation and characterization of a spontaneously immortalized human breast epithelial cell line, MCF-10. Cancer Res. 50, 6075-6086.

Spandidos, D.A., Pintzas, A., Kakkanas, A., Yiagnisis, M., Mahera, H., Patra, E., and Agnantis, N.J. (1987). Elevated expression of the *myc* gene in human benign and malignant breast lesions compared to normal tissue. AntiCancer Res. 7, 1299-1304.

Spirin, K.S., Simpson, J.F., Takeuchi, S., Kawamata, N., Miller, C.W., and Koeffler, H.P. (1996). p27/Kip1 mutation found in breast cancer. Cancer Res. 56, 2400-2404.

Stephenson, T.J., Royds, J.A., Silcocks, P.B., Brammer, H.J., Shothouse,A .J., and Underwood, J.C. (1994). Diagnostic associations of p53 immunostaining in fine needle aspiration cytology of the breast. Cytopathology 5, 146-153.

Stern, D.F., and Kamps, M.P. (1988). EGF-stimulated tyrosine phosphorylation of p185neu: A potential model for receptor interactions. EMBO Journal 7, 995-1001.

Stratton, M.R., Collins, N., Lakhani, S.R., and Sloane, J.P. (1995). Loss of heterozygosity in ductal carcinoma in situ. J. Pathol. 175, 195-201.

Sun, C.C., Lenoir, G., Lynch, H., and Narod, S.A. (1996). In situ breast cancer and BRCA1. Lancet 348, 408.

Tait, L., Dawson, P.J., and Miller, F.R. (1996a). Detection of micro invasion during progression in preneoplastic human breast xenograft model. Int. J. Oncol. 8, 1113-1117.

Tait, L., Dawson, P.J., Wolman, K., and Miller, F.R. (1996b). Multipotent human breast stem cell line MCF10AT. Int. J. Oncol. 9, 263-267.

Tamura, G., Maesawa, C., Suzuki, Y., Kashiwaba, M., Ishida, M., Saito, K., and Satodate, R. (1994). Improved detection of loss of heterozygosity at retinoblastoma gene locus in human breast carcinoma. Pathol. Int. 44, 34-38.

T'Ang, A., Varley, J.M., Chakraborty, S., Murphree, A.L., and Fung, Y.-K. (1988). Structural rearrangement of the retinoblastoma gene in human breast carcinoma. Science 242, 263-266.

Tang, C.K., Perez, C., Grunt, T., Waibel, C., Cho, C., and Lupu, R. (1996). Involvement of heregulin-β2 in the acquisition of the hormone-independent phenotype of breast cancer cells. Cancer Res. 56, 3350-3358.

Tavassoli, F.A., and Norris, H.J. (1990). A comparison of the results of long-term follow-up for atypical intraductal hyperplasia and intraductal hyperplasia of the breast. Cancer 65, 518-529.

Taylor-Papadimitriou, J., Stampfer, M., Bartek, J., Lewis, A., Boshell, M., Lane, E.B., and Leigh, I.M. (1989). Keratin expression in human mammary epithelial cells cultured from normal and malignant tissue: Relation to in vivo phenotypes and influence of medium. J. Cell Sci. 94, 403-413.

Thompson, A.M., Morris, R.G., Wallace, M., Wyllie, A.H., Steel, C.M., and Carter, D.C. (1993). Allele loss from 5q21 (APC/MCC) and 18q21 (DCC) and DCC mRNA expression in breast cancer. Br. J. Cancer 68, 64-68.

Thor, A., Ohuchi, N., Hand, P.H., Callahan, R., Weeks, M.O., Theillet, C., Lidereau, R., Escot, C., Page, D.L., Vilasi, V., and Schlom, J. (1986). *ras* gene alterations and enhanced levels of ras p21 expression in a spectrum of benign and malignant human mammary tissues. Lab. Invest. 55, 603-615.

Toi, M., Tominaga, T., Osaki, A., and Toge, T. (1994). Role of epidermal growth factor receptor expression in primary breast cancer: Results of a biochemical study and an immunohistochemical study. Breast Cancer Res. Treat. 29, 51-58.

Tsuda, H., Fukutomi, T., and Hirohashi, S. (1995). Pattern of gene alterations in intraductal breast neoplasms associated with histologic type and grade. Clinical Cancer Research 1, 261-267.

Tsutsumi, Y., Naber, S.P., DeLellis, R.A., Wolfe, H.J., Marks, P.J., McKenzie, S.J., and Yin, S. (1990). neu oncogene protein and epidermal growth factor receptor are independently expressed in benign and malignant breast tissues. Hum. Pathol. 21, 750-758.

Tzahar, E., Levkowitz, G., Karunagaran, D., Yi, L., Peles, E., Lavi, S., Chang, D., Liu, N., Yayon, A., Wen, D., and Yarden, Y. (1994). ErbB-3 and ErbB-4 function as the respective low and high affinity receptors of all Neu differentiation factor/heregulin isoforms. J. Biol. Chem. 269, 25226-25233.

Umekita, Y., Takasaki, T., and Yoshida, H. (1994). Expression of p53 protein in benign epithelial hyperplasia, atypical ductal hyperplasia, and noninvasive and invasive mammary carcinoma: An immunohistochemical study. Virchows Archives (Pathology and Anatomy) 424, 491-494.

Unemori, E.N., Ways, N., and Pitelka, D.R. (1984). Metastases of murine mammary tumor lines from the mammary gland and ectopic sites. Br. J. Cancer 49, 603-614.

Vaage, J. (1988). Metastasizing potentials of mouse mammary tumors and their metastases. Int. J. Cancer 41, 855-858.

Varley, J.M., Armour, J., Swallow, J.E., Jeffreys, A.J., Ponder, B.A.J., T'Ang, A., Fung, Y.T., Brammar, W.J., and Walker, R.A. (1989). The retinoblastoma gene is frequently altered leading to loss of expression in primary breast tumours. Oncogene 4, 725-729.

Weidner, N., Semple, J.P., Welch, W.R., and Folkman, J. (1991). Tumor angiogenesis and metastasis-correlation in invasive breast carcinoma. N. Engl. J. Med. 324, 1-8.

Weidner, N., Folkman, J., Pozza, F., Bevilacqua, P., Allred, E.N., Moore, D.H., Meli, S., and Gasparini, G. (1992). Tumor angiogenesis: A new significant and independent prognostic indicator in early-stage breast carcinoma. J. Natl. Cancer Inst. 84, 1875-1887.

Weinstat-Saslow, D., Merino, M.J., Manrow, R.E., Lawrence, J.A., Bluth, R.F., Wittenbel, K.D., Simpson, J.F., Page, D.L., and Steeg, P.S. (1995). Overexpression of cyclin D mRNA distinguishes invasive and in situ breast carcinomas from nonmalignant lesions [see comments]. Nat. Med. 1, 1257-1260.

Wellings, S.R., Jensen, H.M., and Marcum, R.G. (1975). An atlas of subgross pathology of the human breast with special reference to possible precancerous lesions. J. Natl. Cancer Inst. 55, 231-273.

Wick, W., Petersen, I., Schmutzler, R.K., Wolfarth, B., Lenartz, D., Bierhoff, E., Hummerich, J., Muller, D.J., Stangl, A.P., Schramm, J., Wiestler, O.D., and von Deimling, A. (1996). Evidence for a novel tumor suppressor gene on chromosome 15 associated with progression to a metastatic stage in breast cancer. Oncogene 12, 973-978.

Wolf, C., Rouyer, N., Lutz, Y., Adida, C., Loriot, M., Bellocq, J.P., Chambon, P. ,and Basset, P. (1993). Stromelysin 3 belongs to a subgroup of proteinases expressed in breast carcinoma fibroblastic cells and possibly implicated in tumor progression. Proc. Natl. Acad. Sci. U. S. A. 90, 1843-1847.

Wooster, R., Neuhausen, S.L., Mangion, J., Quirk, Y., Ford, D., Collins, N., Nguyen, K., Seal, S., Tran, T., Averill, D., Fields, P., Marshall, G., Narod, S., Lenoir, G.M., Lynch, H., Feunteun, J., Devilee, P., Cornilisse, C.J., Menko, F.H., Daly, P.A., Ormiston, W., McManus, R., Pye, C., Lewis, C.M., Cannon-Albright, L.A., Peto, J., Ponder, B.A.J., Skolnick, M.H., Easton, D.F., Goldgar, D.E., and Stratton, M.R. (1994). Localization of a breast cancer susceptibility gene, *BRCA2*, to chromosome 13q12-13. Science 265, 2088-2090.

Xu, C.F., and Solomon, E. (1996). Mutations of the *BRCA1* gene in human cancer. Seminars in Cancer Biology 7, 33-40.

Yaremko, M.L., Recant, W.M., and Westbrook, C.A. (1995). Loss of heterozygosity from the short arm of chromosome 8 is an early event in breast cancers. Genes, Chromosomes, and Cancer 13, 186-191.

Yoon, K., Cole-Strauss, A., and Kmiec, E.B. (1996). Targeted gene correction of episomal DNA in mammalian cells mediated by a chimeric RNA-DNA oligonucleotide. Proc. Natl. Acad. Sci. (USA) 93, 2071-2076.

Younes, M., Lebovitz, R.M., Bommer, K.E., Cagle, P.T., Morton, D., Khan, S., and Laucirica, R. (1995). p53 accumulation in benign breast biopsy specimens. Hum. Pathol. 26, 155-158.

Zelada-Hedman, M., Werer, G., Collins, P., Backdahl, M., Perez, I., Franco, S., Jimenez, J., Cruz, J., Torroella, M., Nordenskjold, M., Skoog, L., and Lindblom, A. (1994). High expression of the EGFR in fibroadenomas compared to breast carcinomas. AntiCancer Res. 14, 1679-1688.

Zhang, S.Y., Caamano, J., Cooper, F., Guo, X., and Klein-Szanto, A.J. (1994). Immunohistochemistry of cyclin D1 in human breast cancer. Am. J. Clin. Pathol. 102, 695-698.

Zhao, Y., Escot, C., Maudelonde, T., Puech, C., Rouanet, P., and Rochefort, H. (1993). Correlation between mannose-6-phosphate/IGFII receptor and cathepsin D RNA levels by in situ hybridization in benign and malignant mammary tumors. Cancer Res. 53, 2901-2905.

Zhuang, Z., Merino, M.J., Chuaqui, R., Liotta, L.A., and Emmert-Buck, M.R. (1995). Identical allelic loss on chromosome 11q13 in microdissected in situ and invasive human breast cancer. Cancer Res. 55, 467-471.

Zukerberg, L.R., Yang, W.I., Gadd, M., Thor, A.D., Koerner, F.C., Schmidt, E.V., and Arnold, A. (1995). Cyclin D1 (PRAD1) protein expression in breast cancer: Approximately one-third of infiltrating mammary carcinomas show overexpression of the cyclin D1 oncogene. Mod. Pathol. 8, 560-567.

ESTROGEN RECEPTOR VARIANTS IN EARLY BREAST DISEASE AND BREAST CANCER PROGRESSION

Adrian V Lee, Steffi Oesterreich, and
Suzanne A. W. Fuqua

Advances in Oncobiology
Volume 2, pages 117-135.

ISBN: 0-7623-0388-3

THE ROLE OF HORMONES IN NORMAL BREAST DEVELOPMENT

Although much is known about the architecture of the breast, the development of this complicated organ is still poorly understood. Three stages of breast development occur in females. In the prepubertal stage, maternal hormones cause an interaction between ductal epithelium and fat tissue. During puberty, ductal elongation and growth occur, and in the final stage during pregnancy, ductal differentiation occurs. The whole process, performed over several years, is controlled by a number of different hormones and growth factors that act in an endocrine, paracrine, autocrine, and intracrine manner, with considerable synergism between the factors.

The main hormones involved in the process are estrogens and progestins (Anderson and Battersby, 1989). Estradiol, the most potent estrogen, is needed both for breast development through puberty, but also for lobule formation during pregnancy. Rodent studies have indicated that estrogen has a positive effect on ductal formation. However, studies in patients have revealed that gonadotrophins also play a critical role; thus, patients with gonadal dysgenesis will only develop normal breasts upon estradiol administration if the hypothalamic–pituitary axis is functioning (Pertzelan et al., 1982). Although estrogens are clearly implicated in the growth of the normal breast, the highest mitotic index, in epithelial cells, occurs during the luteal phase of the menstrual cycle where estrogen levels are at the lowest and progestin levels at their highest (Ferguson and Anderson, 1981). This therefore suggests that either the estrogen effect is mediated by a number of other factors or that progestins are mitogenic. The importance of estrogen in breast development and growth led to immense research into the mechanism whereby estrogen exerts its effects and the eventual elucidation and cloning of the ER (Toft and Gorski, 1966).

ESTROGEN RECEPTOR—BACKGROUND

History

The first indication that estrogen effects were mediated via a receptor came from the observation by Jensen (1960) that rat uterine tissue could bind radio-labeled estradiol. A solubilized nuclear receptor was subsequently isolated by centrifugation of homogenized uterine tissue (Toft and Gorski, 1966). We now know that the ER is a member of a large family of nuclear transcription regulators that includes receptors for androgens, progestins, thyroid hormone, retinoids, and a number of other members whose ligands have yet to be isolated (termed orphan receptors). The detection of ER in both the cytoplasm and the nuclear fraction of uterine tissue led Jensen and colleagues (1968) to propose the two-step mechanism for ER action; unoccupied ER in the cytoplasm-bound estrogen translocated to the nucleus where it then bound to acceptor sites in the genome and thereby affected transcription of specific genes. However, the generation of monoclonal antibodies to the ER (Greene, 1984) immmunolocalized the ER to the nucleus, and enucleation studies (Welshons et al., 1988) further indicated that the majority of unoccupied ER resides in the nucleus.

Domains of the ER

The cloning of ER from the breast cancer cell line MCF-7 led to a rapid understanding of its role in mediating the diverse effects of estrogen. ER mRNA is 6322 nucleotides long, encoding a protein of 595 amino acids that has a relative molecular mass of 66 kDa. The ER sequence is highly homologous between species, and these similarities have led to the identification of seven regions within the ER (termed A through F) that perform the functions of ligand binding (region E), DNA binding (region C), and transcriptional activation (regions A and E; Figure 1). The hormone-binding domain (HBD), domain E, is 250 amino acids in length; a number of studies have delineated the regions required for binding estrogen. Summarizing the data, it is apparent that the entire region from amino acids 302 to 552 is important, with specific studies using the estrogen koetonestrol arizidine, indicating that cysteine 530 may be the most important. The DNA-binding domain (DBD), region C, is 66 amino acids in length and appears to form a zinc finger DNA-binding do-

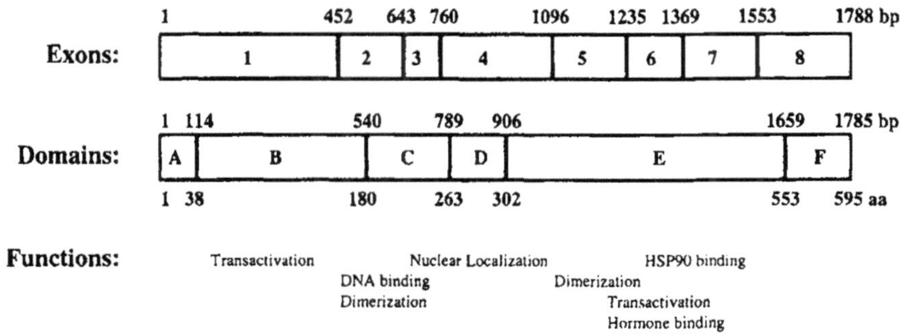

Figure 1. Functional domains of the ER. At the top is the exon structure of ER and below are the functional protein domains. AA–amino acid. BP–base pair

main (Lee et al., 1989). Domain D contains a nuclear localization sequence which, when transferred to another protein, may confer nuclear localization activity (Picard et al., 1990). The transcriptional activating function of ER is contained within domains A and E. The first element, activating function 1 (AF1), which is contained in domains A and B, is constitutively active and hormone-independent whereas activating function 2 (AF-2), found in the E domain, is responsive to hormone.

Receptor Transformation

Unoccupied ER migrates as an 8S form on a low salt–sucrose gradient, but in high salt conditions (>300 mM), transforms to a 4S form or a 5S dimer (Miller et al., 1985). This transformation can also be performed by estrogen and results in activation of the receptor so that it has a higher affinity for DNA, chromatin, and nuclei. Dialysis, ammonium sulphate precipitation, and an increase in temperature also cause transformation, which has led to the hypothesis that it was not simple dimerization of the receptor but that conversion from the 8S to 5S dimer form proceeded via the loss of proteins such as heat shock protein 90 (hsp90). Whether hsps actually prevent dimerization or are able to mask the DBD, which is important for activity, is not known, but mutated receptors lacking the hormone-binding region do not bind hsp90 (Chambraud et al., 1990) and are able to activate transcription, albeit at a lower efficiency than full length ER (Kumar et al., 1987).

Estrogen Response Elements (EREs) and DNA Binding

ER binds to DNA at a region called the estrogen responsive element (ERE), which is usually a palindromic sequence located 5' to the promoter of estrogen-responsive genes (Beato, 1989). A number of EREs have been functionally identified in the promoter regions of estrogen-responsive genes, and from these, a consensus ERE has been defined that is identical to the ERE of the *Xenopus laevis* vitellogenin gene (5'-GGTCA NNN TGACC-3'; Klein-Hitpass et al., 1988).

Many naturally-occurring estrogen-responsive genes contain multiple copies of either consensus EREs or variant forms (imperfect EREs) of this inverted repeat. Whereas there is a more general agreement that progesterone receptor (PgR) and glucocorticoid receptor can bind cooperatively to multiple hormone-responsive elements, there are conflicting results concerning synergism between ERs bound to multiple EREs. Catherino and Jordan (1995) observed that the number of EREs in a reporter plasmid is correlated with the level of estrogen-induced transcription. Similarly, the estrogen response of the *Xenopus* vitellogenin B1 gene is mediated through two closely spaced EREs that have, respectively, one and two nucleotide sequence variations from the consensus response element. In transient transfection assays, the activity of two EREs together was much greater than that of the sum of either alone, suggesting a synergistic action of the two EREs (Martinez and Wahli,

1989). However, others did not find any evidence for cooperative binding of the ER to either consensus or imperfect EREs (Klein-Hitpass et al., 1989; Ponglikitmongkol et al., 1988). A possible answer might lay in a more complex combination of number, spacing, sequence, and last but not least the ERE flanking sequences in the promoter.

It has been shown that the agonist/antagonist balance of antiestrogens may depend on the specific promoter content harboring the ERE. For instance, 4-hydroxy tamoxifen (4-OH TAM), a nonsteroidal antiestrogen, is unable to activate transcription in MCF-7 cells when there is a single ERE in the promoter region, but if there are two or three EREs present, 4-OH TAM becomes an effective agonist (Catherino and Jordan, 1995). Using different estrogen-responsive reporter gene constructs (Fujimoto and Katzenellenbogen, 1994) it was also found that the agonist/antagonist activity of TAM is determined by the ongoing action of EREs within the promoter.

EREs are usually functionally defined as sequences that exhibit homology to the consensus ERE, display specific binding to the ER in gel-retardation assays or footprinting assays, and confer estrogen-inducibility on reporter plasmids in transient transfections. A large number of studies have characterized EREs in various genes, and different methods have been described in order to identify EREs (McDonnell et al., 1991). However, to date, only a very limited number of estrogen-responsive genes has been described. This is somehow surprising since estrogen-stimulation of ER-positive cells leads to a cascade of important phenotypic changes. The identification of more estrogen-responsive genes in the coming years might provide a greater understanding of the role of the ER in normal breast development and breast cancer progression.

Interaction of the ER with the Basal Transcription Machinery

The mechanisms by which the two ER transactivation domains contribute to transcriptional activation is unknown, but by analogy with better understood viral activators, it is now thought that interaction with specific target proteins of the basal transcriptional machinery is important in the transcriptional response (Figure 2). The first finding of an interaction of ER with a member of the basal transcriptional machinery was almost accidental in that purification of chicken ovalbumin upstream promoter-transcription factor (COUP-TF) from HeLa cells yielded co-purification of a second factor, which was also required for activation of the ovalbumin promoter (Ing et al., 1992). Cloning and sequencing of this factor revealed its identity to TFIIB, and consequently these investigators examined the interactions of TFIIB with other steroid receptors. They found that the ER, but not a truncated ER containing only the AF-1 domain, interacted with TFIIB. AF-2s involvement in the interaction with the basal transcription machinery was also found by Jacq et al. (1994), who were able to show that the AF-2-containing region E of the ER binds to $TAF_{II}30$, a TATA-binding protein (TBP) associated protein.

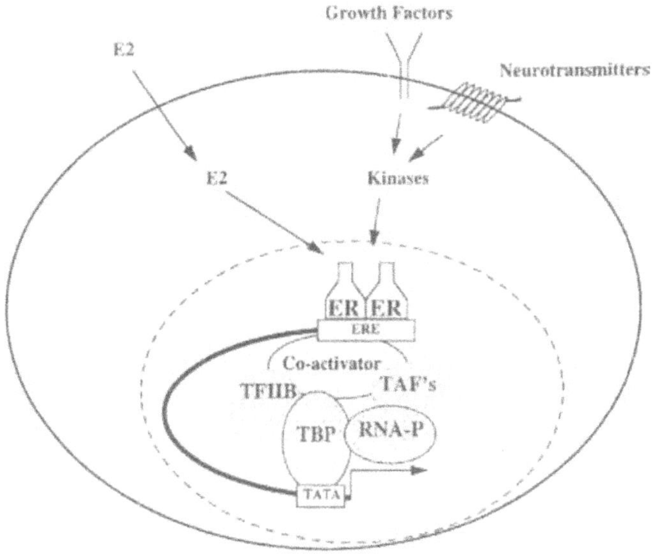

Figure 2. A model for transcriptional activation of the ER. Estradiol (E2) diffuses into the nucleus, binds to the ER and causes dimerization and binding to specific palindromic sequences in DNA termed estrogen response elements (ERE). Here the ER can then interact with a number of co-activators, co-repressors, and members of the basal machinery complex (TBP-TATA–Binding Protein) to increase transcription via RNA polymerase II (RNA-P).

It was recently shown that overexpression of TBP potentiates ER activation of estrogen-regulated reporter genes (Sadovsky et al., 1995). In contrast to the interactions with the other members of the basal transcription machinery, each of the transcriptional activation domains of the ER, AF-1 as well as AF-2, need to be involved in this interaction. Interestingly, other activators such as NF-1 and SP-1 became less potent when TBP was overexpressed, excluding the potential criticism that TBP potentiates all activators.

Further research will presumably indicate interaction of the ER with a number of other components of the basal transcription machinery, and then the effect of each of these interactions can be related to ER function. Although there are clearly multiple interactions between different ER-activating domains and distinct components of the basal transcription, these interactions may not be independent from each other but may result in synergistic effects on transcription. Precise mapping of the exact regions of the ER interacting with these factors, as well as an evaluation of the relevance of these interactions in vivo, is necessary to explain how these contacts between the ER on the one hand, and TAFs and basal transcription factors on the other hand, result in dramatically increased levels of initiation of polymerase II transcription.

ER Accessory Proteins

Whereas the basal machinery complex undoubtedly contributes to the promoter-specific activity of ER, another set of proteins recently discovered, termed ER accessory proteins, may also provide promoter- and or tissue-specific activity by controlling interaction of the ER with the basal transcription machinery (Figure 2). Since the discoveries that the SWI/SNF family of yeast proteins are needed for ER activity in yeast and that human homologs are able to affect ER transcriptional activity (Chiba et al., 1994), many ER accessory proteins have been recently identified (reviewed in Horwitz et al., 1996).

Two proteins, ERAP160 and RIP140, were found to interact with the HBD of ER in a yeast two-hybrid assay, binding to the ER being induced by estrogens, and inhibited by antiestrogens. ERAP160 (Halamachi et al., 1994), like many of the other co-activators, interacts with a number of different members of the steroid receptor family, and it may be a member of a family of co-activators that include SRC-1 (the C-terminus of ERAP160 has 88% homology with SRC-1(Onate et al., 1995)). RIP140 interacts with the AF-2 domain of ER but does not bind to members of the basal machinery such as TBP or TFIIB (Cavailles et al., 1995).

If human homologs are found for the yeast proteins that can interact with the ER and affect its transcriptional activity, then it is clear that an extremely complicated scenario for activation of the ER exists. It is further possible that variants of the ER may interact with a number of different accessory proteins or have altered affinity for those that bind with the ER thus leading to increased or decreased activity of the variant receptor. Although many groups have been instrumental in identifying the factors binding to the ER, very little data exist concerning the actual in vivo biological role of these proteins. In vivo experiments will no doubt lead to a greater understanding of the complex response of a cell or tissue to an estrogen or antiestrogen.

Ligand-Independent Activation of the ER

Since the discovery that progesterone receptor (PgR) can be activated by non-progestins such as dopamine (Power et al., 1991), there has been intense research into ligand-independent activation of steroid hormone receptors, and in the case of the ER, many new activators have been discovered representing other hormones, growth factors, phorbol esters, activators of protein kinase A, and many others (Figure 2). Katzenellenbogen and Norman (1990) showed that expression of PgR is regulated by many factors including estradiol and IGF-I in breast cancer cells and that this occurred at the level of gene transcription (Cho et al., 1994). Further studies in uterine cells indicated transcriptional activation and phosphorylation of the ER by IGF-I (Aronica and Katzenellenbogen, 1993); however, the increase in phosphorylation also occurred with the pure antiestrogen ICI 164,384 (ICI) suggesting that overall phosphorylation did not necessarily correlate with transcriptional activation. The ER can also be activated by other growth factors (Ignar-Trowbridge et al., 1996) and protein kinase activators (Aronica and Katzenellenbogen, 1993; Kato et al., 1995; Patrone et al., 1996). The ability to increase the expression of estrogen-inducible genes such as pS2 and PgR in vitro indicates that cross-talk between growth factor pathways and nuclear steroid hormone receptors can occur within cells that normally express ER, and may have a physiological role in vivo. While it is clear that nonestrogenic ligands can activate the ER, it is also possible that putative variant ER mRNAs that have constitutive transcriptional activity (e.g., the ERΔ5 deletion variant - to be discussed in detail later), along with ligand-independent activity of the receptor, may presumably have a profound effect on downstream targets of the ER.

The mechanism of ligand-independent activation of the ER remains unclear as several reports give contradictory results. The ER is phosphorylated upon stimulation of uterine cells with IGF-I (Aronica and Katzenellenbogen, 1993); however phosphorylation of the ER does not necessarily directly correlate with transcriptional activation. Kato and colleagues (1995) have shown that phosphorylation of the ER on a serine residue at position 118 is important in mitogen-activated protein kinase (MAPK) activation of the ER suggesting that the N-terminal AF-1 region is important. Supporting this, Ignar-Trowbridge and colleagues (1996) have shown that this same domain is important in activation of the ER by a number of growth factors; however, some activation was directed by the carboxy-terminal AF-2 domain. Conversely, ras and protein kinase C seem to be involved in ligand-independent activation of the ER through the AF-2 domain as dominant–negative inhibitors of ras inhibit insulin activation of the ER stably expressed in neuroblastoma cells (Patrone et al., 1996). These conflicting results probably represent the cell type and promoter-dependent activity of ER (Tzukerman et al., 1994) with some of the data arising from transient or stable expression of ER in cells that are normally ER-negative.

Cellular Effects of Estrogen

A greater understanding of the action of estrogen came from the establishment of breast cancer cell lines, many of which contain detectable levels of ER and are growth-responsive or dependent upon the hormone. Estrogen can have both genomic and nongenomic actions in the cell with nongenomic effects occurring in the absence of receptors usually at pharmacological doses of hormone (Duval et al., 1983). Genomic effects occur at physiological concentrations and, in ER-positive breast cancer cell lines, result in a number of diverse effects including increased expression of DNA-synthesizing proteins (Bronzert et al., 1981), regulation of epidermal growth factor, insulinlike growth factor and fibroblast growth factor family members (Clarke et al., 1991), increased expression of receptors (PgR, laminin), proteases (cathepsin D, plasminogen, collagenase), heat-shock protein 27 (Adams et al., 1983), pS2 (Westley and Rochefort, 1980), and many other proteins. The ability of the ER to regulate expression of a combination of growth factors and proteases suggests that aberrant expression of the ER, or other putative variant ER mRNAs may result in altered expression of these pathways and therefore be implicated in breast cancer development and progression.

THE ROLE OF ESTROGENS AND THE ER IN BREAST CANCER

While estrogens are involved in the growth and development of the normal breast, they are also clearly implicated in both the generation and progression of breast cancer. In animal studies, rats treated with the carcinogen 7,12-dimethylbenzanthracene develop hormone-dependent tumors, which will not grow if the rats are ovariectomized (Dao, 1972). Tumors will form however if mice are treated with pharmacological doses of estrogen (Huggins et al., 1961). Epidemiological evidence also implicates estrogen in the progression of cancer. Firstly, daughters of women who took diethylstilbestrol in the 1950s to prevent aborted pregnancies, have a higher incidence of vaginal adenocarcinoma (Herbst et al., 1971). Secondly, early forms of estrogen replacement therapy, which involved high doses of unopposed estrogens, showed that users had a higher incidence of endometrial cancer (Smith et al., 1975), and this was also observed with early forms of the contraceptive pill (Henderson et al., 1983).

The majority of risk factors for breast cancer can be simply viewed as representing the total exposure of the breast to estrogen, that is, the longer the exposure to estrogen, the higher the risk of developing breast cancer. Thus, high risk factors include early menarche, late first pregnancy, late menopause and nulliparity, obesity, and estrogen replacement therapy. In contrast, protective factors include early first pregnancy, lactation, and physical activity—all elements that reduce the exposure of the breast to estrogen (Henderson and Bernstein, 1996). Furthermore, breast cancer occurs in women at a rate of about 100 times that seen in men even though the male breast contains the same epithelial component. Additionally, breast cancer frequency rises in male transvestites that take estrogen replacement therapy.

Substantiating the importance of estrogen in the growth and progression of breast cancer was the finding that reducing estrogen exposure to the breast can initiate breast tumor regression. Thus, Beatson performed bilateral ovariectomy on a patient with inoperable breast cancer that resulted in tumor regression (1896). Although he did not know the mechanism of this effect, the elucidation and characterization of the molecular structure of the ER have since led to a greater understanding of the role of endocrine treatment in breast cancer treatment. The relatively low success rate of chemotherapy and surgery has led to the search for therapies that were less toxic and reversible, the most widely used now being antiestrogens. The most successful antiestrogen to date is tamoxifen (ICI 46,474; Nolvadex), a nonsteroidal antiestrogen that has minimal toxicity, can inhibit growth of hormone-responsive advanced breast cancer (Furr and Jordan, 1984), and can prolong survival if given adjuvantly (Early Breast Cancer Trialists Collaborative Group, 1992). Interestingly, tamoxifen not only causes regression in ER-positive tumors (Lippman and Chabner, 1986) but also in a small number of ER-negative tumors (Nolvadex Adjuvant Trial Organization, 1988). The beneficial effects of tamoxifen therapy are lessened by the fact that most estrogen-responsive tumors will eventually return in a hormone-independent state and require a more toxic therapy.

ER VARIANTS AND BREAST CANCER

The structure of the ER gene is commonly altered in both normal and malignant breast tissue. A number of different alterations have been discovered, generally representing truncations and exon skipping, with rare cases of point mutations and exon duplication. The number of changes discovered and the rare occurrence of these events within other steroid hormone receptors may simply indicate that the ER gene is a hot spot for such alterations, or it may give a clue as to the importance of the ER in breast growth and development. A large number of variants have also been either created in vitro or discovered in cell lines treated with various drug regimes. This review will concentrate on those naturally occurring ER variants found in normal and breast cancer tissue and the role they may play in breast cancer growth and progression.

Early Breast Disease

If a significant risk factor for developing breast cancer is the total exposure of the breast to estrogen, this implies that the ER may be a important risk factor for disease progression as shown by the clinical use of antiestrogens in breast cancer treatment. Abnormal expression of the ER in the wild-type or variant form may account for an increase in the epithelial component of the breast and may therefore provide more proliferating cells susceptible to an oncogenic hit; alternatively, increased ER expression may drive proliferation of those cells that have already received an initial insult (Figure 3).

There is accumulating data indicating that the wild-type ER is frequently overexpressed in early breast disease. The ER is normally expressed at low levels in normal breast tissue (Ricketts et al., 1991), but levels change throughout the menstrual cycle; thus, this is a difficult issue to quantitate. ER expression is higher in premalignant or malignant lesions as compared with adjacent normal epithelial tissue (van Agthoven et al., 1994), and this expression correlates with the intensity of hyperplasia in some lesions (Jacquemier et al., 1982). Furthermore, Khan and colleagues (1994) have shown that ER expression in benign breast epithelium is a significant risk factor for breast cancer, although the number of samples used was relatively small in this case control study (51 breast cancer vs. 69 benign breast tissue).

We have recently hypothesized that expression of ER variants early in breast disease may also play a role in breast cancer development. We have screened early

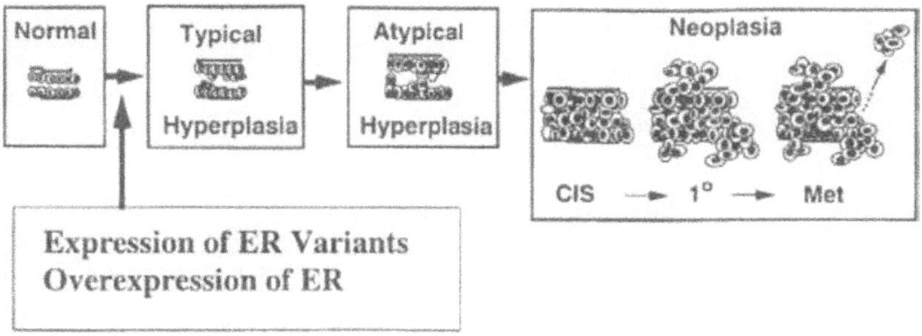

Figure 3. A model proposing a role for increased expression of the ER or ER variants in early breast tumor progression. Elevated receptor levels may provide an increase in the number of cells that potentially could receive an oncogenic hit or promote the growth of cells that have already received an insult.

breast lesions for ER variants by PCR and isolated a variant ER that contained an alteration (Lys 303 Arg) in the beginning of the hormone-binding domain (Wiltschke et al., 1995). This variant shows increased transcriptional activity and increased sensitivity to estrogen when expressed transiently in an ER-negative breast cancer cell line, however, the receptor binds estrogen and DNA normally. We are currently determining the effect of expression of this variant on breast development and examining whether expression of this variant might play a role early in breast disease.

Leygue and colleagues (1996a) have used a novel form of semiquantitative RT-PCR (RT-triple primer-PCR) to amplify both wild-type and variant ER mRNAs at the same time from breast tissue. They found that there was a significant increase in expression of a truncated ER variant called Clone 4, which contains exons 1 and 2 of the ER but which then contains sequences similar to long-interspersed repetitive elements (LINE-1). There was a significant increase in expression of Clone 4 mRNA between normal breast tissue and breast cancer that had good prognostic features (which may be representative of an early stage of breast cancer), suggesting that expression of this variant may be an early event in breast tumor progression.

Breast Cancer

Although there are few data concerning ER variants in early breast disease, there have been a number of studies examining the presence of ER variants in breast cancer. It should be noted however, that to date, these variants have only been found as variant mRNA species and isoforms, and as yet, no altered proteins have been identified in tumor tissue. Expression of these variants can be detected in breast cancer cell lines in vitro; however, detailed protein studies must first be performed in vivo in human clinical samples before concrete conclusions can be made about the role of specific ER variants in breast cancer progression. Here we will detail those variants that have recently been found in clinical samples and whose altered structure suggest that they may play a role in breast cancer.

Murphy and Dotzlaw (1989) found by Northern analysis of human breast tumor biopsies small ER mRNAs (2.4 and 3.8 kb) that were expressed in conjunction with full-length wild type ER (6.5 kb). Cloning and expression of these variant mRNAs indicated that they represented aberrantly spliced ER mRNA. Two of these variants called clones 5 and 24 represented abnormal splicing at exon 3 resulting in a protein that would contain the DBD, but no HBD; however, they had no transcriptional activity when expressed in COS cells. The third variant called clone 4 is a variant ER that represents exons 1 and 2 of the wild type ER but then contains sequences similar to long-interspersed repetitive sequence elements (LINE-1). Again, this ER variant is inactive in transient transactivation assays. Further investigation of the existence of these variants by RT-PCR has indicated the presence of variant ERs lacking exon 2, exon 3, exons 2–3, exon 5, and exon 7 (Leygue et al., 1996b) in both normal and breast tumor samples. The expression of a number of variants along with wild-type ER complicates drawing any simple model for a role of these variants in breast tumor progression.

Recently, Murphy et al. (1996), have extended these initial observations by the use of RT-PCR in detecting large forms of ER mRNA in 9.4% of human breast cancer tissue samples. The most common variant was a duplication of exon 6 (7.5% of samples) with detection of a duplication of exons 3 and 4 in a single tumor and a 69-bp insertion between exons 5 and 6 in three tumors. No functional data is yet available on the altered ER proteins that are predicted to be transcribed from these variant mRNAs, however the exon 6 duplication could potentially be translated into a protein of 51.4 kDa that would lack the HBD, and thus may act as a dominant–positive ER. Interestingly, an 80-kDa, immunoreactive ER variant was recently cloned from an MCF-7 cell line that is unresponsive to estradiol growth stimulation, this variant ER containing a complete duplication of exons 6 and 7 (Pink et al., 1996). Although it has been difficult to detect protein products from putative variant ER mRNAs that have been reported, the existence of an immunoreactive, 80-kDa ER variant in MCF-7 human breast cancer cells suggests that attempts to make antibodies specific to the variants may eventually be successful.

Point Mutations in the ER

The detection of point mutations within the ER from human breast tumor samples is much less common than the reports of exon splicing. Karnik and colleagues (1994) have identified a 42-bp insertion and a single bp deletion in two patients with tamoxifen-resistant breast cancer. Both of these variants would be predicted to encode truncated ERs, however functional data on these are not yet available. We have recently screened 30 patients who had received tamoxifen therapy for at least 6 months prior to recurrence and found three different point mutations within the ER in these samples (Zhang et al., 1996). Functional analysis of one of these variants with a point mutation in exon 8 has identified a variant of ER that has constitutive transactivation, which has been increased to that seen with wild-type ER in the presence of estrogen. Additionally, this high level of constitutive transcriptional activity is not affected, or is only minimally, by antiestrogens, depending on the promoter utilized. We hypothesize that this variant may play a role in breast cancer progression by interruption of normal ER action.

The Exon 5 ER Deletion Variant

We originally identified a variant form of ER mRNA that completely lacks exon 5 (called ERΔ5) from human breast tumors that were ER-negative but PgR-positive. We had originally hypothesized that in these samples there might be a variant receptor that would not bind ligand but was transcriptionally active and thereby responsible for the discordant expression of PgR in these tumors. Indeed, the ERΔ5 variant does represent this type of receptor; however, its transactivating potential appears to be promoter- and cell-dependent, similar to that seen with the activity of wild-type ER in different cells (Tzukerman et al., 1994). We showed that the ERΔ5 variant has 10–15% of the activity of wild-type ER in yeast cells, 40–50% in another type of yeast background, and 75% activity in MDA-MB-231 human breast cancer cells (Fuqua et al., 1991). Rea and Parker (1996) have shown that the ERΔ5 variant has transcriptional activity in the absence of hormone in chicken embryo fibroblasts but only weak activity in MCF-7 breast cancer cells. The differences seen are in accordance with the fact that ERΔ5 only contains the AF-1 transactivation domain, which has previously been shown to have promoter- and cell-dependent activity (Tzukerman et al., 1994).

Although we originally identified ERΔ5 in ER-negative /PgR-positive tumors, further analysis indicates that it is also expressed at the RNA level in the majority of ER-positive tumors; thus ERΔ5 variant is almost always co-expressed along with wild-type ER RNA (Zhang et al., 1993). This variant was not found in any ER-negative /PgR-negative tumors. The expression of two forms of ER RNA makes drawing conclusions about the cellular effects of ERΔ5 especially difficult. Thus, if ERΔ5 is less active than wild-type ER, it could conceivably reduce the effect of estrogen within a cell, whereas if it has activity similar to or greater than wild-type ER, then it might increase the overall estrogen effect.

As detailed above, simple expression of the ERΔ5 variant may not be a good indicator of the net effect of the variant in cells. This may be one of the reasons for the discrepancy between our studies and those of Rea and Parker (1996). In our laboratory subline of MCF-7 cells, the ERΔ5 variant is transcriptionally active in the absence of hormone and, when constitutively expressed, leads to estrogen-independent growth that is unaffected by tamoxifen (Fuqua et al., 1995). Rea and Parker (1996) found that ERΔ5 was only slightly active in their MCF-7 cells and that stable expression resulted in no significant change in estrogen-regulated growth or growth inhibition by antiestrogens. Perhaps this result simply reflects the differences that have been noted several times between MCF-7 sublines held in different laboratories (Osborne et al., 1987). Supporting this, Klotz and colleagues (1995) have shown that expression of varying levels of this variant in sublines of MCF-7 cells from different laboratories was associated with different responses to estrogen (higher ERΔ5 expression led to a decrease in estrogen responsiveness). However, it must be noted that a number of other ER mRNA variants were seen to vary in these cell lines, suggesting that the difference may not only be due to expression of the ERΔ5 variant alone.

Studies have also been performed analyzing the expression of the ERΔ5 variant in normal breast tissue. Leygue and colleagues (1996b) detected a number of ER variants in all 9 normal breast tissue samples examined (clone 4 variant, and variants with deletions in exon 2, exon 3, exons 2-3, and exon 7). Of particular interest was the fact that ERΔ5 mRNA expression was found to be significantly lower ($p<0.001$) relative to wild-type ER in normal breast tissue compared to that in breast tumor tissue. Pfeffer and colleagues (1996) also detected expression of a number of ER variants in a single normal breast tissue specimen, and higher levels specifically of ERΔ5 were found in one tumor sample. However, this study was limited to a single normal and tumor sample.

Daffada and colleagues (1995) have examined ERΔ5 mRNA expression in primary breast tumors and tamoxifen-resistant breast cancer. While there was no significant difference between ERΔ5 mRNA expression between these two tumor groups, it was noted that tumors that expressed PgR and pS2, in the absence of detectable ER, had significantly elevated levels of ERΔ5 mRNA. This therefore suggests that this variant may account for the hormone-independent expression of ER-regulated genes and that the tumor now may have ligand-independent activation of the ER. Further studies have indicated that, while ER levels are significantly reduced in tamoxifen-resistant breast tumors, when compared with breast tumors that have not seen tamoxifen, pS2 levels remain unchanged (Johnston et al., 1995). This again suggests that, although there seems to be little or no ER by immunohistochemical analysis, estrogen-responsive genes are still expressed. This may be accounted for by the expression of ERΔ5, suggesting a role for this variant in tamoxifen resistance of breast cancer. It must be noted, however, that many other factors can alter pS2 expression indicating that multiple pathways for tamoxifen resistance assuredly exist.

FUTURE DIRECTIONS

It is clear from both in vitro and in vivo studies on wild-type ER activity that the simple measurement of ER variant mRNA expression will not be an indicator of the effect of that variant within specific cells or tissues. The ER is an extremely complicated nuclear transcription factor being regulated by a number of different ligands but also being controlled in a promoter- and tissue-specific manner by a number of co-activators and co-repressors. Complicating analysis of the role of variant ERs is the fact that, in nearly all cases, wild-type ER is co-expressed. It is hypothesized that a different sets of genes will be activated by variant ERs with the ability to interact with a different affinity with activators or repressors. Elucidation of the genes induced by different variant receptors may eventually provide a blueprint for variant ER action just as we use PgR and pS2 as markers for ER action today.

SUMMARY

As a downstream mediator of estrogen action, the ER can have a number of pleiotropic effects on a variety of cell types. The critical importance of estrogen in both normal and breast cancer growth has made the ER a matter of intense study, and still its complete mode of action is unclear. One of the discoveries to come out of recent work is the finding that the ER is often altered at the RNA level in both normal and malignant breast tissue. When the interaction of these variants with the growing number of regulators of wild-type ER activity are studied in detail, we may get a clearer picture of the role of variant ERs in breast cancer disease and progression.

ACKNOWLEDGMENTS

The authors would like to acknowledge grant support from the National Institutes of Health (NIH) CA30195, P30CA54174, and the Department of Defense DAMD17-94-J-4112.

REFERENCES

Adams, D. J., Hajj, H., Edwards, D. P., Bjercke, R. J., and McGuire, W. L. (1983). Detection of a *Mr* 24,000 estrogen-regulated protein in human breast cancer by monoclonal antibodies. Cancer Research 43, 4297-4301.

Anderson, T. J., and Battersby, S. (1989). The involvement of oestrogen in the development and function of the normal breast: Histological evidence. Proceedings of the Royal Society of Edinburgh 95B, 23-32.

Aronica, S., and Katzenellenbogen, B. (1993). Stimulation of estrogen receptor-mediated transcription and alteration in the phosphorylation state of the rat uterine estrogen receptor by estrogen, cyclic adenosine monophosphate, and insulin-like growth factor-I. Molecular Endocrinology 7(6), 743-752.

Beato, M. (1989). Gene regulation by steroid hormones. Cell 56, 335-344.

Beatson, G. T. (1896). On the treatment of inoperable cases of carcinoma of the mamma. Suggestions for a new treatment with illustrative cases. Lancet 2, 104-107.

Bronzert, D. A., Monaco, M. E., Pinkus, L., Aitken, S., and Lippman, M. E. (1981). Purification and properties of estrogen responsive cytoplasmic thymidine kinase from human breast cancer. Cancer Research 41, 604-610.

Catherino, W. H., and Jordan, V. C. (1995). Increasing the number of tandem estrogen response elements increases the estrogenic activity of a tamoxifen analogue. Cancer Letters 92(1), 39-47.

Cavailles, V., Dauvois, S., L'Horset, F., Lopez, G., Hoare, S., Kushner, P., and Parker, M. (1995). Nuclear factor RIP140 modulates transcriptional activation by the estrogen receptor. EMBO-J 14(14), 3741-3751.

Chambraud, B., Berry, M., Redeuilh, G., Chambon, P., and BAlieu, E. E. (1990). Several regions of human estrogen receptor are involved in the formation of receptor-heat shock protein 90 complexes. Journal of Biological Chemistry 265, 20686-20691.

Chiba, H., Muramatsu, M., Nomoto, A., and Kato, H. (1994). Two human homologues of *Sacharomyces cerevisiae* SNF2/SW12 and *Drosophila brahma* are transcriptional coactivators cooperating with the estrogen receptor and the retinoic acid receptor. Nucleic Acids Research 22, 1815-1820.

Cho, H., Aronica, S., and Katzenellenbogen, B. S. (1994). Regulation of progesterone receptor gene expression in MCF-7 breast cancer cells: A comparison of the effects of cyclic adenosine 3'-5'-monophosphate, estradiol, insulin-like growth factor-I and serum factors. Endocrinology 134, 658-664.

Clarke, R., Dickson, R. B., and Lippman, M. E. (1991). The role of steroid hormones and growth factors in the control of normal and malignant breast. In: *Nuclear Hormone Receptors*. (Parker, M., Ed.), Academic Press, London.

Daffada, A. A. I., Johnston, S. R. D., Smith, I. E., Detre, S., King, N., and Dowsett, M. (1995). Exon 5 deletion variant estrogen receptor messenger RNA expression in relation to tamoxifen resistance and progesterone receptor/pS2 status in human breast cancer. Cancer Research 55, 288-293.

Dao, T. L. (1972). Ablation therapy for hormone dependent tumors. Annual Review of Medicine 23, 1-18.

Duval, D., Durant, S., and Homo-delarche, F. (1983). Non-genomic effects of steroids. Biochimica et Biophysica Acta 737, 409-442.

Early Breast Cancer Trialists Collaborative Group. (1992). Systemic treatment of early breast cancer by hormonal, cytotoxic, or immune therapy: 133 randomized trials involving 31000 recurrences and 24000 deaths among 75000 women. Lancet 339, 1-15, 71-85.

Ferguson, D. J. P., and Anderson, T. J. (1981). Morphological evaluation of cell turnover in relation to the menstrual cycle in the nesting human breast. British Journal of Cancer 44, 177-181.

Fujimoto, N., and Katzenellenbogen, B. S. (1994). Alteration in the agonist/antagonist balance of antiestrogens by activation of protein kinase A signaling pathways in breast cancer cells: Antiestrogen selectivity and promoter dependence. Molecular Endocrinology 8, 296-304.

Fuqua, S., Wiltschke, C., Castles, C., Wolf, D., and Allred, D. (1995). A role for estrogen-receptor variants in endocrine resistance. Endocrine-Related Cancer 2, 19-25.

Fuqua, S. A. W., Fitzgerald, S. D., Chamness, G. C., Tandon, A. K., McDonnell, D. P., Nawaz, Z., O'Malley, B. W., and McGuire, W. L. (1991). Variant human breast tumor estrogen receptor with constitutive transcriptional activity. Cancer Research 51(1), 105-109.

Furr, B. J. A., and Jordan, V. C. (1984). The pharmacology and clinical uses of tamoxifen. Pharmacology and Therapeutics 25, 127-205.

Greene, G. L., Sobel, N., King, W. J., and Jensen, E. J. (1984). Immunochemical Studies of Estrogen Receptors. Journal of Steroid Biochemistry 20(1), 51-56.

Halamachi, S., Marden, E., Martin, G., MacKay, H., Abbondanza, C., and Brown, M. (1994). Estrogen receptor-associated proteins: Possible mediators of hormone-induced transcription. Science 264, 1455-1458.

Henderson, B. E., and Bernstein, L. (1996). Endogenous and exogenous hormonal factors. In: *Diseases of the breast.* (Harris, J. R., Lippman, M. E., Morrow, M., and Hellman, S., Eds.), pp. 185-200. Lippincott-Raven, Philadelphia, PA.

Henderson, B. E., Casagrande, J. T., and Pike, M. C. (1983). The epidemiology of endometrial cancer in young women. British Journal of Cancer 47, 749-756.

Herbst, A. L., Ulfelder, H., and Poskanzer, D. C. (1971). Adenocarcinoma of the vagina: Association of maternal stibestrol therapy with tumor appearance in young women. New England Journal of Medicine 284, 878-881.

Horwitz, K. B., Jackson, T. A., Bain, D. L., Richer, J. K., Takimoto, G. S., and Tung, L. (1996). Nuclear receptor coactivators and corepressors. Molecular Endocrinology 10, 1167-1177.

Huggins, C., Grand, L. C., and Brillantes, F. P. (1961). Mammary cancer induced by a single feeding of polynuclear hydrocarbons, and its suppression. Nature 189, 204-207.

Ignar-Trowbridge, D. M., Pimentel, M., Parker, M. G., McLachlan, J. A., and Korach, K. S. (1996). Peptide growth factors cross-talk with the estrogen receptor requires the A/B domain and occurs independently of protein kinase C or estradiol. Endocrinology 137, 1735-1744.

Ing, N. H., Beekman, J. M., Tsai, S. Y., Tsai, M.-J., and O'Malley, B. W. (1992). Members of the steroid hormone receptor superfamily interact with TFIIB (S300-II). Journal of Biological Chemistry 267(5), 17617-17623.

Jacq, X., Brou, C., Lutz, Y., Davidson, I., Chambon, P., and Tora, L. (1994). Human TAF(II)30 is present in a distinct TFIID complex and is required for transcriptional activation by the estrogen receptor. Cell 79(1), 107-117.

Jacquemier, J. D., Rolland, P. H., Vague, D., Lieutaud, R., Spitalier, J. M., and Martin, P. M. (1982). Relationships between steroid receptor and epithelial cell proliferation in benign fibrocystic disease of the breast. Cancer 49, 2534-2536.

Jensen, E. V. (1960). Studies of growth phenomenon using tritium labelled steroids. Proceedings of the International Congress of Biochemists 15, 119.

Jensen, E. V., Suzuki, T., Kawashima, T., Stumpf, W. E., Jungblut, P. W., and DeSombre, E. R. (1968). A two-step mechanism for the interaction of estradiol with rat uterus. Proceedings of the National Academy of Science of the United States of America 59, 632-638.

Johnston, S., Saccani-Jotti, G., Smith, I., Newby, J., and Dowsett, M. (1995). Discordance between pS2 and ER expression in tamoxifen-resistant breast cancer. Breast Cancer Research and Treatment 37S, 33.

Karnik, P. S., Kulkarni, S., Liu, X.-P., Budd, G. T., and Bukowski, R. M. (1994). Estrogen receptor mutations in tamoxifen-resistant breast cancer. Cancer Research 54(2), 349-353.

Kato, S., Endoh, H., Masuhiro, Y., Kitamoto, T., Uchiyama, S., Sasaki, H., Masushige, S., Gotoh, Y., Nishida, E., Kawashima, H., Metzger, D., and Chambon, P. (1995). Activation of the estrogen receptor through phosphorylation by mitogen-activated protein kinase. Science 270, 1491-1494.

Katzenellenbogen, B., and Norman, M. J. (1990). Multihormonal regulation of the progesterone receptor in MCF-7 human breast cancer cells: Interrelationships among insulin/insulin-like growth factor-I, serum, and estrogen. Endocrinology 126, 891-898.

Khan, S. A., Rogers, M. A. M., and Tamsen, A. (1994). Estrogen receptor expression of benign breast epithelium and its association with breast cancer. Cancer Research 54, 993-997.

Klein-Hitpass, L., Ryffel, G. U., Heitlinger, E., and Cato, A. C. B. (1988). A 13-bp palindrome is a functional estrogen responsive element and interacts specifically with estrogen receptor. Nucleic Acids Research 16, 647-664.

Klein-Hitpass, L., Tsai, S. Y., Greene, G. L., Clark, J. H., Tsai, M.-J., and O'Malley, B. W. (1989). Specific binding of estrogen receptor to the estrogen response element. Molecular and Cellular Biology 9(1), 43-49.

Klotz, D. M., Castles, C. G., Fuqua, S. A. W., Spriggs, L. L., and Hill, S. M. (1995). Differential expression of wild-type and variant ER mRNAs by stock MCF-7 breast cancer cells may account for differences in estrogen responsiveness. Biochemical and Biophysical Research Communications 210(2), 609-615.

Kumar, V., Green, S., Stack, G., Berry, M., Jin, J.-R., and Chambon, P. (1987). Functional domains of the human estrogen receptor. Cell 51, 941-951.

Lee, M., Gippert, G., Soman, K., Case, D., and Wright, P. (1989). Three-dimensional solution structure of a single, zinc-finger, DNA-binding domain. Science 245, 635-637.

Leygue, E., Murphy, L., Kuttenn, F., and Watson, P. (1996a). Triple primer polymerase chain reaction. A new way to quantify truncated mRNA expression. American Journal of Pathology 148(4), 1097-1103.

Leygue, E. R., Watson, P. H., and Murphy, L. C. (1996b). Estrogen receptor variants in normal human mammary tissue. Journal National Cancer Institute 88, 284-290.

Lippman, M. E., and Chabner, B. A. (1986). Editorial overview: Proceedings of the NIH concensus development conference on adjuvant chemotherapy and endocrine therapy for breast cancer. NCI Monograms 1, 5-10.

Martinez, E., and Wahli, W. (1989). Cooperative Binding of Estrogen Receptor to Imperfect Estrogen-Responsive DNA Elements Correlates with Their Synergistic Hormone-Dependent Enhancer Activity. EMBO Journal 8(12), 3781-3791.

McDonnell, D. P., Nawaz, Z., and O'Malley, B. W. (1991). In situ distinction between steroid receptor binding and transactivation at a target gene. Mol. Cell. Biol. 11, 4350-4355.

Miller, M. A., Mullick, A., Greene, G. L., and Katzenellenbogen, B. S. (1985). Characterization of the subunit nature of nuclear estrogen receptors by chemical cross-linking and dense amino acid labelling. Endocrinology 117, 515-522.

Murphy, L. C., and Dotzlaw, H. (1989). Variant estrogen receptor mRNA species detected in human breast cancer biopsy samples. Molecular Endocrinology 3, 687-693.

Murphy, L. C., Wang, M., Coutt, A., and Dotzlaw, H. (1996). Novel mutations in the estrogen receptor messenger RNA in human breast cancers. Journal of Clinical Endocrinology and Metabolism 81(4), 1420-1427.

Nolvadex Adjuvant Trial Organization. (1988). Controlled trial of tamoxifen as a single adjuvant agent in the management of early breast cancer. British Journal of Cancer 57, 608-611.

Onate, S., Tsai, S., Tsai, M.-J., and O'Malley, B. (1995). Sequence and characterization of a coactivator for the steroid hormone receptor superfamily. Science 270, 1354-1357.

Osborne, C. K., Hobbs, K., and Trent, J. M. (1987). Biological differences among MCF-7 human breast cancer cell lines from different laboratories. Breast Cancer Research and Treatment 9, 111-121.

Patrone, C., Ma, Z. Q., Pollio, G., Agrati, P., Parker, M. G., and Maggi, A. (1996). Cross-coupling between insulin and estrogen receptor in human neuroblastoma cells. Molecular Endocrinology 10(5), 499-507.

Pertzelan, A. L., Yalon, L., Kauli, R., and Laron, Z. (1982). A comparative study of the effect of oestrogen substitution therapy on breast development in girls with hypo- and hypergonadotrophic hypogonadism. Clinical Endocrinology 16, 359-368.

Pfeffer, U., Fecarotta, E., Arena, G., Forlani, A., and Vidali, G. (1996). Alternative splicing of the estrogen receptor primary transcript normally occurs in estrogen receptor positive tissues and cell lines. Journal of Steroid Biochemistry and Molecular Biology 56, 99-105.

Picard, D., Khursheed, B., Garabedian, M. J., Fortin, M. G., Lindquist, S., and Yamamoto, K. R. (1990). Reduced levels of hsp90 compromise steroid receptor action in vivo. Nature 348, 166-168.

Pink, J., Wu, S., Wolf, D., Bilimoria, M., and Jordan, V. (1996). A novel 80-kDa human estrogen receptor containing a duplication of exons 6 and 7. Nucleic Acids Research 24(5), 962-969.

Ponglikitmongkol, M., Green, S., and Chambon, P. (1988). Genomic organization for the human oestrogen receptor gene. EMBO Journal 7, 3385-3388.

Power, R., Mani, S., Cadina, J., Conneely, O., and O'Malley, B. (1991). Dopaminergic and ligand-independent activation of steroid hormone receptors. Science 254, 1639-1639.

Rea, D., and Parker, M. G. (1996). Effect of an exon 5 variant of the estrogen receptor in MCF-7 human breast cancer cells. Cancer Research 56(7), 1556-1563.

Ricketts, D., Turnbull, L., Ryall, G., Bakhshi, R., Rawson, N. S. B., Gazet, J.-C., Nolan, C., and Coombes, R. C. (1991). Estrogen and progesterone receptors in the normal female breast. Cancer Research 51 (April 1), 1817-1822.

Sadovsky, Y., Webb, P., Lopez, G., Baxter, J. D., Fitzpatrick, P. M., Gizang-Ginsberg, E., Cavailles, V., Parker, M. G., and Kushner, P. J. (1995). Transcriptional activators differ in their reposes to overexpression of TATA-Box-Binding protein. Molecular and Cellular Biology 15, 1554-1563.

Smith, D. C., Prentice, R., Thompson, D. J., and Herrmann, W. L. (1975). Association of exogenous estrogen and endometrial carcinoma. New England Journal of Medicine 293, 1164-1167.

Toft, D., and Gorski, J. (1966). A receptor molecule for estrogens: Isolation from the rat uterus and preliminary characterization. Proceedings of the National Academy of Sciences of the United States of America 55, 1574.

Tzukerman, M., Esty, A., Santiso-Mere, D., Danielian, P., Parker, M., Stein, R., Pike, J., and McDonnell, D. (1994). Human estrogen receptor transactivational capacity is determined by both cellular and promoter context and mediated by two functionally distinct intramolecular regions. Molecular Endocrinology 8, 21-30.

van Agthoven, T., Timmermans, M., Foekens, J. A., Dorssers, L. C. J., and Henzen-Logmans, S. C. (1994). Differential expression of estrogen, progesterone, and epidermal growth factor receptors in normal, benign, and malignant human breast tissues using dual staining immunohistochemistry. American Journal of Pathology 144 (June 6), 1238-1246.

Welshons, W. V., Cormier, E. M., Wolf, M. F., Williams, P. O., and Jordan, V. C. (1988). Estrogen receptor distribution in breast cancer cell lines using cell enucleation. Endocrinology 122, 2379-2386.

Westley, B., and Rochefort, H. (1980). A secreted glycoprotein induced by estrogen induced in human cancer cell lines. Cell (Cambridge, Mass.) 20, 353-362.

Wiltschke, C., Wolf, D., Castles, C., O'Connell, P., Allred, D., and Fuqua, S. (1995). An estrogen receptor variant with increased transcriptional activity isolated from premalignant breast lesions. Proceedings of the American Association for Cancer Research, 255.

Zhang, Q.-X., Wolf, D. M., Borg, A., Oesterreich, S., and Fuqua, S. A. W. (1996). Human estrogen receptor tyrosine 537 mutant identified in a metastatic breast tumor confers strong transactivation independent of hormone binding. Cancer Research 57, 1244-1249.

Zhang, Q. X., Borg, A., and Fuqua, S. A. W. (1993). An exon-5 deletion variant of the estrogen receptor frequently coexpressed with wild-type estrogen receptor in human breast cancer. Cancer Research 53(24), 5882-5884.

BIOLOGICAL BASIS OF GENETIC PREDISPOSITION TO BREAST CANCER

Shanaz H. Dairkee and Helene S. Smith*

* *Deceased*

Advances in Oncobiology
Volume 2, pages 137-158.

ISBN: 0-7623-0388-3

INTRODUCTION

The paradigm of the retinoblastoma gene, originally identified in childhood tumors, is the best-documented example of the two-hit model of cancer progression (Knudson, 1971). According to this model, familial tumors result from germ-line inactivation of one allele of a tumor suppressor gene; the inactivated allele is, therefore, present in all cells of the body. A high incidence of malignancy occurs in these families because only one additional inactivation event causes loss of suppressor gene function. Sporadic cancer arises due to the inactivation of both alleles in the target somatic cells; therefore, as expected, the requisite mutations in two alleles of the same gene occur at a much lower frequency than in familial disease.

Inherited mutations at several genetic loci have been implicated in familial breast cancer. It is often hypothesized that these familial genes function in a manner similar to the retinoblastoma gene and that the same genes are relevant for the 90% of breast cancers that are not inherited. While this may be true for some cases, the Knudson model alone may not be sufficient to account for many, if not all, breast cancers. Breast cancer differs from other types of cancers in being very heterogeneous both in biological behavior and in the accumulation of genetic aberrations. There is evidence for various recessive genes (putative tumor suppressor genes) that are involved in sporadic breast cancer that have not been implicated in familial disease. There is also evidence for involvement of a number of different gain-of-function genes (possibly new oncogenes).

A number of molecular studies have been done on colon cancer and have led to a model for the genesis of other epithelial malignancies such as breast cancer. Colon cancer is known to display a histologic continuum from normal colonic mucosa through increasingly aggressive adenomas to invasive carcinoma. Specific genetic lesions are associated with each of these histologic entities and appear to accumulate in the more aggressive lesions in a manner consistent with a specific pathway of malignant progression (Kinzler and Vogelstein, 1996).

Although, a morphological continuum from atypical hyperplasia through carcinoma in situ to invasive disease is observed in breast cancer, the pattern of molecular changes and biologic behavior suggests a more complex etiology than with colon cancer. Unlike the specific sequence of genetic lesions associated with colon cancer progression, specific genetic alterations are seen in some breast lesions at early preinvasive stages, whereas, in other tumors, they may not occur until a late stage of metastatic dissemination (reviewed in Dairkee and Smith, 1996).

Heterogeneity in the sequential acquisition of specific genetic lesions, as well as widespread genetic heterogeneity per se, have led to a stochastic model of breast cancer progression (Smith et al., 1993). According to this model, phenotypes essential for malignancy, such as dysregulated growth, invasion, angiogenesis, or metastasis, can be acquired in more than one way. Different breast tumors may therefore result from independent pathways of malignant progression. Furthermore, an initiated cell can accrue genetic lesions stochastically in any order instead of following a set sequence of events. Consequently, if a tumor acquires the ability to invade prior to dysregulated cell proliferation, the invasive tumor will not be preceeded by an in situ stage. Much work still needs to be done to establish with certainty which molecular changes seen only during sporadic breast cancer progression are causal and which are consequential.

GERM LINE MUTATIONS OBSERVED IN FAMILIAL BREAST CANCER

Familial or hereditary breast cancer is operationally defined as the occurrence of two or more affected relatives within a pedigree. Clinical features known to differ markedly between familial and sporadic breast cancer include early age of onset (Lynch et al, 1976) and bilaterality (Harris et al.,1978). Epidemiological studies have suggested that up to 10% of all breast cancer is due to the inheritance of predisposing genes (Claus et al.,1991).

In recent years, mutations in a number of genes, presumably tumor suppressor genes, have been implicated in increasing susceptibility to familial breast cancer. These genes differ considerably in terms of the degree of risk they confer, the breast cancer incidence they account for, and the other cancers and pathological phenotypes with which they are associated. Individuals who inherit one mutant allele of the gene are at an increased risk for breast cancer, and the tumors they develop arise from somatic cells that have lost the wild-type allele (and therefore the tumor-suppressing ability). This loss can occur by any mechanism causing loss of function, such as heterozygous deletion, genomic imprinting (Sapienza, 1990), or increased degradation of gene product (Scheffner et al., 1990). Alternately, loss of gene function can result from dominant–negative mechanisms (Herskowitz, 1987). A number of susceptibility genes such as the *p53* tumor suppressor gene, the androgen receptor gene, and the ataxia telangiectasia gene, were initially identified due to their association with syndromes that encompass a wide spectrum of malignancies, only one of which is breast cancer.

In contrast to other familial cancer genes, breast cancer susceptibility genes confer a predisposition primarily towards breast cancer. The lifetime risk of breast cancer may approach 80–90% in women who have germ-line mutations in these genes, commonly known as *BRCA-1* and *BRCA-2* (Easton et al., 1993; Wooster et al., 1994). Tumors of individuals from *BRCA-1*– and *BRCA-2*–linked families show consistent loss of the wild-type allele, suggesting a tumor suppressor function for these genes (Smith et al.,1992; Collins et al., 1995). However, it is possible that other genetic and environmental factors can abrogate the predisposing influence of mutated susceptibility genes.

The *p53* Tumor Suppressor Gene

Inactivation of the *p53* tumor suppressor gene constitutes one of the most common gene alterations in human tumors, which includes breast cancer (Hollstein et al., 1991). The 53-kDa product of this gene is a phosphoprotein involved in a variety of cellular processes, particularly in assessment of DNA damage and cell cycle checkpoint control, which lead to transactivation of genes involved in G1 arrest, DNA repair, and programmed cell death (Kinzler and Vogelstein, 1996). The vast majority of the *p53* mutations (>98%), are somatically acquired and generally occur as missense mutations in exons 5 through 8, an evolutionarily conserved region of the gene in which the DNA-binding core of the protein is located (Hollstein et al., 1991).

Germ-line mutations in the *p53* gene are reportedly prevalent in individuals with the Li-Fraumeni (LFS) and Li-Fraumeni-like (LFL) syndromes characterized by familial childhood sarcoma, leukemia, and early onset breast cancer (Malkin et al., 1990; Srivastava et al., 1990). Families are designated as LFS on the basis of a well-defined set of criteria that include the proband having been diagnosed with sarcoma before 45 years of age, a first degree relative having cancer before 45 years of age, and another first or second degree relative in the pedigree having any cancer diagnosed up to this age or having a sarcoma occurring at any age (Birch et al.,1994). Families with features suggesting LFS, but not strictly conforming to the definition, are referred to as LFL. The reported frequency of germ-line *p53* mutations in LFS and LFL families varies depending on the methods used for detecting mutations as well as on stringency of the criteria for defining the syndrome. For example, Eeles and colleagues (1994) report a frequency of 7% in LFL families while Varley and colleagues (1997) have observed a frequency of 22% in another series of LFL families and 71% in LFS families. A comparison of the characteristics of cancers occurring in LFS families with *p53* mutations against LFS families in general demonstrates that the two groups are remarkably similar (Birch et al., 1994).

Germ-line *p53* mutations have a distribution pattern within the gene that is similar to that of somatically acquired *p53* mutations (Birch et al., 1994: Frebourg et al., 1995). The diversity of tumor types in the carriers of germ-line *p53* mutations suggests that some tissues may be more susceptible to transformation in the presence of one wild-type *p53* allele. As noted by Malkin and colleagues (1990), the frequency of distribution by site of malignancy in LFS patients is somewhat different from the general population. For example, in LFS families, breast cancer is much more common than colon cancer, whereas the frequency of *p53* mutations in breast and colon cancers in the general population is quite comparable. Based on this finding, it can be speculated that in breast carcinogenesis, inactivation of the *p53* gene product may be a more critical rate-limiting step.

In a few cases of germline *p53* mutations, retention of the wild-type *p53* allele has been observed in the primary breast tumor. It is possible that the mutant form of the protein in the cells of these individuals interferes with the wild-type *p53* function in a dominant–negative manner in some tissues. Alternately, the *p53* germline mutation in these cases may not be relevant to the disease, or it may confer a subtle influence on tumor progression (Prosser et al., 1992; Sun et al., 1996).

The Ataxia-Telangiectasia Gene (*ATM*)

Ataxia-telangiectasia (AT) is an autosomal, recessive syndrome characterized by immunological, neurological, and developmental defects and an increased risk of cancer. Although AT is a rare disease (1 in 40,000), the frequency of the heterozygosity status is high (about 1.4%). The literature suggests that not only homozygotes but also AT heterozygotes show a predisposition to many types of cancers, particularly breast cancer for which there is a fivefold increased risk (Swift et al., 1987). It is estimated that as many as 20% of breast cancer patients may be *ATM* heterozygotes (Swift et al., 1991).

The *ATM* gene is located at 11q22-11q23, a region of frequent loss of heterozygosity in sporadic breast cancer (Carter et al., 1994; Negrini et al., 1995). However, studies on breast tumors with LOH at 11q have not found any mutations in the *ATM* gene (Vorechovsky et al., 1996).

Recent observations have demonstrated that cells from AT patients exhibit radiation-resistant DNA synthesis and show a reduced or delayed gamma radiation–induced increase in p53 protein levels (Birrell and Ramsay, 1995). In addition, the *ATM* gene has been shown to have extensive homology to several cell cycle checkpoint genes in other organisms (Savitsky et al., 1995; Morrow et al., 1995). A model based on these findings proposes that in *ATM* homozygotes, the absence of damage-sensitive cell cycle checkpoints and damage-induced repair leads to genetic instability and cancer (Meyn, 1995).

The Androgen Receptor (*AR*) Gene

In males, androgens are primarily synthesized in the testes and secreted as testosterone, whereas in females, the major source is the adrenal cortex, which secretes dihydroepiandosterone (Wilson et al., 1981). Androgens are involved in many regulatory processes in breast epithelium, but their role in the development of breast cancer is poorly understood.

The androgen receptor belongs to a family of nuclear receptors and contains three functional domains: a carboxy-terminal, hormone-binding region; a central DNA-binding region; and an amino terminal region involved in the expression of androgen-related genes (Jenster et al., 1991). In the developing male fetus, the androgen receptor acts as a ligand-dependent DNA transcription factor that binds androgens that cause masculinization. Abnormalities of receptor function result in the failure of normal male differentiation (Grumbach and Conte, 1985). Androgen receptors have been demonstrated immunohistochemically in the nuclei of normal breast epithelium and in 79% of primary breast tumors. A significant association was found between the expression of androgen and estrogen receptors (Isola, 1993).

Molecular studies of androgen receptor gene structure have recently provided new insights toward defining a genetic basis for the pathology associated with breast carcinoma affecting middle-aged and older men. Specific germ-line mutations affecting the steroid-binding domain of the androgen receptor gene have been observed in males suffering from partial androgen insensitivity (AIS). These mutations have also been reported in rare cases of male breast cancer (Lobaccaro et al., 1993; Wooster et al., 1992). It is speculated that such mutations could account for the development of male breast cancer by the loss of the protective effect of androgens on these cells (Lobaccaro et al., 1993).

The *BRCA-1* Gene

The breast cancer susceptibility gene, *BRCA-1*, is located on chromosome 17q21 (Miki et al., 1994). Over 100 distinct mutations of *BRCA-1* scattered throughout the coding region have been identified in early onset breast and ovarian cancers resulting from hereditary predisposition (Shattuck-Eidens et al., 1995). These mutations may account for one half of all familial breast cancers (Easton et al., 1993).

The majority of the mutations are predicted to result in the loss of a *BRCA-1* transcript or in the formation of a truncated *BRCA-1* protein of 2%–88% of the expected normal length (Serova et al., 1996). Gene alterations have included frameshift insertions or deletions, nonsense mutations, splice acceptor site mutations, and so forth. (Johannson et al., 1996).

In one study, quantitative mRNA in situ hybridization analysis was performed on archival paraffin-embedded tumor specimens from 25 patients with known germline *BRCA-1* mutations. The BRCA-1 mRNA levels were invariably low in these tumors whereas the normal breast epithelium surrounding these tumors showed higher mRNA levels suggesting that the low levels in tumor cells may be due to somatic inactivation of the wild-type *BRCA-1* allele (Kainu et al., 1996). Expression levels in sporadic breast tumors are significantly higher; therefore, low BRCA-1 expression has been proposed as a means of identifying patients with BRCA-1–linked breast cancer (Kainu et al., 1996).

In contrast to the wide variety of germ line mutations in BRCA-1 that have been identified among the general population of high-risk families, a single *BRCA-1* mutation in exon 2, 185delAG, has been noted in approximately 20% of Ashkenazi Jewish women with early onset breast cancer (Fitzgerald et al., 1996). Less commonly, three other loss-of-function mutations have also been reported in this population, including an 11-bp deletion in exon 2, 188del11, and a 5382insC mutation in exon 20 (Berman et al., 1996). A recent large-scale population study in Ashkenazi Jewish individuals unselected for breast cancer has determined a carrier frequency of 1.09% for the 185delAG mutation (Roa et al., 1996).

Analyses to determine the correlation between the presence of germ-line mutations in BRCA-1 and the histoprognostic grade of the tumor have revealed a statistically significant increase in the prevalence of high-grade tumors in BRCA-1–associated breast cancers compared to breast tumors from the general population (Eisinger et al., 1996). A determination of the mitotic index in a series of breast cancers from 20 BRCA1-associated families has demonstrated a prevalence of highly proliferative tumors when mutations occur in the most conserved domains located in the terminal regions of the gene (Sobol et al., 1996). These findings suggest that these regions must play an important role in the growth suppression of breast epithelium. Determining the biological relevance of mutations in other areas of the gene and how they affect breast cancer development are of obvious importance.

The biological function of BRCA-1 in normal development and in malignancy remains unknown. An individual with a family history of breast cancer whose parents were both heterozygous for the same BRCA1 was found to be homozygous for this mutation in her somatic cells (Boyd et al., 1995). This finding indicates that the presence of this genetic aberration is not a lethal event in fetal development, nor is it critical for normal cell functioning prior to malignant progression. In order to verify the biological implications of such a finding, it will be important to identify additional cases with similar alterations. In the case of a *BRCA-1* mouse homozygous knockout, fetal lethality has been observed with multiple developmental abnormalities of the central nervous system (Gowen et al., 1996).

There is some controversy regarding the role of BRCA-1 in sporadic breast cancer. As expected of a tumor suppressor gene product, preliminary studies have found that the putative 220-kDa BRCA-1 protein localizes to the nucleus of nonmalignant breast epithelial cultures (Chen et al., 1995; Scully et al., 1996). Chen and colleagues (1995) characterized 50 specimens of sporadic primary breast cancer and found the normal pattern of nuclear localization in only 16% of the cases. The remainder of cases predominantly showed aberrant subcellular localization while absence of BRCA-1 protein was seen in 4% of the cases. Based on these findings, Chen and colleagues (1995) have suggested that BRCA-1 aberrations may be involved in the pathogenesis of sporadic as well as familial breast cancer. Scully and colleagues (1996) also observed nuclear localization of the *BRCA-1* gene product but found no differences between malignant and nonmalignant cells. Further increasing the controversy, Jensen and colleagues (1996) reported that *BRCA-1* encodes a 190-kDa protein that localizes in secretory vesicles and bears close sequence homology and biochemical analogy to the granin protein family. In the light of the latter report, BRCA-1 appears to function by a novel mechanism as yet not described for tumor suppressor gene products. The discrepancy in the findings on *BRCA-1* gene product may be due to the use of relatively uncharacterized and nonspecific antibody reagents. Furthermore, until a comprehensive demonstration of BRCA-1 localization in multiple samples of normal breast as well as in other normal tissues is completed, the significance of these results remains questionable.

The *BRCA-2* Gene

Another highly penetrant, autosomal dominant gene that confers an increased risk of early onset breast cancer is the *BRCA2* gene, which has been localized to chromosome

13q12-q13. Germline mutations in BRCA2 are thought to account for as much as 70% of inherited breast cancers not linked to BRCA-1 (Wooster et al., 1994). Additionally, BRCA-2 mutations were detected in 14% of male breast cancer (Couch et al., 1996).

The mutations have been shown to cause considerable disruption of the open reading frame of the transcriptional unit (Wooster et al., 1994). In some studies, a wide diversity of cancers such as ovarian, colon, esophageal, pancreas, stomach, and lymphatic system have also been observed in BRCA-2 mutant families (Berman et al., 1996; Phelan et al., 1996; Thorlacius et al., 1996). These examples are reminiscent of the multiple cancer syndromes and suggest that BRCA-2 inactivation may lead to a general tendency for chromosomal abnormalities and genetic instability throughout the body. The risk of ovarian cancer in BRCA-2 mutation carriers, although not as high as in BRCA-1 carriers, is higher than the general population (Stratton, 1996).

Recent studies have demonstrated specific mutations to be more prevalent within population subgroups. For example, a 185delAG mutation in *BRCA-1* and a 6174delT frameshift mutation in *BRCA-2* together may account for over 25% of all early-onset breast cancer in Ashkenazi Jewish women with a family history of breast and ovarian cancer (Neuhausen et al., 1996). Similarly, studies on Icelandic high risk breast cancer families have identified positive BRCA-2 linkage in which all members carry a 999del5 mutation (Arason et al., 1993; Gudmundsson et al., 1996; Thorlacius et al., 1996; Johannesdottir et al., 1996).

The *BRCA-1* and *BRCA-2* genes have several structural and functional similarities. Both genes have an unusually large exon 11 (>5 kb), exhibit high levels of transcript in the testis, and have a hydrophobic signal sequence in the granin domain unlike other molecules containing granin sequences (reviewed in Stratton, 1996). Both genes also occur in chromosomal locations at which loss of heterozygosity (LOH) is often observed in sporadic breast cancer (Dairkee and Smith, 1996), particularly of the wild type allele in familial breast cancer (Smith et al., 1992; Collins et al., 1995). However, inactivating mutations have not been reported in either gene in sporadic breast cancer.

In order to obtain reliable data on gene penetrance and for interpreting the correlation between a given genotype and phenotype in familial diseases, the following are essential: (1) germ-line mutations that have not been previously described are confirmed as mutations and not polymorphisms by screening several normal alleles, and (2) as many individuals as possible within a family are screened for the presence of germ-line mutations. Particularly if presymptomatic testing is to be offered within families, confirmation of the mutation in a first-degree relative is important.

SOMATIC GENETIC LESIONS OBSERVED IN SPORADIC BREAST CANCER

Although germline mutations lead to only a small percentage of breast malignancies, it is widely anticipated that the identification of these will lead to a better understanding of the vast majority of sporadic breast cancers. However, mutations in the genes associated with familial breast cancer have not yet been observed in sporadic breast cancer. Evidence in sporadic breast cancers for alterations in other recessive genes that have not been found in familial disease comes from studies on the identification of loss of heterozygosity (LOH) in specific chromosomal regions (reviewed in Dairkee and Smith, 1996). Other types of analyses have also identified putative tumor suppressor genes such as *Brush-1, maspin*, and *nm23* (Schott et al., 1994; Gilles et al., 1991; Leone et al., 1991; Pemberton et al., 1995; Stahl et al.,1991; and Zou et al.,1994). Moreover, there is evidence for dominant-acting genetic lesions in sporadic breast cancers. Besides dominant and recessive genetic alterations, a variety of morphological and phenomenological alterations have been observed in sporadic breast disease whose genetic basis remains to be defined.

Common Sites of Loss of Heterozygosity (LOH)

In breast tumors, LOH represents the most frequent type of genetic aberration (for reviews, see Devilee and Cornelisse, 1990; Mackay et al., 1990; Callahan et al., 1992). Using standard methods of DNA analysis to compare normal and tumor DNA, sites of putative tumor suppressor genes are revealed as LOH of one of a pair of polymorphic alleles in the tumor sample. This finding is commonly interpreted as either a physical loss of one allele, or a recombination event that results in two copies of the same allele.

The baseline level of LOH for any randomly selected probe is approximately 5% (Chen, 1992). LOH has been reported at significantly higher levels in a varying proportion of breast cancers at specific chromosomal regions such as 1p, 1q, 3p, 6q, 7p, 11p, 13q, 16q, 17p, 17q, 18p, 18q, and 22q (Devilee et al.,1991a; Takita et al., 1992; Sato et al., 1990; Ali et al. 1987; Beiche et al., 1990; Lundberg et al., 1987; Futreal et al., 1992; Mackay et al., 1988; Coles et al., 1990; Devilee et al., 1991b; Chen et al., 1986; Chen et al., 1991; Gendler et al., 1990; Merlo et al., 1989; Cornelisse, 1992; Ali et al., 1989; Genuardi et al., 1989; Devilee et al., 1989b; Devilee et al., 1991c; Thorlacius et al., 1991; Andersen et al., 1992; Sato et al., 1991). The incidence of loss varies for different regions with the most frequent loss being on 3p, 6q, 7p, 16q, and 17p (40–60%), while losses in the other regions are seen in 15–20% tumor cases.

For three chromosomal locations, namely 3p (Ali et al., 1989; Sato et al., 1991; Chen et al., 1994), 13q (Devilee et al., 1989), and 17p (Coles et al., 1990; Chen et al, 1991), more than one region of deletion has been identified suggesting that several genes in that region may be important in disease progression. However, it is also possible that LOH is merely incidental and is a reflection of the overall genomic instability within rapidly proliferating, aneuploid tumor cells, thus raising the question of cause or effect (Smith 1990; McGuire and Naylor 1989). An argument against this possibility is provided by observations that aneuploidy and rapid proliferation correlate with LOH at specific sites (i.e., 17p) rather than with the overall incidence of LOH (Chen et al., 1991).

Other Putative Recessive Gene Aberrations Involved in Sporadic Breast Cancer

Brush-1

This is a recently identified candidate suppressor gene located at 13q12-q13 proximal to the RB gene. *Brush-1* encodes a 4.7kb mRNA that is expressed at high levels in normal breast epithelium. In primary tumors, LOH at this site results in markedly decreased levels of *Brush-1* mRNA while maintaining the normal levels for RB (Schott et al., 1994).

Maspin

This product of a candidate tumor suppressor gene is related to the serpin family of protease inhibitors. It is expressed in normal breast cells but not in breast cancer cell lines. Analysis of malignant breast tissue has revealed that loss of maspin expression occurs most frequently in advanced cancers (Zou et al., 1994). A recent report suggests maspin is more closely related to ovalbumin and angiotensinogen and that its tumor suppressor activity is independent of a latent or intrinsic trypsinlike serine proteinase–inhibitory activity (Pemberton et al., 1995).

nm23

Two closely related metastasis–suppressor genes *nm23-1*(*NME1*) and *nm23-2* (*NME2*), have been implicated in the control of metastatic potential (Leone et al,1991; Stahl et al, 1991). It has been demonstrated that the nm23 gene product displays nucleoside diphosphate kinase (NDPK) activity (Gilles et al, 1991).

Dominant-Acting Genetic Lesions

This class of alterations are generally operative through the activation of specific genes that increase malignancy (by increasing tumor growth or invasiveness) due to either mutations or other regulatory changes, which result in increased gene expression. The most common class of dominant-acting genetic lesions are referred to as proto-oncogenes. The most frequent approach to discovering specific genes amplified in breast cancer has been to search for amplification of oncogenes frequently found in model systems. This approach has led to the identification of c-*erb*B-2 (17q12), c-*myc* (8q24), and *prad* 1/ cyclin D (11q13) amplification in 20–30% of breast cancers. In isolated reports, *flg* (8p12), *bek* (10q24), and IGFR-1/*fes* (15q24-q25) were shown to be amplified in a smaller percentage of breast cancers

(Slamon et al., 1987; Machotka et al.,1989; Lammie et al.,1991; Adnane et al., 1991; Berns et al., 1992a,b).

Another approach for identifying dominant-acting genes is by determining the target gene within regions of chromosomal amplification. A number of amplified chromosomal regions have been found in a proportion of breast cancers. Kallioniemi and colleagues (1994) reported that two-thirds of the tumors examined showed increases in copy number affecting a total of 26 chromosomal subregions. Sequences originating from 17q22-q24 and 20q13 showed the highest frequency of amplification (Kallioniemi et al., 1994).

It remains a difficult problem to identify the target gene of any given amplicon. One problem is that the amplified regions are large chromosomal segments containing many different genes. Because of the increased copy number of these genes, many can be overexpressed even though they do not contribute to the malignant phenotype. We have suggested (Dairkee and Smith, 1996) that the target gene within a given amplicon should always be present whenever that chromosomal region is amplified. In contrast, irrelevant extraneous genes residing near the target gene might not always be in the amplicon since the amplicons of various tumors differ in length. Additionally, the product of the target gene should sometimes be overexpressed even if the gene is not amplified; this is because other mechanisms besides gene amplification are known to exist and to lead to overexpression.

Several proto-oncogenes code for products that are either growth factors, growth factor receptors, signal transducers, protein kinases, or transcriptional activators. Since peptide growth factors are important regulators of human mammary epithelial proliferation and differentiation, it is expected that amplification and overexpression of these growth factors and their receptors would provide a growth advantage to the tumors expressing them. Indeed this has been observed for the Type 1 Growth Factor Receptor (GFR) Family (tyrosine kinase receptors including the epidermal growth factor receptor, EGFR; c-*erb*B1), c-*erb*B2 (*HER*2, neu), c-*erb*B3 (*HER*3), and c-*erb*B4 (*HER*4) receptors. Unlike EGFR and c-*erb*B2, c-*erb*B3 is expressed at moderately high levels in normal breast tissue. This level is further increased in 13–22% primary tumors (Gasparini et al., 1994; Lemoine et al.,1992). The c-*erb*B4 protein is the most recently characterized member of this family. Although the level of protein expression in primary tumors is not yet known, its mRNA is expressed at high levels in normal breast tissue as well as some established cell lines (Plowman et al.,1993).

Amplification of the c-*myc* oncogene, located at 8q, has been reported in 20% of primary breast tumors and implicated in the pathogenesis of aggressively growing breast tumors (Berns et al., 1992b). Some studies have demonstrated an unusually high proliferative index accompanying c-*myc* amplification in breast tumors (Kreipe et al., 1993). Despite this circumstantial data, there has been no conclusive study demonstrating that c-myc rather than some other nearby gene is the target for amplification at chromosome 8q.

Amplification of chromosomal region, 11q13, has been reported in 5–23% of breast cancers (Peters et al.,1995). This chromosomal region also encompasses several candidate target genes including fgf3/*wnt*-2 , fgf4/*hst*-1, and ccnd1/*prad*1 (which encodes cyclin D1). Neither fgf3 nor fgf4 proteins were expressed in any breast cancers with this amplification. In contrast, expression of the ccnd1/*prad*1 gene product at elevated levels was seen in all cases of 11q13 amplification. In some cases, increased expression of the gene is found in tumors where there is no evidence for DNA amplification or gross gene rearrangement, further implicating *prad*-1 as the target for the 11q13 amplicon (Gillett et al.,1994).

Histological Aberrations Associated with Breast Cancer

A number of biologic and histologic aspects of breast cancer have been described in the literature. Our current level of understanding the biologic basis for genetic predisposition to breast cancer is just beginning to consider the importance of analyzing these phenomena at the molecular level. Both stromal and epithelial alterations in breast cancer progression need to be addressed at the molecular level to determine which phenomena are genetic and which are epigenetic in etiology.

Stromal Alterations

Although the branching mammary tree is mostly comprised of epithelium, the stroma that surrounds the epithelium constitutes more than three-fourths of the resting breast (Drife, 1986), and in malignancy, it may constitute >90% of the tumor (Dvorak, 1986). Complex interactions occur between the epithelial and mesenchymal components of the gland. We have previously hypothesized that early events in malignancy involve a field effect causing subtle changes within a defined region of the breast (Deng et al., 1996). The changes may not only be limited to epithelial cells that directly give rise to breast cancer but may also involve stromal cells, such as fibroblasts (Smith et al., 1987).

Many components and characteristics of the tumor stroma are found to be altered in the invasive tumor (reviewed in Ronnov-Jessen et al.,1996). Such changes are clearly important in facilitating tumor invasion and dissemination. The "field effect" hypothesis predicts that abnormal stroma may initiate the malignant process by presenting inappropriate signals to normal epithelial cells. Possible mechanisms include the synthesis of inappropriate (fetal) proteins and trophic factors.

In primary breast tumors, expression of IGF-I and IGF-II in stromal fibroblasts depends on their location in relation to the tumor. IGF-I expression is favored over IGF-II in the stroma of macroscopically normal breast areas adjacent to the tumor, whereas in tumor stroma, IGF-II expression is more prominent (Singer et al., 1995). Since both ligands are mitogenic for malignant breast epithelium, the preferential expression of IGF-II in the tumor stroma may be related to additional effects that enhance tumor survival.

Some studies suggest that the abnormal "field effect" seen with stromal fibroblasts can extend throughout the body in a manner analogous to germline genetic changes. It has been reported that skin fibroblasts from breast cancer patients display abnormalities in growth characteristics and in the ability to migrate through collagen gels (Azzarone et al., 1984; Schor et al., 1985). Migration is facilitated by the secretion of a soluble migration-stimulating factor (MSF), which is produced by fetal fibroblasts and breast cancer fibroblasts but not made by normal adult cells (Grey et al., 1989).

Epithelial Alterations

Differentiation Status of Lobules. The branching mammary tree in the adult human female breast consists of several thousand hormone-sensitive, potentially milk-secreting structures called lobules. The secretions of these lobules are collected through terminal ducts that in turn drain into the lumen of larger ducts. A given lobule and its terminal duct are referred to as a terminal–ductal lobular unit (TDLU). The seminal studies of Wellings and colleagues (1975) on the microarchitecture of a large series of normal and cancerous human breasts has demonstrated that cancerous breast tissue displays morphological alterations predominantly in the TDLU. Based on this evidence and as originally suggested by Parks (1959), it is generally accepted that breast cancer originates in the TDLU.

Epidemiological data indicate that breast cancer incidence is greater in nulliparous women whereas early parity confers protection (MacMahon et al., 1970). Investigations to determine what basic differences exist between the parous and the nulliparous womens' breasts at a histologic level have been pioneered by Russo and colleagues (1990). It is hypothesized that, morphologically, the lobular structures of the breast generally reflect different stages of development. Type I lobules (Lob 1) are the most undifferentiated. Type 2 lobules(Lob 2), which evolve from type I, are composed of a higher number of ductular structures per lobule. They progress to lobule types 3 and 4 (Lob 3 and Lob 4), which are present in the breast during pregnancy and lactation. In the breast of nulliparous women, the predominant structure present is Lob 1. In parous women free of cancer, the breast contains a greater percentage of Lob 3 and a moderately increased number of Lob 2 with a concomitant reduction in Lob 1. Parous women with breast cancer, on the other hand, exhibit a greater percentage of Lob 1 and fewer Lob 3 than the noncancerous group, approaching the percentages found in nulliparous women (Russo et al., 1994; Russo and Russo, 1995).

Atypical Hyperplasia. Women who have been diagnosed with benign breast disease are known to have an increased risk of breast cancer (Ernster, 1981; Love et al., 1982). In an attempt to subdivide the histologically heterogeneous group of benign lesions into prognostically relevant categories, a large, retrospective study with long-term follow-up of women who had undergone breast biopsies for benign breast disease was carried out (Dupont and Page, 1985). A major finding of this study was that biopsies lacking proliferative disease did not identify women at increased risk of cancer. Proliferative lesions without atypical hyperplasia conferred a risk of cancer 1.9 times greater than the risk in women with nonproliferative lesions. The presence of atypical hyperplasia increased the risk five times that of nonproliferative lesions. The greatest risk (11 times greater than nonproliferative lesions) was observed in women with atypical hyperplasia and a family history of breast cancer.

Increased risk in all of the above subcategories is not limited to the breast in which the lesion was first detected. Investigations directed at determining the risk of bilateral breast disease by examining contralateral breast tissue suggest that at least 20% of patients with invasive breast cancer have an atypical proliferative lesion in the opposite breast (Tulusan et al., 1985). The incidence of this lesion was 3.6% in women without breast cancer who had presented with benign breast disease (Dupont and Page, 1985).

Ductal Carcinoma In Situ *(DCIS).* The noninvasive malignant lesion commonly known as DCIS continues to be enigmatic, particularly from the perspective of patient management. A recent study by Cady and colleagues (1996) predicts that within the next decade the proportion of breast lesions diagnosed as DCIS will approach 33% (the present rate is 14–18%). Due to the proximity of its association with the invasive lesion, it is widely speculated that DCIS may be a direct precursor of invasive carcinoma. However, the role of DCIS as an obligatory prerequisite in the development of invasive cancer is uncertain and whether all DCIS become invasive is unknown.

To shed light on the relationship between various morphologically distinct lesions at the molecular level, comparative studies have been undertaken between invasive carcinoma and adjacent atypias and DCIS for the incidence of loss of heterozygosity (LOH) at various chromosomal loci known to be frequently associated with breast tumors (Andersen et al.,1992; Radford et al.,1993; Allred et al., 1994; Koreth et al.,1995; Zhuang et al.,1995).

We have examined several independent cases for LOH at various chromosomal loci in invasive and DCIS components of the same carcinoma. In 20% of the cases studied, the DCIS components had some, but not all of the LOH's observed in the invasive tumor. In the majority of cases examined, the DCIS was identical in LOH profile to the invasive tumor (Deng and Smith unpublished). This result is consistent with previous studies in which invasive breast cancers and associated DCIS were found to have similar frequencies of LOH at many different loci. All of the accumulated data therefore supports the assumption that the DCIS and the invasive lesion are closely related.

Immunohistochemical comparisons of noninvasive and invasive breast cancer for the product of the c-*erb*B-2 oncogene have demonstrated that high levels of c-*erb*B2 expression in high grade, comedo-type DCIS is relatively common (80%)

as compared with invasive cancers (25–30%; Gusterson et al., 1988a; Gusterson et al., 1988b; Allred et al., 1992). This finding suggests that many DCIS may not progress to invasive cancer. Alternately, it is possible that some invasive tumors may develop without passing through a high grade DCIS stage. Yet another possibility is that those DCIS cases that are high grade but lack expression of c-*erb*B2 are the ones most likely to become invasive.

The similarities observed between the two lesions could be interpreted as molecular evidence that DCIS is a closely related precursor of the invasive component. However, it is not entirely unlikely that foci of DCIS adjacent to the invasive tumor in fact represent intraductal spread of malignant cells that are equally invasive in other areas of the tumor; these cells, however, are like DCIS in appearance because the tumor cells have not yet invaded the basement membrane.

SUMMARY

In summary, there have indeed been major breakthroughs in the identification of multiple genetic aberrations in breast cancer. These include germ-line mutations in familial disease, genes that confer increased susceptibility to many different cancers including breast cancer, and genetic aberrations observed only in breast cancers. The large number of aberrations reported may seem perplexing. However, strategies which identify how they fit into models for the etiology and evolution of this disease and how familial genes may be involved in sporadic breast cancer are beginning to appear and will continue to be a challenge for the future.

ACKNOWLEDGMENTS

Support for the authors was provided in part by NIH grants PO1-CA44768, 1P59-CA58207, and R01-CA6698.

REFERENCES

Adnane, J., Gaudray, P., Dionne, C. A., Crumley, G., Jaye, M., Schlessinger, J., Jeanteur, P., Birnbaum, D., and Theillet, C. (1991). BEK and FLG, two receptors to members of the FGF family, are amplified in subsets of human breast cancers. Oncogene. 6, 659-663.

Ali, I. U., Lidereau, R., Theillet, C., and Callahan, R. (1987). Reduction to homozygosity of genes on chromosome 11 in human breast neoplasia. Science. 238, 185-188.

Ali, I.U., Lidereau, R., Theillet, C., and Callahan, R. (1989). Presence of two members of c-*erb*A receptor gene family (c-*erb*Ab and c-*erb*A2) in smallest region of somatic homozygosity on chromosome 3p21-p25 in human breast carcinoma. J. Natl. Cancer Inst. 81, 1815-1820.

Allred, D.C., Clark, G.M., Molina, R., Tandon, A.K., Schnitt, S.J., Gilchrist, K.W., Osborne, C.K., Tormey, D.C., and McGuire, W.L. (1992). Overexpression of *HER-2/neu* and its relationship with other prognostic factors change during the progression of *in situ* to invasive breast cancer. Hum. Pathol. 23, 974-979.

Allred, D.C., O'Connell, P., Fuqua, S.A.W., and Osborne, C.K. (1994). Immunohistochemical studies of early breast cancer. Br. Cancer Res. Treat. 32, 13-18.

Andersen, T. I., Gaustad, A., Ottestad, L., Farrants, G.W., Nesland, J.M., Tveit, K.M., and Borresen, A.L. (1992). Genetic alterations of the tumor suppressor gene regions 3p, 11p, 13q, 17p, and 17q in human breast carcinomas. *Genes, Chromosomes, and Cancer.* 4, 113-121.

Arason, A., Barkardottir, R.B., and Egilsson, V. (1993). Linkage analysis of chromosome 17q markers and breast-ovarian cancer in Icelandic families, and possible relationship to prostatic cancer. Am. J. Hum. Genet. 52, 711-717.

Azzarone, B., Mareel, M., Billard, C., Scemama, P., Chaponnier, C., and Macieira-Coelho, A. (1984). Abnormal properties of skin fibroblasts from patients with breast cancer. Int. J. Cancer 33, 759-764.

Beiche, I., Champeme, M.H., Merlo, G. , Larsen, C-J., Callahan, R., and Lideraeau, R. (1990). Loss of heterozygosity of the L-myc oncogene in human breast tumors. Human Genet. 85, 101-105.

Berman, D.B., Wagner-Costalas, J., Schultz, D.C., Lynch, H.T., Daly, M., and Godwin, A.K. (1996). Two distinct origins of a common BRCA1 mutation in breast-ovarian cancer families: A genetic study of 15 185delAG-mutation kindreds. Amer. J. Human Genet. 58, 1166-1176.

Berns, E.M., Klijn, J.G., van Staveren, I.L., Portengen, H. and Foekens, J.A. (1992a). Sporadic amplification of the insulinlike growth factor 1 receptor gene in human breast tumors. Cancer Res. 52, 1036-1039.

Berns, E.M., Klijn, J.G.M., Van Putten, W.L.J., Van Staveren, I.L., Portengen, H., and Foekens, J.A. (1992b). c-*myc* amplification is a better prognostic factor than *HER2/neu* amplification in primary breast cancer. Cancer Res 52, 1107-1113.

Birch, J.M., Hartley, A.L., Tricker, K.J., Prosser, J., Condie, A., Kelsey, A.M., Harris, M., Morris Jones, P.H., Binchy, A., Crowther, D., Craft, A.W., Eden, O.B., Evans, D.G.R., Thompson, E., Mann, J.R., Martin, J., Mitchell, E.L.D., and Santibanez-Koref, M.F. (1994). Prevalence and diversity of constitutional mutations in the *p53* gene among 21 Li- Fraumeni families. Cancer Res. 54, 1298-1304.

Birrell, G.W., and Ramsay, J.R. (1995). Induction of p53 protein by γ radiation in lymphocyte lines from breast cancer and ataxia telangiectasia patients. Brit. J. Cancer 72, 1096-1101.

Boyd, M., Harris, F., McFarlane, R., Davidson, H.R., and Black, D.M. (1995). A human *BRCA1* gene knockout. Nature 375, 541-542.

Cady, B., Stone, M.D., Schuler, J.G., Thakur, R., Wanner, M.A., and Lavin, P.T. (1996). The new era in breast cancer. Invasion, size, and nodal involvement dramatically decreasing as a result of mammographic screening. Archives of Surgery 131, 301-308.

Callahan, R., Cropp, C.S., Merlo, G.R., Liscia, D.S., Cappa, A.P., and Lidereau, R. (1992). Somatic mutations and human breast cancer. A status report. Cancer 69, 1582-1588.

Carter, S.L., Negrinin, M., Baffa, R., Gillum, D.R., Rosenberg, A.L., Schwartz, G.F., and Croce, C.M. (1994). Loss of heterozygosity at 11q22-q23 in breast cancer. Cancer Res. 54, 6270-6274.

Chen, L.C., Dollbaum, C., and Smith, H.S. (1986). Loss of heterozygosity on chromosome 1q in human breast cancer. PNAS USA. 86, 7204-7207.

Chen, L.C., Neubauer, A., Kurisu, F., Waldman, Ljung, B-M., Goodson, W., Goldman, E., Moore, D., Liu, E., Mayall, B., and Smith, H.S. (1991). Loss of heterozygosity on the short arm of chromosome 17 is associated with high proliferative capacity and DNA aneuploidy in primary human breast cancer. PNAS USA. 88, 3847-51.

Chen, L.C., Kurisu, W., Ljung, B.M., Goldman, E.S., Moore, D., and Smith, H.S. (1992). Heterogeneity for allelic loss in human breast cancer. JNCI. 84, 506-510.

Chen, L.C., Matsumara, K., Deng, G., Kurisu, W., Ljung, B.M., Lerman, M., Waldman, F.M., and Smith, H.S. (1994). Deletion of two separate regions on chromosome 3p in breast cancrs. Cancer Res. 54, 3021-3024.

Chen, Y., Chen, C.F., Riley, D.J., Allred, D.C., Chen, P.L. Von Hoff, D., Osborne, C.K., and Lee, W.H. (1995). Aberrant subcellular localization of BRCA-1 in breast cancer. Science 270, 789-791.

Claus, E.B., Risch, N., and Thompson, D. (1991). Genetic analysis of breast cancer in the cancer and steroid hormone study. Am. J. Hum. Gen. 48, 232-242.

Coles, C., Thompson, A.M., Elder, P.A., Cohen, B., Mackenzie, I.M., Steel, C.M. Cranston, G., Chetty, U., Mackay, J., Macdonald, M., and Nakamura, Y. (1990). Evidence implicating at least two genes on chromosome 17p in breast carcinogenesis. The Lancet. 336, 761-763.

Collins, N., McManus, R., Wooster, R., Mangion, J., Seal, S., Lakhani, S.R., Ormiston, W., Daly, P.A., Ford, D., Easton, D.F., and Stratton, M.R. (1995). Consistent loss of the wild type allele in breast cancers from a family linked to the *BRCA-2* gene on chromosome 13q12-13. Oncogene 10, 1673-1675.

Cornelisse, C.J. (1992). Loss of heterozygosity, chromosome 7q, and breast cancer. The Lancet 339, 1423.

Couch, F.J., Farid, L.M., Deshano, M.L., Tavtigian, S.V., Calzone, K., Campeau, L., Peng, Y., Bogden, B., Chen, Q., Neuhausen, S., Shattuck-Eidens, D., Godwin, A.K., Daly, M., Radford, D.M., Sedlacek, S., Rommens, J., Simard, J., Garber, J., Merajver, S., and Weber, B.L. (1996). BRCA2 germline mutations in male breast cancer cases and breast cancer families. Nature Genet. 13, 123-125.

Dairkee, S.H. and Smith, H.S. (1996). Genetic analysis of breast cancer progression. J. Mammary Gland Biol. and Neoplasia 1, 139-151.

Deng, G., Lu, Y., Zlotnikov, G., Thor, A.D., and Smith, H.S. (1996). Loss of heterozygosity in normal tissue adjacent to breast carcinomas. Science 274, 2057-2059.

Devilee, P., Van Den Broek, M., Kulpers-Dukshoorn, N., Kolluri, R., Khan, P.M., Pearson, P.L., and Cornelisse, C.J. (1989). At least four different chromosomal regions are involved in loss of heterozygosity in human breast carcinoma. Genomics 5, 554-560.

Deville, P. and Cornelisse, C.J. (1990). Genetics of human breast cancer. Cancer Surv. 9, 605-630.

Devilee, P., van Vliet, M., Bardoel, A., Kievits, T., Kuipers-Diujkshoorn, N., Pearson, P.L., and Cornelisse, C.L. (1991a). Frequent somatic imbalance of marker alleles for chromosome 1 in human primary breast carcinoma. Cancer Res. 51, 1020-1025.

Devilee, P., van Vliet, M., van Sloun, P., Dijkshoorn, N., Hermans, J., Pearson, P.L., and Cornelisse, C.L. (1991b). Allelotype of human breast carcinoma: A second major site for loss of heterozygosity is on chromosome 6q. Oncogene. 6, 1705-1711.

Devilee, P., van Vliet, M., Kuipers-Dijkshoorn, N., Pearson, P.L., and Cornelisse, C.J. (1991c). Somatic genetic changes on chromosome 18 in breast carcinomas: Is the DCC gene involved? Oncogene. 6, 311-315.

Drife, J.O. (1986). Breast development in puberty. In: *Annals of the New York Academy of Sciences. Endocrinology of the Breast: Basic and Clinical Aspects.*(Angeli, A., Bradlow, H.L., and Dogliotti, L., Eds.) pp 58-65. NY Acad. Sci., New York.

Dupont, W.D. and Page, D.L. (1985). Risk factors for breast cancer in women with proliferative breast disease. N. Engl. J. Med. 312, 146-151.

Dvorak, H.F. (1986). Tumors: Wounds that do not heal. Similarities between tumor stroma generation and wound healing. N. Engl. J. Med. 315, 1650-1659.

Easton, D.F., Bishop, D.T., Ford, D., Cockford, G.P., and the Breast Cancer Linkage Consortium. (1993). Genetic linkage analysis in familial breast and ovarian cancer. Am. J. Hum. Genet. 52, 718-722.

Eeles, R.A., Stratton, M.R., Goldgar, D.E., and Easton, D.F. (1994). The genetics of familial breast cancer and their practical implications. Euro. J. Cancer 30, 1383-1390.

Eisinger, F., Stoppa-Lyonnet, D., Longy, M., Kerangueven, F., Noguchi, T., Bailly, C., Vincent-Salomon, A., Jacquemier, J., Birnbaum, D., and Sobol. H. (1996). Germline mutation at BRCA1 affects the histoprognostic grade in hereditary breast cancer. Cancer Res., 56, 471-474.

Ernster, V.L. (1981). The epidemiology of benign breast disease. Epidemiol. Rev. 3, 184-202.

Fitzgerald, M.G., MacDonald, D.J., Krainer, M., Hoover, I., O'Neil, E., Unsal, H., Silva-Arrieto, S., Finkelstein, D.M, Beer-Romero, P., and Englert, C. (1996). Germ-line BRCA-1 mutations in Jewish and non-Jewish women with early onset breast cancer. N. Engl. J. Med., 334, 143-149.

Frebourg, T., Barbier, N., Yan, Y.X., Garber, J.E., Dreyfus, M., Fraumeni, J., Li, F.P., and Friend, S.H. (1995). Germ-line p53 mutations in 15 families with Li-Fraumeni syndrome. Am. J. Hum. Genet. Mar, 56 (3), 608-615.

Futreal, P.A., Soderkvist, P., Marks, J.R., Iglehart, J.D., Cochran, C., Barrett, J.C., and Wiseman, R.W. (1992). Detection of frequent allelic loss on proximal chromosome 17q in sporadic breast carcinoma using microsatellite length polymorphisms. Cancer Res. 52, 2624-2627.

Gasparini, G., Gullick, W.J., Maluta, S., Palma, P.D., Caffo, O., Meli, S., Leonardi, E., Boracchi, P., Pozza, F., Lemoine, N.R., and Bevilacqua, P. (1994). c-*erb*B3 and c-*erb*B2 gene expression in node-negative breast carcinoma. Eur. J. Cancer. 30A, 16-22.

Gendler, S., Cohen, E., Craston, A., Duhig, T., Johnstone, G., and Barnes, D. (1990). The locus of the polymorphic epithelial mucin (pem) tumour antigen on chromosome 1q21 shows a high frequency of alteration in primary human breast tumours. Int. J. Cancer. 45, 431-435.

Genuardi, M., Tsihira, H., Anderson, D.E., and Saunders, G.F. (1989). Distal deletion of chromosome 1p in ductal carcinoma of the breast. Am. J. Hum. Genet. 45, 73-82.

Gilles, A., Presecan, E., Vonica, A., and Lascu, I. (1991). Nucleoside diphosphate kinase from human erythrocytes. J. Biol. Chem. 266, 8784-8789.

Gillett, C., Fantl, V., Smith, R., Fisher, C., Bartek, J., Dickson, C., Barnes, D., and Peters, G. (1994). Amplification and overexpression of cyclin D1 in breast cancer detected by immunohistochemical staining. Cancer Res. 54, 1812-1817.

Gowen, L.C., Johnson, B.L., Latour, A.M., Sulik, K.K., and Koller, B.H. (1996). Brca-1 deficiency results in early embryonic lethality characterized by neuroepithelial abnormalities. Nature Genet. 12, 191-194.

Grey, A-M., Schor, A.M., Rushton, G., Ellis, I., and Schor, S.L. (1989). Purification of the migration stimulating factor produced by fetal and breast cancer patient fibroblasts. Proc. Natl. Acad. Sci. USA. 86, 2438-2442.

Grumbach, M.M., Conte, F.A. (1985). Disorders of sexual differentiation. In: *Textbook of Endocrinology*. (Wilson, J.D., Foster, D.W., Eds.), pp 312-401. Saunders, Philadelphia.

Gudmundsson, J., Johannesdottir, G., Arason, A., Bergthorsson, J.T., Ingvarsson. S., Egilsson, V., and Barkardottir, R.B. (1996). Frequent occurrence of BRCA2 linkage in Icelandic Breast cancer amilies and segregation of a common BRCA2 haplotype. Am. J. Hum. Genet., 58, 749-756.

Gusterson, B.A., Machin, L.G., Gullick, W.J., Gibbs, N.M., Powles, T.J., Price, P., McKinna, A., and Harrison, S. (1988a). Immunohistochemical distribution of c-*erb*B-2 in infiltrating and *in situ* breast cancer. Int. J. Cancer 42, 842-845.

Gusterson, B.A., Machin, L.G., Gullick, W.J., Gibbs, N.M., Powles, T.J., Elliot, C., Ashley, S., Monaghan, P., and Harrison, S. (1988b). c-*erb*B-2 expression in benign and malignant breast disease. Br. J. Cancer 58, 453-457.

Harris, R.E., Lynch, H.T., and Guirgis, H.A. (1978). Familial breast cancer: Risk to the contralateral breast. JNCI. 60, 955-960.

Herskowitz, I. (1987). Functional inactivation of genes by dominant negative mutations. Nature 329, 219-222.

Hollstein, M., Sidransky, D., Vogelstein, B. and Harris, C. (1991). p53 mutations in human cancers. Science 253, 49-53.

Isola, J.J. (1993). Immunohistochemical demonstration of androgen receptor in breast cancer and its relationship to other prognostic factors. J. Pathol. 170, 31-35.

Jensen, R.A., Thompson, M.E., Jetton, T.L., Szabo, C.I., van der Meer, R., Helou, B., Tronick, S.R., Page, D.L., King, M-C., and Holt, J. (1996). BRCA-1 is secreted and exhibits properties of a granin. Nature Genetics 12, 303-308.

Jenster, G., Van der Korput, H.A.G.M., Van Vroonhoven, C., Van der Kwast, T.H., Trapman, J., and Brinkmann, A.O. (1991). Domains of the human androgen receptor involved in steroid binding, transcriptional activation, and subcellular localization. Mol. Endocrinol. 5, 1396-1404.

Johannesdottir, G., Gudmundsson, J., Bergthorsson, J.T., Arason, A., Agnarsson, B.A., Eiriksdottir, G., Johannsson, O.T., Borg, A., Ingvarsson, S., Easton, D., Egilsson, V., and Barkardottir, R.B. (1996). High prevalence of the 999del5 mutation in Icelandic breast and ovarian cancer patients. Cancer Res. 56, 3663-3665.

Johannson, O., Ostermeyer, E.A., Hakansson, S., Friedman, L.S., Johansson, U., Sellberg, G., Brondum-Nielsen, K., Sele, V., Olsson, H., King, M.C., and Borg, A. (1996). Founding BRCA1 mutations in hereditary breast and ovarian cancer in southern Sweden. Amer. J. Human Genet. 58, 441-450.

Kainu, T., Kononen, J., Johansson, O., Olsson, H., Borg, A., and Isola, J. (1996). Detection of germline BRCA1 mutations in breast cancer patients by quantitative messenger RNA *in situ* hybridization. Cancer Research 56, 2912-2915.

Kallioniemi, A., Kallioniemi, O.P., Piper, J., Tanner, M., Stokke, T., Chen, L., Smith, H.S., Pinkel, D., Gray, J.W., and Waldman, F. (1994). Detection and mapping of amplified DNA sequences in breast cancer by comparative genomic hybridization. PNAS, 91, 2156-2160.

Kinzler, K.W. and Vogelstein, B. (1996). Lessons from hereditary colorectal cancer. Cell 87, 159-170.

Knudson, A.G. (1971). Mutation and cancer: Statistical study of retinoblastoma. Proc. Natl. Acad. Sci. USA 68, 820-823.

Koreth, J., Bethwaite, P.B., and McGee, J.O'D. (1995). Mutation at chromosome 11q23 in human non-familial breast cancer: A microdissection microsatellite analysis. J. Path. 176, 11-18.

Kreipe, H., Feist, H., Fischer, L., Felgner, J., Heidorn, K., Mettler, L, and Parwaresch, R. (1993). Amplification of c-*myc* but not of c-*erb*B2 is associated with high proliferative capacity in breast cancer. Cancer Res. 53, 1956-1961.

Lammie, G.A. and Peters, G. (1991). Chromosome 11q13 abnormalities in human cancer. Cancer Cells. 3, 413-420.

Lemoine, N.R., Barnes, D.M., Hollywood, D.P., Hughes, C.M., Smith, P., Dublin, E., Prigent, S.A., Gullick, W.J., and Hurst, H.C. (1992). Expression of *erb*B3 gene product in breast cancer. Br. J. Cancer. 66, 1116-1122.

Leone, A., McBride, O.W., Weston, A., Wang, M.G., Anglard, P., Cropp, C.S., Goepel, J.R., Lidereau, R., Callahan, R., Linehan, W.M., Rees, R.C., Harris, C.C., Liotta, L.A., and Steeg, P.S. (1991). Somatic allelic deletion of nm23-H1 in human cancer. Cancer Res. 51, 2490-2493.

Lobaccaro, J.M., Lumbroso, S., Belon, C., Galtier-Dereure, F., Bringer, J., Lesimple, T., Namer, M., Cutuli, B.F., Pujol, H., and Sultan, C. (1993). Androgen receptor gene mutation in male breast cancer. Human Mol. Genet. 2, 1799-1802.

Love, S.M., Gelman, R.S., and Silen, W. (1982). Fibrocystic "disease" of the breast-a nondisease? N. Engl. J. Med. 307, 1010-1014.

Lundberg, C., Skoog, L., Cavenee, W., and Nordenskjold, M. (1987). Loss of heterozygosity in human ductal breast tumors indicates a recessive mutation on chromosome 13. Proc. Natl. Acad. Sci. 84, 2372-2376.

Lynch, H.T., Guirgis, H.A., Brodkey, F., Maloney, K., Lynch, P.M., Rankin, L., and Lynch, J.F. (1976). Early age of onset in familial breast cancer: Genetic and cancer control implications. Arch. Surg., 111, 126-131.

Machotka, S.V., Garret, C.T., Schwartz, A.M., and Callahan, R. (1989). Amplification of the proto-oncogenes *int*-2, c-*erb* B-2, and c-*myc* in human breast cancer. Clin. Chim. Acta. 4, 207-218.

Mackay, J., Elder, P.A., Steel, C.M., and Forrest, A.P.M. (1988). Allele loss on short arm of chromosome 17 in breast cancers. The Lancet 1384-1385.

Mackay, J., Thompson, A.M., Coles, C., and Steel, C.M.. (1990). Molecular lesions in breast cancer. Int. J. Cancer Suppl. 5, 47-50.

MacMahon, B., Cole, P., Lin, T.M., Lowe, C.R., Mirra, A.P., Ravnihar, B., Salber, E.J., Valaoras, V.G., and Yuasa, S. (1970). Age at first birth and breast cancer risk. Bull. WHO 43, 209-211.

Malkin, D., Li, F.P., Strong, L.C., Fraumeni, J.F., Nelson, C.E., Kim, D.H., Kassel, J., Gryka, M.A., Bischoff, F.Z., Tainsky, M.A., and Friend, S.H. (1990). Germ line p53 mutations in a familial syndrome of breast cancer, sarcomas, and other neoplasms. Science, 250 1233-1238.

McGuire, W., and Naylor, S.L. (1989). Loss of heterozygosity in breast cancer: Cause or effect? JNCI 81, 1764-1765.

Merlo, G., Siddiqui, J., Cropp, C., Liscia, D., Lidereau, R., Callahan, R., and Kufe, D. (1989). Frequent alteration of the DF3 tumor-associated antigen gene in primary human breast carcinomas. Cancer Res. 49, 6966-6971.

Meyn, M.S. (1995). Ataxia-telangiectasia and cellular responses to DNA damage. Cancer Research 55, 5991-6001.

Miki, Y., Swensen, J., Shattuck Eidens, D., Futreal, P.A., Harshman, K., Tavtigian, S., Liu, Q, Cochran, C., Bennett, L.M., Ding, W., Bell, R., Rosenthal, J., Hussey, C., Tran, T., McClure, M., Frye, C., Hattier, T., Phelps, R., Haugen-Strano, A., Katcher, H., Yakumo, K., Gholami, Z., Shaffer, D., Stone, S., Bayer, S., Wray, C., Bogden, R., Dayananth, P., Ward, J., Tonin, P., Narod, S., Bristow, P.K., Norris, F.H., Helvering, L., Morrison, P., Rosteck, P., Lai, M., Barrett, J.C., Lewis, C., Neuhausen, S., Cannon-Albright, L., Goldgar, D., Wiseman, R., Kamb, A., and Skolnick, M.H. (1994). A strong candidate for the breast and ovarian cancer susceptibility gene BRCA1. Science 266, 66-71.

Morrow, D.M., Tagle, D.A., Shiloh, Y., Collins, F., and Hieter, P. (1995). TELI, an *S. cerevisiae* homolog of the human gene mutated in ataxia-telangiectasia is functionally related to the yeast checkpoint gene MECI. Cell 82, 823-829.

Negrini, M., Rasio, D., Hampton, G.M., Sabbioni, S., Rattan, S., Carter, S.L., Rosenberg, A.L., Schwartz, G.F., Shiloh, Y., Cavenee, W.K., and Croce, C.M. (1995). Definition and refinement of chromosome 11 regions of loss of heterozygosity in breast cancer: Identification of a new region at 11q23.3. Cancer Res. 55, 3003-3007.

Neuhausen, S., Gilewski, T., Norton, L., Tran, T., Mcguire, P., Swensen, J., Hampel, H., Borgen, P., Brown, K., Skolnick, M. Shattuck-Eidens, D., Jhanwar, S., Godgar, D., and Offit, K. (1996). Recurrent BRCA2 617delT mutations in Ashkenazi Jewish women affected by breast cancer. Nature Genet. 13, 126-128.

Parks, A.G. (1959). The micro-anatomy of the breast. Ann. R. Coll. Surg. Engl. 24, 235-240.

Pemberton, P.A., Wong, D.T., Gibson, H.L., Kiefer, M.C, Fitzpatrick, P.A., Sager, R., and Barr, P.J. (1995). The tumor suppressor maspin does not undergo the stressed to relaxed transition or inhibit trypsinlike serine proteases. Evidence that maspin is not a protease inhibitory serpin. J. Biol. Chem. 270, 15832-15837.

Peters, G., Fantl, V., Smith, R., Brookes, S., and Dickson, C. (1995). Chromosome 11q13 markers and D-type cyclins in breast cancer. Breast Cancer Research and Treatment 33, 125-135.

Phelan, C.M., Lancaster, J.M., Tonin, P., Gumbs, C., Cochran, C., Carter, R., Ghadirian, P., Perret, C., Moslehi, R., and Dion, F. (1996). Mutation analysis of the *BRCA2* gene in 49 site-specific breast cancer families. Nature Genetics 13 (1), 120-122.

Plowman, G.D., Culouscou, J.M., Whitney, G.S., Green, J.M., Carlton, G.W., Foy, L., Neubauer, M.G., and Shoyab, M. (1993). Ligand specific activation of HER4/p180 *erb*B4, a fourth member of the epidermal growth factor receptor family. PNAS 90, 1746-1750.

Prosser, J., Porter, D., Coles, C., Condie, A., Thompson, A.M., Chetty, U., Steel, C.M., and Evans, H.J. (1992). Constitutional p53 mutation in a non-Li-Fraumeni cancer family. Brit. J. Cancer Apr, 65 (4), 527-528.

Radford, D.M., Fair, K., Thompsom, A.M., Ritter, J.H., Holt, M., Steinbrueck, T., Wallace, M., Wells, S.A., and Donis-Keller, H.R. (1993). Allelic loss on chromosome 17 in Ductal carcinoma *in situ* of the Breast. Cancer Res. 53, 2947-2950.

Roa, B.B., Boyd, A.A., Volcik, K., and Richards, C.S. (1996). Ashkenazi Jewish population frequencies for common mutations in BRCA1 and BRCA2. Nature Genet. 14, 185-187.

Ronnov-Jessen, L., Petersen, O.W., and Bissell, M.J. (1996). Cellular changes involved in the conversion of normal to malignant breast: Importance of the stromal reaction. Physiological Rev. 76, 69-125.

Russo, J., Gusterson, B.A., Rogers, A.E., Russo, I.H., Wellings, S.R., and van Zwieten, M.J. (1990). Biology of Disease. Comparative study of human and rat mammary tumorigenesis. Lab. Inves. 62, 244-277.

Russo, J., Romero, A.L., and Russo, I.H. (1994). Architectural pattern of the normal and cancerous breast under the influence of parity. Cancer Epidemiology, Biomarkers and Prevention 3, 219-224.

Russo, J. and Russo, I.H. (1995). The etiopathogenesis of breast cancer prevention. Cancer Letters 90, 81-89.

Sapienza, C. (1990). Genome imprinting, cellular mosaicism, and carcinogenesis. Mol. Carcinog. 3, 118-121.

Sato, T., Tanigami, A., Yamakawa, K., Akiyama, F., Kasumi, F., Sakamoto, G., and Nakamura, Y. (1990). Allelotype of breast cancer: Cumulative allele losses promote tumor progression in primary breast cancer. Cancer Res. 50, 7184-7189.

Sato, T., Akiyama, F., Sakamoto, G., Kasumi, F., and Nakamura, Y. (1991). Accumulation of genetic alterations and progression of primary breast cancer. Cancer Res. 51, 5794-5799.

Savitsky, K., Sfez, S., Tagle, D.A., Ziv, Y., Sartiel, A., Collins, F.S., Shiloh, Y., and Rotman, G. (1995). The complete sequence of the coding region of the ATM gene reveals similarity to cell cycle regulators in different species. Hum. Mol. Genet. 4, 2025-2032.

Serova, O., Montagna, M., Torchard, D., Narod, S.A., Tonin, P., Sylla, B., Lynch, H.T., Feunteun, J., and Lenoir, G.M. (1996). A high incidence of BRCA1 mutations in 20 breast-ovarian cancer families. An. J. Hum. Genet. 58, 42-51.

Scheffner, M., Werness, B.A., Huibregtse, J.M., Levine, A.J., and Howley, P.M. (1990). The E6 oncoprotein encoded by human papillomavirus type 16 and 18 promotes the degradation of p53. Cell 63, 1129-1136.

Schor, S.L., Schor, A.M., Durning, P., and Rushton, G. (1985). Skin obtained from cancer patients displays fetal-like behaviour in collagen gels. J. Cell Sci. 73, 235-244.

Schott, D.R., Chang, J.N., Deng, G., Kurisu, W., Kuo, W-L, Gray, J. and Smith, H.S. (1994). A candidate tumor suppressor gene in human breast cancers. Cancer Res. 54, 1393-1396.

Scully, R., Ganesan, S., Brown, M., DeCaprio, J., Cannistra, S., Feunteun, J., Schnitt, S., Livingston, D. (1996). Location of BRCA1 in human breast and ovarian cancer cells. Science 272, 123-126.

Shattuck-Eidens, D., McClure, M., Simard, J., Labrie, F., Narod, S., Couch, F., Hoskins, K., Weber, B.L., Erdos, M., Brody, L., Friedman, L., Ostemeyer, E., Szabo, C., King, M-C., Jhanwar, S., Offit, K., Norton, L., Gilewski, T., Lubin, M., Osborne, M., Black, D., Boyd, M., Steel, M., Ingles, S., Haile, R., Lindblom, A., Olsson, H., Borg, A., Bishop, T., Solomon, H., Radice, P., Spatti, G., Gayther, S., Ponder, B., Warren, W., Stratton, M., Liu, Q., Fujimura, F., Lewis, C., Skolnick, M., and Goldgar, D.E. (1995). A collaborative survey of 80 mutations in the BRCA1 breast and ovarian cancer susceptibility gene. JAMA. 273, 535-541.

Singer, C., Rasmussen, A., Smith, H.S., Lippman, M.E., Lynch, H.T., and Cullen, K.J. (1995). Malignant breast epithelium selects for insulin-like growth factor II expression in breast stroma: Evidence for paracrine function. Cancer Res. 55, 2448-2454.

Slamon, D.J., Clark, G.M., Wong, S.G., Levin, W.J., Ullrich, A., and McGuire, W.L. (1987). Human Breast Cancer: Correlation of relapse and survival with amplification of the HER-2/neu oncogene. Science. 235, 177-182.

Smith, H.S., Dairkee, S.H., Ljung, B-M., Mayall, B., Sylvester, S.S., and Hackett, A.J. (1987). Cellular manifestations of human breast cancer. In: *Cellular and Molecular Biology of Breast Cancer*.(Medina, D., Kidwell, W., Heppner, G., and Anderson E., Eds.) pp. 437-452. Plenum Publishing Corp, New York.

Smith, H.S. (1990). Stochastic model for interpreting the data on loss of heterozygosity in breast cancer. JNCI. 82, 793-794.

Smith, H.S., Lu, Y., Deng, G., Martinez, O., Krams, S., Ljung, B-M., Thor, A., and Lagios, M. (1993). Molecular aspects of early stages of breast cancer progression. J. Cell. Biochem. 17G, 144-152.

Smith, S.A., Easton, D.F., Evans, D.G.R., and Ponder, B.A. (1992). Allele losses in the region 17q12-21 in familial breast and ovarian cancer involve the wild type chromosome. Nat. Genet. 2, 128-131.

Sobol, H., Stoppa-Lyonnet, D., Bressac-de-Paillerets, B., Peyrat, J-P., Kerangueven, F., Janin, N., Noguchi, T., Eisinger, F., Guinebretiere, J-M., Jacquemier, J., and Birnbaum, D. (1996).

Truncation at conserved terminal regions of BRCA1 protein is associated with highly proliferating hereditary breast cancers. Cancer Res., 56, 3216-3219.

Srivastava, S., Zou, Z., Pirollo, K., Blattner, W., and Chang, E.H. (1990). Germ-line transmission of amutated *p53* gene in a cancer-prone family with Li-Fraumeni syndrome. Nature 348, 747-749.

Stahl, J.A., Leone, A., Rosengard, A.M., Potter, L., King, C.R., and Steeg, P.S. (1991). Identification of a second human nm23 gene, *nm23-H2*. Cancer Res. 51, 445-449.

Stratton, M.R. (1996). Recent advances in understanding of genetic susceptibility to breast cancer. Human Mol. Genet. 5, 1515-1519.

Sun, X-F., Johannsson, O., Hakansson, S., Sellberg, G., Nordenskjold, B., Olsson, H., and Borg, A. (1996). A novel p53 germline alteration identified in a late onset breast cancer kindred. Oncogene 13, 407-411.

Swift, M., Reitnauer, P.J., Morrell, D., and Chase, C.L. (1987). Breast and other cancers in families with ataxia telangiectasia. N. Engl. J. Med. 316, 1289-1294.

Swift, M., Morrell, D., Massey, R.B., and Chase, C.L. (1991). Incidence of cancer in 161 families affected by ataxia telangiectasia. N. Engl. J. Med. 325, 1831-1836.

Takita, K., Sato, T., Miyagi, M., Watatani, M., Akiyama, F., Sakamoto, G., Kasumi, F., Abe, R., and Nakamura, Y. (1992). Correlation of loss of alleles on the Short arm of chromosome 11 and 17 with metastasis of primary breast cancer to lymph nodes. Cancer Res. 52, 3914-3917.

Thorlacius, S., Jonasdottir, O., and Eyfjord, J.E. (1991). Loss of heterozygosity at selective sites on chromosomes 13 and 17 in human breast carcinoma. Anticancer Res. 11, 1501-1507.

Thorlacius, S., Olafsdottir, G., tryggvadottir, L., Neuhausen, S., Jonasosn, J.G., Tavtigan, S.V., Tulinius, H., Ogmundsdottir, H.M., and Eyfjord, J.E. (1996). A single BRCA2 mutation in male and female breast cancer families from Iceland with varied cancer phenotypes. Nat.genet. 13, 117-119.

Tulusan, A.H., Ronay, G., Egger, H., and Willgeroth, F. (1985). A contribution to the natural history of breast cancer. Arch. Gynecol. 237, 85-91.

Varley, J.M., McGown, G., Thorncroft, M., Santibanez-Koref, M.F., Kelsey, A.M., Tricker, K.J., Evans, G.R., and Birch, J.M. (1997). Germline mutations of TP53 in Li-Fraumeni families: An extended study of 39 families. Cancer Res. 57, 3245-3252.

Vorechovsky, I., Rasio, D., Luo, L., Monaco, C., Hammarstrom, L., Webster, D.B., Zaloudik, J., Barbanti-Brodano, G., James, J., Russo, G., Croce, C.M., and Negrini, M. (1996). The ATM gene susceptibility to breast cancer: Analysis of 38 breast tumors reveals no evidence for mutation. Cancer Res. 56, 2726-2732.

Wellings, S.R., Jensen, H.M., Marcum, R.G. (1975). An atlas of subgross pathology of the human breast with special reference to possible precancerous lesions. JNCI 55, 231-238.

Wilson, J.D., Griffin, J.E., Leshin, M., and George, F.W. (1981). Role of gonadal hormones in development of the sexual phenotypes. Hum Genet. 58, 78-84.

Wooster, R., Mangion, J., Eeles, R., Smith, S., Dowsett, M., Averill, D., Barrett-Lee, P., Easton, D.F., Ponder, B.A., and Stratton, M.R. (1992). A germline mutation in the androgen receptor gene in two brothers with breast cancer and Reifenstein syndrome. Nature Genetics 2, 132-134.

Wooster, R., Neuhausen, S.L., Mangion, J., Quirk, Y., Ford, D., Collins, N., Nguyen, K., Seal, S., Tran, T., Averill, D., Fields, P., Marshall, G., Narod, S., Lenoir, G.M., Lynch, H., Feunteun, J., Devilee, P., Cornelisse, C.J., Menko, F.H., Daly, P.A., Ormiston, W., McManus, R., Pye, C., Lewis, C.M., Cannon-Albright, L.A., Peto, J., Ponder, B.A., Skolnick, M.H., Easton, D.F., Goldgar, D.E., and Stratton, M.R. (1994). Localization of breast cancer susceptibility gene *BRCA2* to chromosome 13q12-13. Science 265, 2088-2090.

Zhuang, Z., Merino, M.J., Chuaqui, R., Liotta, L., and Emmert-Buck, M.R. (1995). Identical allelic loss on chromosome 11q13 in microdissected in situ and invasive human breast cancer. Cancer Res. 55, 467-471.

Zou, Z., Anisowicz, A., Hendrix, M.J., Thor, A., Neveu, M., Sheng, S., Rafidi, K., Seftor, E., and Sager, R. (1994). Maspin, a serpin with tumor-suppressing activity in human mammary epithelial cells. Science. 263, 526-529.

THE CLINICOPATHOLOGIC SIGNIFICANCE OF GENETIC INSTABILITY IN BREAST CARCINOMA PROGRESSION

Daniel W. Visscher, and Susan M. Wykes

INTRODUCTION

Formal criteria for "genetic instability" are difficult, perhaps impossible, to enumerate. In the context of solid tumors such as breast carcinoma, the concept of genetic instability arose from the observation that individual tumors were often

Advances in Oncobiology
Volume 2, pages 159-176.

ISBN: 0-7623-0388-3

characterized by a high degree of karyotypic diversity which, in some cases, continued to evolve upon serial passage in cell culture. The striking complexity and heterogeneity of observed cytogenetic aberrations implied that inability to sustain genomic organization constituted a biologically important, possibly causative, feature of neoplasia. Genetic instability, then, may be viewed either as a process inherent to, or as a consequence of, neoplastic growth.

It has become axiomatic that structural genetic alterations are at least partially responsible for neoplastic phenotype and, furthermore, that they are acquired in a stepwise manner over protracted time intervals. Moreover, the degree of clinical tumor progression (as determined by characteristics such as invasion and metastasis) is widely perceived to reflect the cumulative number, or complexity, of cellular genetic lesions. Genetic instability, by virtue of fueling spontaneous acquisition of mutational events, would theoretically provide the impetus for selection and differential growth of neoplastic subpopulations having progressively more dysregulated growth characteristics. This model of neoplastic pathophysiology has driven the search to identify and study the specific genes that perform three functions: orchestration of cellular growth and differentiation, involvement in pathologic events such as invasion and metastasis, and maintainence of genomic organization.

A reasonable corollary to this scenario of tumor progression is that the acquisition of mutations in neoplastic populations involves specific genes in a random as opposed to an ordered, or sequential, manner. Some randomly acquired mutations, moreover, may be functionally silent thereby resulting in "genetic noise". Other genetic alterations may directly facilitate tumor progression by leading to abnormal expression of growth regulatory genes thereby resulting in clonal expansion. This concept is one factor that has limited investigation into the identification of consistent patterns by which cellular genetic alterations are acquired during the evolution of neoplasia. By the term "patterns", we are referring either to the identification of mutations that represent "early" versus "late" events, or permissive versus rate-limiting events. The measurement of cellular parameters that reflect aggregate, or accumulated, genetic pathology would perhaps represent another strategy to explore the relationship between genetic instability and tumor progression.

There are also biological factors that preclude systematic analyses of aggregate genomic destabilization, particularly in breast neoplasia. Most significant, perhaps, is the inability to perform sequential analysis of individual human neoplasms, particularly with samples straddling biologically critical events such as the transition from pre-invasive to invasive growth. Lengthy disease natural history, difficulty in clinical detection of early lesions, and the intimate microscopic admixture of benign, pre-malignant, pre-invasive, and invasive neoplastic cells within aliquots of human breast tumors all further complicate attempts to discern unifying patterns of genetic pathophysiology in breast neoplasia. Finally, the well described clinical, epidemiologic, and pathologic diversity of human breast neoplasia is compelling evidence that multiple pathophysiologically distinctive disease subsets may exist.

Basic scientists, therefore, have been engaged largely with specialized investigations involving the regulation of expression, or cellular effects of expression, for specific genes that regulate growth and differentiation. The translational researchers, on the other hand, have mostly performed clinicopathologic analyses comparing the status of individual genes to parameters such as stage or disease outcome. Given the impressive complexity and diversity of structural genetic alteration in breast neoplasia, though, it hardly comes as a surprise that the informative content of single genes such as *ERBB-2* or *p53* have limited clinical value in the care of individual patients.

In this chapter, we shall summarize an admittedly small body of literature that attempts to correlate measures of aggregate genetic deviation—such as cellular DNA content, karyotype, chromosome aneuploidy, or allelotype—to phenotypic measures of tumor progression, including pathologic evidence of malignant transformation, host invasion, or metastasis. Implicit to the discussion will be the assumption that the tendency to accumulation of mutations—or genetic instability—is an intrinsic variable in clinical breast neoplasia, albeit difficult to quantify. Although attempts will also be made to discern sequences, or patterns, of genomic pathology that evolve from genetic instability, their presence remains speculative at this time.

MODELS OF GENETIC PROGRESSION—CLINICOPATHOLOGIC CORRELATIONS

Patterns of Cellular Level Genetic Instability

By correlating karyotypes with flow cytometric DNA content on large numbers of clinically derived breast carcinoma samples, Dutrillaux and colleagues (1991) have constructed a plausible model of cellular-level genetic evolution in breast neoplasia. Their data suggest that, initially, an accumulation of multiple unbalanced chromosomal rearrangements leads to a small net loss of genomic DNA via segmental deletions. These events would potentially contribute to altered gene expression through losses of heterozygosity that "unmask" recessive alleles at tumor suppressor loci. The number of observed chromosomal rearrangements, as should be noted, varies considerably among individual tumors, reflecting either inherent differences in rate of mutation or differences in sampling with respect to the length of time or number of cell doublings through which individual tumors have evolved. Genetic rearrangements in breast carcinomas may affect up to 80% of cellular chromosome compliments, but overall they are involved in an average of 30%. Comparison of karyotypes derived from tumors having few rearrangements with those having many demonstrates that shared, and therefore *important* or *early*, alterations involve chromosomes 1, 8, 11, 16, and 17 (Dutrillaux et al., 1990; Dutrillaux et al., 1991). Structural alterations involving 1q or 16p are especially common, with an observed prevalence of up to 60%.

During the evolution of a particular neoplasm, genetic rearrangements develop in conjunction with diminished total chromosome number, which is sometimes considerable, by virtue of acquired monosomies. Chromosome loss, though, is generally out of proportion to cellular DNA losses since formation of large chromosomes occurs (Remvikos et al., 1988). Although chromosome number in near diploid tumors may fall to about 35, the net loss of cellular DNA is limited to about 15% of the normal total complement. About one-third of human breast carcinomas will indefinitely maintain karyotypic profiles having near diploid cellular DNA content, whereas the remainder will eventually develop aneuploid "stemlines", presumed to be clonal, of cells having increased total DNA content (1.2–3 times normal). Tumors having combined diploid range and aneuploid karyotypic modes have been shown by Dutrillaux and colleagues to share common aberrations. They are also estimated to contain quantitatively similar amounts of DNA as "stemlines" observed in corresponding flow cytometric DNA histograms. Thus, cytophotometrically, DNA aneuploid stemlines, which occur in the majority of breast carcinomas, are theorized to develop via endoreduplication ("tetraploidization") of near-diploid progenitor cells, followed by clonal expansion.

Approximately one-half of tumors harboring aneuploid stemlines will retain the near-diploid progenitor stemline, which is presumed to continue "evolving". At least two lines of evidence, however, suggest that tumors that harbor aneuploid stemlines represent more genetically "evolved", or unstable, neoplastic populations than tumors with diploid range populations only. Firstly, the probability of endoreduplication, at least insofar as generation of a cytophotometrically detectable aneuploid population was concerned, was shown to be greater in neoplasms that had acquired either more chromosomal rearrangements or greater decreases in chromosome number (Dutrillaux et al., 1991). Secondly, synthesis phase fraction is usually greater in DNA aneuploid- versus diploid-range stemlines (Remvikos et al., 1992). The latter association is in keeping with a greater degree of "tumor progression" because proliferative autonomy is widely agreed to reflect a clinically aggressive (generally estrogen receptor–negative) phenotype, which directly or indirectly facilitates invasive growth.

In addition to presence of shared or "marker" chromosomes in karyotypic stemlines, the patterns of DNA content in corresponding stemlines analyzed by flow cytometry are compatible with the theory that polyploidization represents the dominant mechanism by which aneuploid populations are generated. This is illustrated in Table 1. Note that the DNA indices of aneuploid stemlines from clinically derived cytophotometrically multiclonal breast neoplasms typically represent near multiples of one another. Additionally, the preponderance of DNA aneuploid human breast carcinomas have DNA indices in the hypotetraploid range (i.e., DNA index 1.6–1.8), which is fully compatible with net loss of cellular DNA occurring prior to and/or following genomic endoreduplication (Visscher et al., 1995b)

The molecular pathogenesis of "polyploidization" in vivo is poorly defined. Neither is it entirely clear whether the observed association with abnormal cell cycling

Table 1. Flow Cytometric DNA Indices of Hypodiploid Carcinomas Having Accompanying Hyperdiploid Stemlines

DNA INDEX	
Hypodiploid Peak	*Hyperdiploid Peak*
0.8	1.6
0.8	1.5
0.8	1.6
0.8	1.5/1.7
0.9	1.8
0.9	1.9
0.9	1.7
0.9	1.7
0.9	1.7

(either proliferative fraction or kinetics) represents a cause or an effect of aneuploid stemline generation. Observations made from experimental polyploidization of malignant cell lines in vitro suggest these events result from genetic instability occurring during mitosis (Sennerstam and Auer, 1993). Relatively short cell cycles, moreover, predispose cells to endoreduplication events. These data would seem to implicate dysfunction of genes that regulate genomic organization during DNA synthesis and cytokinesis as critical events in breast carcinoma progression. The mechanism(s) of maintaining (or failing to maintain) genomic integrity, particularly during cell division, remain(s) largely undefined in human breast neoplasia. The list of genes potentially involved with genomic instability, though, is lengthy (Volpe et al., 1988). Perhaps in this context it is worth mentioning that one of the genes known to maintain genomic organization—*p53*—is less often mutated in breast neoplasia than in most other adult adenocarcinomas. However, *p53* mutation is associated with DNA aneuploidy in breast carcinoma (Schmitt et al, 1995) as well as with frequency of allelic loss (Eyfjörd et al, 1995). Similarly, microsatellite instability (so-called mutator phenotype) is observed, but with variable frequency, in breast carcinomas (Shaw et al., 1996). Other authors (Shay et al., 1993; Odagiri et al., 1994) have noted that breast carcinomas are characterized by loss of telomeric DNA, a region which is believed responsible for genomic organization.

According to Dutrillaux and colleagues (1991), chromosomal alterations involving either the near diploid or triploid range stemline may continue to accumulate following the endoreduplication event that generates a cytophotometrically aneuploid stemline. Cytogenetically, however, DNA losses following endoreduplication events more frequently involve deletion of whole chromosomes as opposed to unbalanced rearrangements. Progressive DNA losses from aneuploid stemlines initially having DNA indices of 1.6–1.8 are theorized to eventually generate clonal populations exhibiting DNA indices of 1.3–1.4. Thus, it is the character of DNA aneuploidy, as opposed to the absolute deviation of DNA content from normal, that best reflects degree of cellular genetic evolution.

Phenotypic parameters of human breast carcinomas would appear to support this model of dynamic genetic progression. It has been reported, for example, that DNA tetraploid breast carcinomas (i.e., those having DNA indices of 1.9–2.1) are characterized by patient survival similar to or slightly worse than diploid-range tumors, but better than triploid range cases having DNA indices of 1.3–1.7 (Joensuu et al., 1992). These findings are presumably explained by the inference that peritetraploid DNA content reflects an endoreduplication event that was neither preceded, nor followed, by appreciable DNA content loss or, by analogy, chromosomal rearrangement. Other empirical data also support this theory of genetic pathophysiology. Synthesis phase fractions of breast carcinomas, for example, progressively increase as DNA index decreases from 2.0 to 1.3 (Remvikos et al., 1992). Furthermore, DNA aneuploid stemlines may undergo a second round of genomic endoreduplication. Polyploidization of triploid-range stemlines, if followed by clonal expansion, gives rise to hypertetraploid stemlines with DNA indices between 2.6 and 2.8. As might be expected, such cases are phenotypically high grade (i.e., poorly differentiated) and clinically aggressive (Fallenius et al., 1988). Hypertetraploid stemlines, though, are relatively uncommon in series of flow cytometrically analyzed breast carcinomas, accounting for only 10–15% of cases. They are rarer still in karyotypes.

From the standpoint of cellular DNA content and karyotype then, breast carcinomas are characterized by three major genetic subgroups with respect to dominant clonal population: (i) purely near-diploid, (ii) purely hypotetraploid / triploid, and (iii) a combined diploid-range to aneuploid, or heterogeneous, pattern. These patterns of cellular level clonal genetic alteration have been observed in other forms of human epithelial neoplasia, such as colorectal carcinoma, in which karyotypically and cytophotometrically peridiploid or endoreduplication-derived, triploid-range DNA stemlines exist alone or concurrently (Giaretti, 1994).

Readers familiar with flow cytometric studies of human breast carcinomas will note that the mechanism of cytogenetic evolution proposed by Dutrillaux and colleagues appears to reconcile the apparent ineffectiveness of ploidy status as a highly reliable prognostic index. Cytophotometrically diploid-range breast carcinomas, notably, are clinically very heterogeneous with at least some characterized by metastatic phenotype and therefore, an "advanced" degree of tumor progression. Such aggressive cases, it may be hypothesized, could have developed extensive chromosomal rearrangements leading to metastatic phenotype in the absence of stable polyploidization. Although published studies correlating karyotype complexity to disease outcome in near-diploid breast carcinomas are lacking, it is certainly the case that diploid-range carcinomas may also be characterized by considerable diversity in proliferative fraction.

Potential insights into the heterogeneous pathobiology of near-diploid breast carcinomas have also been provided by investigators who employ image cytophotometric (as opposed to flow cytometric) DNA analysis. Studies of cellular DNA content using these methods demonstrate that approximately 18% of diploid-range breast carcinomas harbor small numbers (less than 2% of total cells) of highly aneuploid cells, often with hypertetraploid DNA content (Siitonen et al., 1993). By virtue of limited representation, such cells would not form a visible population in flow cytometric DNA histograms nor would they likely be observed in karyotypes. Their presence in small but quantifiable numbers, though, is compatible with the notion that endoreduplication events occur with significant frequency in some tumors in keeping with the notion of variable intrinsic genetic instability. However, for reasons that remain unclear, they do not necessarily give rise to polyploid stemlines through clonal expansion. Interestingly, the presence of isolated hypertetraploid cells is associated with elevated proliferative fraction—additional evidence linking genetic instability to cell cycling (Siitonen et al., 1993). Detection of highly aneuploid cells also correlates with adverse outcome, albeit in tumors with either diploid-range or aneuploid DNA content. Although the functional significance of highly aneuploid cells in tumor progression remains unclear, their apparent clinical relevance is compatible with the notion that markers of inherent genetic instability per se have clinical relevance.

Compared to flow cytometry, image cytophotometric analyses reveal another, qualitatively different, pattern of cellular DNA content aberration that is in keeping with subsets of breast carcinoma characterized by a high degree of intrinsic genetic instability. Auer and colleagues (1980), among others, have emphasized the presence of breast carcinomas having so-called Type IV DNA histograms. Cytophotometric DNA content analyses of these cases reveal numerous aneuploid cells with many hypertetraploid events, but without dominant (clonal) stemlines. Although flow cytometric DNA content analyses rarely, if ever, reveal such "diffuse" patterns of DNA aneuploidy, cell-by-cell karyotype analyses of some individual cases confirms the existence of highly diverse, "unstable" tumors. Even among cytophotometrically homogeneous aneuploid tumors, though, chromosome counts using fluorescence in situ hybridization demonstrate an unsettling degree of genetic heterogeneity in human breast neoplasia (Visscher et al., 1995c). Counts for individual chromosomes obtained by fluorescence hybridization techniques are often characterized by marked intercellular variability such that individual chromosome complement is best characterized by a modal value. The consistent and somewhat paradoxical finding that such tumors nonetheless produce discrete DNA "stemlines" in flow cytometric DNA histograms remains to be fully explained, although it is confirmed by metaphase analyses, which usually demonstrate total chromosome count heterogeneity within apparently dominant stemlines (Bell et al., 1990). It is perhaps the case that individual variabilities in chromosome aneuploidy tend to "cancel one another out", thereby producing discrete, quantitatively-definable stemlines with "stable" DNA content. Thus, even within the predominant group of cytophotometrically homogeneous breast cancers, the relatively common presence of highly aneuploid "outlier" cells as well as marked variability in individual chromosome number nevertheless confirms a variable, but often high degree of structural genetic diversity compatible with genomic instability.

The observed intratumoral and intertumoral variabilities of genetic aberration in human breast neoplasia have long represented an issue of importance and a source of consternation. As already noted, metaphase spreads of breast carcinoma cells are often highly complex with rearrangements often involving a majority of chromosomes. However, the variety of consistent or specific anomalies (i.e., peculiar to breast neoplasia) is limited. The extent to which genotypic complexity is a reflection of "random" genetic alteration (i.e., genetic instability) resulting in genetic "noise" or the existence of multiple genetically-distinct breast carcinoma subsets (as defined by characteristic translocations, deletions, etc... possibly corresponding to distinct clinical–epidemiologic groups of patients) remains unclear. At present, developing a "genetic classification" for breast carcinoma is frustrated by the inability to evaluate the large number of cases required to account for both the variety of clinical subsets and genetic aberrations.

Genetic Instability in Early Breast Neoplasia

Cultivating an understanding of "random", as opposed to "progression-related" genetic alterations, as implied by the clinical relevance of "rare" aneuploid events, has potentially enormous biological significance since genetic instability per se would, arguably, precede clonal expansion. For this reason, investigators have recently focused attention on the supposed precursor lesions of breast carcinoma—the hyperplasias—in an attempt to identify evidence that "early" stages of neoplastic progression are characterized by genetic instability. These studies unfortunately, are limited in number, lacking in clinical follow-up, and quite variable with respect to methodology, which includes cytophotometry, FISH, and molecular-level assays of allelic loss (loss of heterozygosity, or LOH). Although these studies are reasonably consistent in demonstrating the presence of genetic abnormalities in many "proliferative" lesions, they provide somewhat differing pictures of genetic instability among these breast epithelial populations. Several groups, for example, have reported the presence of cytophotometrically detectable DNA aneuploidy or abnormal p53 expression in florid or atypical ductal hyperplasias (Crissman et al., 1990; Younes et al., 1995). To the extent that cytophotometrically detectable aneuploid stemlines represent a clonal, or "genetically advanced," anomaly (at least in terms of the model proposed by Dutrillaux et al., 1991), these data would appear wholly compatible with the notion that at least some "hyperplastic" breast lesions represent neoplasms, perhaps akin to colorectal adenoma. The histologic resemblance of atypical hyperplasia (AH) to well differentiated forms of intraductal breast carcinoma is further evidence of such a relationship.

More sensitive in situ hybridization methods reveal that many hyperplastic lesions (and virtually all hyperplastic lesions with morphologic "atypia") are characterized by numerical alterations (aneuploidy) of at least one chromosome (Micale et al., 1994). Several observations from this admittedly incompletely developed literature are particularly revealing about the nature of genetic instability in "early" breast neoplasia. Firstly, chromosome aneuploidies in hyperplastic lesions are more or less randomly distributed through the genome, without apparent pattern. Sneige and colleagues (1996) for example, evaluated 25 proliferative and pre-invasive breast lesions for aneuploidies of chromosomes 7–12, 17, 18, and X. The frequency of individual chromosome affected by numerical abnormality varied between 39% (for chromosome 11) and 73% (for chromosome 7). Secondly, chromosome aneuploidies were, in general, limited to a relatively small minority of cells in the lesion. Chromosomal gain in AH, for example, involved a mean of 8.5% of counted nuclei (range 3–37%), a gain that hardly seems sufficient to result in DNA shifts sufficient to be predictably detected as DNA aneuploidy by cytophotometry. There are, unfortunately, no studies that correlate aneuploidy of individual chromosomes to karyotype or cytophotometric aneuploidy in proliferative breast lesions. It is thus not possible to determine whether all of these findings reflect "random" events involving isolated cells or possibly the presence of a small subpopulation of cells that have resulted (either individually or via clonal expansion) from endoreduplication. It is interesting, in this context, to note that numerical chromosome gains were more frequently observed by Sneige and colleagues than were numerical losses. These data, in any event, seem at least compatible with the notion that putatively early stages in the evolution of breast neoplasia may be characterized by cellular level genetic alterations. The extent to which such alterations reflect genetic instability, genetic evolution or both remains speculative however.

Would the presence of a focal and seemingly disorderly pattern of genetic anomalies, as observed using interphase cytogenetics, however, necessarily exclude the synchronous occurrence of clonal or *progression-related* events in so-called premalignant lesions? A preliminary answer to this question has been forwarded by O'Connell and colleagues (1994), who compared proliferative lesions to synchronous pre- invasive and invasive breast carcinomas for losses of heterozygosity at four loci. The observed frequency of LOH, for at least one locus, was 63% in proliferative lesions with atypia. This is similar to the frequency (63%) in ductal carcinoma in situ (DCIS). Perhaps more significant, though, was the finding that 50% of proliferative lesions shared LOH patterns with concurrent but more advanced lesions (i.e., invasive carcinoma in the same breast). Concurrent foci of pre-invasive and invasive carcinoma shared LOH patterns in an even greater proportion of cases (80%). Apart from adding support to the concept that at least some *hyperplastic* breast lesions actually represent neoplastic proliferations, the data from this study imply that histologically recognizable *stages* of breast carcinomas evolve, at least partially, in a *serial* manner. That is, the presence of shared genetic lesions is compatible with a direct *clonal evolution* of cells from a hyperplastic lesion into a phenotypically more advanced pre-invasive carcinoma, and so on. This contrasts with the competing "parallel" model of breast carcinoma evolution in which proliferative lesions accompany but develop independently from co- existing malignant lesions.

It should be noted, of course, that studies of this type are preliminary. The study by O'Connell and colleagues (1994) included only 60 cases all of which had advanced to invasive ductal carcinoma. The data also fail to exclude a mixed serial and parallel evolution pattern because only partial homology was observed between proliferative and malignant lesions. Given the considerable epidemiologic, clinical, and pathologic heterogeneity of breast carcinoma, moreover, it is hardly inconceivable that multiple evolutionary pathways of tumor progression may exist.

In summary then, the perhaps sparse data would seem to suggest that breast tissues are characterized by evidence of genetic instability before the appearance of histologically-detectable malignant transformation. Proliferative breast disease, as such, would appear to represent a marker of sorts for genetic instability. The high prevalence of PBD in relation to breast carcinoma, though, suggests either that genetic instability alone is an insufficient condition for progression to malignant neoplasia or that PBD is biologically heterogeneous. The lengthy and complex natural history of human breast neoplasia in addition to the focal anatomic distribution of hyperplastic lesions represent unfortunate but significant obstacles to an improved understanding of the early events in breast tumor progression.

The Transition from Preinvasive to Invasive Disease

In contrast to so-called proliferative (hyperplastic) lesions, karyotypic and DNA content aberrations in pre-invasive breast carcinomas, are well described. Although there are notable differences with invasive neoplasia, the data show that in situ carcinomas are generally characterized by a relatively advanced degree of genetic pathology, at least according to models described previously herein. At the karyotypic level, for example, DCIS is uniformly characterized by abnormal metaphases (Nielsen et al., 1987; Nielsen et al.,1989). Among diploid-range DCIS cells, moreover, different metaphases from a given case typically demonstrate a spread, or scattering, of the chromosome number around 46. These data are analogous to previously described results of in situ hybridization analyses of chromosome number in both proliferative and invasive lesions, which were characterized by a surprising cell-to-cell variability in chromosome counts. Numerical chromosome losses, though, are limited in karyotypes of pre-invasive breast carcinomas. The modal (diploid-range) chromosome counts for DCIS reported by one group ranged from 45 to 49 (Nielsen et al., 1989). Invasive lesions studied by the same authors, in contrast, had diploid-range modal chromosome counts between 37 and 40. Thus, significant degrees of numerical chromosome loss appear to be associated with the invasive stages of tumor growth. These findings would seem to link a pathologically recognizable step in tumor progression to the Dutrillaux model of genetic evolution. Although karyotypes of most pre-invasive carcinomas exhibit diploid-range chromosome counts (triploid-range stemlines were observed in approximately 40% of cases), most cases also exhibited varying numbers of polyploid cells compatible with random endoreduplication events as earlier described in image analysis studies. Triploid-range karyotypes in DCIS, moreover, contained greater numbers of chromosomes than metaphases derived from invasive tumors (69–72 vs. 60–69), also in keeping with the Dutrillaux model.

Flow cytometric studies generally confirm the greater frequency of aneuploid stemlines in invasive, as compared with pre-invasive, breast carcinomas. DNA aneuploidy has been reported in about 40% of DCIS cases versus 60–70% of invasive tumors (Aasmundstad and Haugen, 1990; Killeen and Namiki, 1991). Caution should be exercised in interpreting analyses of this type since relatively small DNA aneuploid stemlines in some tumors could be numerically dominated by diploid-range populations and thus not identified in DNA histograms. This is implied by the results of interphase cytogenetic studies, which to date have reported chromosome aneuploidy in a minority of the cells derived from DCIS lesions (20–40%) as compared with virtually all of the cells from invasive neoplasms (>80%) (Sneige et al., 1996; Visscher et al., 1996). The extent to which this is explained by the presence of "small" aneuploid populations, differential admixture of benign hyperplastic populations within pre-invasive lesions, or by lesser degrees of intrinsic genomic instability is unclear at this time. Furthermore, it is arguably true that clinically detected, pre-invasive carcinomas have significant biological differences from areas of in situ carcinomas that represent residual in situ components within overly invasive breast neoplasia. In other words, breast neoplasms that have reached a detectable, sometimes considerable size without evolving to invasive phenotype may be fundamentally different (i.e., not merely an earlier stage) from neoplasms that invade following an undetected and presumably shorter phase of pre-invasive neoplasia. These differences may bias extrapolations of genetic pathophysiologic mechanisms based on presumed analogies.

Finally, the frequency of DNA aneuploidy in DCIS and all breast neoplasms for that matter is highly dependent on nuclear grade, or the degree of nuclear enlargement, hyperchromatism, and pleomorphism. Approximately one half of pre-invasive breast carcinomas exhibit high grade nuclear cytologic features. (Such tumors are often referred to as "comedocarcinomas" owing to the presence of lumenal necrotic debris.) The results of genetic studies that evaluate pre-invasive breast neoplasia are thus rendered vulnerable to potential errors associated with selection bias based on grade. Predictably, DNA aneuploidy is observed in virtually all high-grade (so-called comedo) in situ ductal carcinomas whereas the frequency of aneuploid stemlines in low-grade tumors approaches that of atypical hyperplasia (Crissman et al., 1990). It is perhaps not surprising and hardly a source of comfort to note that many pre-invasive breast carcinomas (at least one-third) are also characterized by a morphologically heterogeneous phenotype, which is compatible with significant (or rapid) genetic evolution during the pre-invasive phase of tumor growth (Lennington et al., 1994). Thus, an important dichotomy in the biology of breast neoplasia, that of grade-dependent behavior, becomes manifest during the pre-invasive stage of tumor growth.

Most would agree that high-grade, and thereby aneuploid, DCIS cases are more likely to evolve to invasive growth (within limited time intervals) than well differentiated DCIS cases. Does this imply that aneuploidy and, by analogy, genetic instability, facilitate invasive growth? In tumors with multiple stemlines, unfortunately, the distribution of diploid and aneuploid stemlines between pre-invasive and invasive disease components is not predictable (Visscher et al., 1993). That the absence of detectable endoreduplication is not necessarily a barrier to invasive phenotype is also implied by the aforementioned 30% frequency of karyotypically and cytophotometrically diploid-range invasive carcinomas.

Possibly more dramatic evidence of genetic divergence between pre-invasive and invasive breast neoplasia is to be found at the molecular level with assays of allelic loss or imbalance (so called allelotype). Aldaz and colleagues (1995) have recently compared the incidence of allelic imbalance (at 20 loci) in 23 pre-invasive and 29 invasive carcinomas. None of the loci was more frequently altered in pre-invasive compared to invasive lesions. However, eight of twenty alleles (40%) were significantly more likely to be altered in the invasive tumors, especially loci on 3p, 16p, 18q, and 22q. These data seem to imply that most genetic events acquired during pre-invasive growth would also provide selective advantage to invasive neoplastic growth and were thereby retained during progression to invasive phenotype. The observed tendency to conserve genetic alterations in the transition from in situ to invasive growth, alternatively, may be interpreted to imply that invasive phenotype reflects additive, or interactive, effects of previously acquired, functionally significant genetic lesions.

Although the data of Aldaz and colleagues (1995) largely substantiate the existence of significant genetic distinctions between pre-invasive and invasive breast neoplasia, they are subject to the potential for case selection bias as previously enumerated. In particular, the inability to assess individual neoplasms at different points in their natural history imposes considerable limitations on interpreting the biologic relevance of genetic alterations to a given neoplasm. As with other aspects of breast carcinoma, the task of analyzing the significance of genetic instability in pre-invasive breast neoplasia is complicated significantly by extreme variability in pre-invasive disease interval as implied by intertumoral heterogeneity in the relative proportions of in situ and invasive disease that comprise breast masses (Silverberg and Chitale, 1973). Some clinically presenting breast tumors are characterized by more or less equal volumes of pre-invasive and invasive neoplasia. In other invasive neoplasms, it may be difficult to identify any residual in situ neoplasm, even with extensive tissue sampling. It is reasonable to hypothesize that length of pre-invasive disease interval reflects the degree of intrinsic genetic instability and thereby the tendency to progress. Careful study of genetic alterations in the context of concurrent invasive and pre-invasive disease extent, then, may provide an alterative to the likely insoluble problem of obtaining longitudinal, or serial, analyses of human breast tumor progression.

A notable such study that has compared pre-invasive to concurrent invasive disease components, however, shows that at least one oncogene, ERBB-2, is paradoxically overexpressed more frequently in pre-invasive that in invasive breast neoplasia. Allred and colleagues (1992) have shown that ERBB-2 immunoreactivity is present in 56% of purely in situ ductal carcinomas versus the only 11% of purely invasive lesions. Tumors with admixed pre-invasive and invasive neoplasia overexpressed ERBB-2 in an intermediate proportion of cases (22%). Apart from the apparent divergence from the study by Aldaz and colleagues, these data are also paradoxical in view of numerous reports that associate ERBB-2 alterations with aggressive clinical behavior in breast neoplasia such as recurrence, metastasis, tumor grade, and DNA aneuploidy. They would seem, at least superficially, to suggest that ERBB-2 facilitates pre- invasive, but not invasive, stages of tumor growth.

In the study of Allred and colleagues (1992) though, presence of ERBB-2 overexpression in the invasive tumor component correlated perfectly with overexpression in co-existing DCIS. This finding is directly analogous to previously cited molecular-level experiments. Interestingly though, the "high-grade" DCIS variants (most of which overexpressed ERBB-2) were significantly less likely to be associated with host invasion than "lower grade" DCIS variants, which typically lacked ERBB-2 overexpression. To the extent that oncogene amplification, genetic instability, and high-grade are correlated, as noted earlier, these data would seem to represent additional data that imply lack of direct causal association between genetic instability and invasive phenotype. In view of these data, moreover, it would seem that the biological significance of ERBB-2 within a given neoplasm would need to be interpreted in the context of other genetic parameters, host factors that impact tumor growth, and the size and growth rate of the neoplasm. The observed association of ERBB-2 with high grade (and DNA aneuploidy) would also be understandable if it were also true that tumors with high levels of intrinsic genetic instability evolve in a manner that differs fundamentally from those with less or differing patterns of genetic instability.

Clonal Heterogeneity

An important theoretical consequence of genetic instability is the potential to generate multiple populations of cells having variable phenotypic traits. The expected consequences of this process may be viewed from at least two mechanistic perspectives which, although divergent, are not necessarily mutually exclusive. Firstly, it may reasonably be hypothesized that selection of genetic variants based on growth advantage would lead to eventual numerical dominance of each tumor by the most aggressive, or "genetically evolved," clonal population. Such models of neoplastic progression minimize the biological significance of productive interactions between populations of neoplastic cells. An alternative model that emphasizes the relevance of cellular interaction, predicts that genetic instability would potentially facilitate tumor progression through synergistic interactions between multiple genetically distinct, functionally-diverse, populations.

Static karyotypic, cytophotometric, or molecular genetic data, of course, can neither prove nor refute either of these alternatives since they cannot measure (or detect) functional cellular interactions. Studies using experimental tumor models, however, have revealed a variety of productive cellular interactions. Furthermore, the literature collectively demonstrates that many, if not most breast carcinomas, harbor at least two genetically distinguishable, co-dominant *populations. Dutrillaux and colleagues (1991) as previously described, reported that about one-third of breast carcinomas contained a mixture of near-diploid and triploid karyotypes. To the extent that single-sample karyotype analyses constitute a limited sampling, they likely represent a conservative estimate of heterogeneity. Cytophotometric or karyotype assays derived from multiple geographically separated areas of neoplastic tissue reveal that a minimum one-half, and possibly as many as all breast carcinomas, contain multiple stemlines (Bonsing et al., 1993; Visscher et al., 1995b). Some authors have even reported the admixture of multiple unrelated clonal populations (Teixeira et al., 1996).

In the absence of assays that detect synergistic, growth-promoting interactions, one must concede that clonal heterogeneity may simply reflect rapid clonal evolution owing to genetic instability (i.e., with some populations in decline, others gain in degree of representation). Some authors have attempted to infer the presence of cooperative growth-promoting relationships between neoplastic subpopulations, therefore, by examining the distribution of stemlines in different areas of primary neoplasm or by comparing stemline heterogeneity in primary tumors with metastases or recurrences. As might be anticipated, the data from such studies lend themselves neither to conclusive nor to straightforward interpretation. Co-dominance of intermingled cells from differing stemlines, for example, would be compatible with growth-promoting cellular interaction (e.g., via paracrine mechanisms). Many tumors do, in fact, contain co-dominant populations within geographically limited areas. Other cases, however, clearly demonstrate geographic isolation of stemlines (Visscher et al., 1995a; Teixeira et al., 1996). Furthermore, the proportions of admixed, versus isolated, genetically co-dominant neoplasms varies between the few published studies. It should perhaps be noted, in view of these data, that the spectrum of potentially synergistic interactions is considerable. It is thus unlikely that all such relationships would be limited by degree of proximity. Furthermore, it is not entirely clear whether geographically isolated clones have necessarily evolved away from, as opposed to towards, interdependence or interaction with other stemlines.

The case for "productive heterogeneity" is strengthened somewhat by studies that compare cytophotometric stemline distribution in primary tumors either to synchronous metastases or to recurrences (Feichter et al., 1989; Beerman et al., 1991; Bonsing et al., 1993; Symmans et al., 1996). Cytophotometric DNA stemline heterogeneity has been observed in the majority of metastatic deposits from clinical breast carcinomas. In most but not all cases, moreover, the primary tumors and metastasis contain DNA stemlines that share similar DNA content (i.e., DNA indices).

Lack of dramatic cellular DNA content shifts between primary and metastatic neoplastic populations implies either considerable homogeneity between the primary and metastatic populations of a tumor or that evolution of metastatic phenotype involves limited genetic progression (i.e., per the model of Dutrillaux et al., 1991). The observed combinations of DNA indices in cytophotometrically defined stemlines, whether in primary or in metastasis, is nonrandom. Typically, one or more DNA aneuploid (triploid-range) stemlines is observed together with a near-diploid stemline as described in karyotype analyses of primary neoplasms. Near-diploid stemlines, in fact, are disproportionately observed in post-therapy recurrences or breast neoplasia, thereby affirming their biological viability (Beerman et al., 1991). This apparent co-evolution of near-diploid with aneuploid stemlines in a large number, possibly a preponderance of breast neoplasms, seems compelling albeit indirect evidence for the significance of stable interactions, however obscure, between clonal populations of neoplastic cells.

Although the cytophotometrists generally fail to discern clear differences in degree of stemline heterogeneity between primary and metastatic lesions, it is usually the case that cytophotometrically homogenous primaries are associated with homogenous metastases. Conversely, cytophotometrically heterogeneous primaries are usually associated with heterogenous metastases. Whether this apparently intrinsic tendency to clonal heterogeneity reflects permissive cellular interactions or merely differential rates of mutation and clonal expansion is not clear. In either case, however, the observation would seem to be yet additional evidence that degree of genetic diversity represents an intrinsic trait that is conserved throughout lengthy periods of a tumor's natural history. This is not unlike traditional phenotypic traits such as histologic subtype or degree of differentiation, which are maintained throughout the natural history of individual breast neoplasms. More recent investigators have also demonstrated concordant proliferative index and ERBB-2 expression between primary neoplasm and metastasis (Goodson et al., 1993).

CONCLUSION

It is presumed, but not proven, that the inability to maintain consistent genomic organization is a driving force behind continuing tumor progression. The current accumulated data suggest that histologic precursor lesions of human breast carcinoma—one or more components of proliferative breast disease—are characterized by DNA pathology in keeping with genetically unstable populations. The degree of genetic aberration at the "hyperplastic stage", though, is poorly defined. Some cases may be limited to widely dispersed cells with chromosome aneuploidy (i.e., without evidence of clonal expansion) compatible with mere presence of genetic instability. Other precursor lesions may, in fact, contain "progression-related" DNA pathology more in keeping with pre-invasive malignancy. Pre-invasive carcinomas, then, most likely evolve directly from "partially transformed" precursor lesions following acquisition of a sufficient (but yet undefined) number of events. Chromosomal rearrangement is clearly an integral component of this process.

In addition to structural karyotypic or molecular-level alterations that impact phenotype, breast neoplasia is characterized by at least three variably quantifiable manifestations of intrinsic genetic instability—clonal cellular DNA content aberration, "rare" highly aneuploid cells, and heterogeneity of chromosome count within karyotypically and cytophotometrically defined "stemlines". In many neoplasms, aneuploid cells derived from genomic endoreduplication have sufficient growth advantage to undergo clonal expansion. This event seems to define the degree of differentiation (or heterogeneity of differentiation) for a neoplasm at the pre-invasive stage. DNA aneuploidy also reflects a greater intrinsic degree of genetic instability by virtue of associations with karyotypic complexity, gene amplification, and *p53* mutation. Other breast tumors seem to be characterized by an inherently limited degree of genetic instability as shown by indefinite maintenance of "low-grade", near-diploid cell populations throughout the disease natural history. At a superficial level, this dichotomy of genetic pathology and progression may account for the clinically apparent division of human breast carcinomas into well differentiated or slow-growing versus poorly differentiated or rapid-growing disease subsets. Breast neoplasia, though, is also characterized by a sizable group of neoplasms characterized by apparently stable clonal stemline heterogeneity, although the clinical correlates of this pattern remain undetermined. In our view, the large number of cases that sustain (in both primary and metastatic lesions) admixed diploid-range with triploid stemlines is compelling evidence that there is functional relevance to this and perhaps other more subtle forms of genetic heterogeneity. Further study of this pattern may illuminate the apparent lack of correlation between "evolved" aneuploid clones and critical events in tumor progression such as invasion and metastasis.

REFERENCES

Aasmundstad, T.A. and Haugen, O.A. (1990). DNA ploidy in intraductal breast carcinomas. Eur. J. Cancer, 26, 956-959.

Aldaz, C. M., Chen, T., Sahin, A., Cunningham, J., and Bondy, M. (1995). Comparative allelotype of in situ and invasive human breast cancer: High frequency of microsatellite instability in lobular breast carcinomas. Cancer Res., 55, 3976-3981.

Allred, D.C., Clark, G.M., Molina, R., Tandon, A.K., Schnitt, S.J., Gilchrist, K.W., Osborne, C.K., Tormey, D.C., and McGuire, W.L. (1992). Overexpression of *HER-2/neu* and its relationship with other prognostic factors change during the progression of in situ t invasive breast cancer. Hum. Pathol., 23, 974-979.

Auer, G.U., Caspersson, T.O., and Wallgren, A.S. (1980). DNA content and survival in mammary carcinoma. Anal. Quant. Cytology, 2, 161-165.

Beerman, H., Bonsing, B.A., van de Vijver, M.J., Hermans, J., Kluin, Ph.M., Caspers, R.J., van de Velde, C.J.H., and Corneliss, C.J. (1991). DNA ploidy of primary breast cancer and local recurrence after breast-conserving therapy. Br. J. Cancer 64, 139-143.

Bell, C., Frost, P., Kerbel, R.S. (1991). Cytogenetic heterogeneity of genetically marked and metastatically competent "dominant" tumor cell clones. Cancer Genet. Cytogenet 54, 153-161.

Bonsing, B.A., Beerman, H., Kuipers-Kijkshoorn, N., Fleuren, G.J., Cornelisse, C.J. (1993). High levels of DNA index heterogeneity in advanced breast carcinomas. Evidence for DNA ploidy differences between lymphatic and hematogenous metastases. Cancer, 71, 382-391.

Crissman, J.D., Visscher, D.W., and Kubus, J. (1990). Image cyto-photometric DNA analysis of atypical hyperplasias and intraductal carcinomas of the breast. Arch. Pathol. Lab. Med. 114, 1249-1253.

Dutrillaux, B., Gerbault-Seureau, M., Remvikos, Y., Zafrani, B., and Prieur, M. (1991). Breast cancer genetic evolution: I. Data from cytogenetics and DNA content. Breast Cancer Res. Treat., 19, 245-255.

Dutrillaux, B., Gerbault-Seureau, M., and Zafrani, B. (1990).

Characterization of chromosomal anomalies in human breast cancer: A comparison of 30 paradiploid cases with few chromosome changes. Cancer Genet. Cytogenet. 49, 203-227.

Eyfjörd, J.E., Thorlacius, S., Steinarsdottir, M., Valgardsdottir, R., Ögmundsdottir, H.M., and Anamthawat-Jonsson, K. (1995). *p53* abnormalities and genomic instability in primary human breast carcinomas. Cancer Res. 5, 646-651.

Fallenius, A.G., Auer, G.U., and Carstensen, J.M. (1988). Prognostic significance of DNA measurements in 409 consecutive breast cancer patients. Cancer, 62, 331-341.

Feichter, G.E., Kaufmann, M., Müller, A., Haag, D., Eckhardt, R., and Goerttler, K. (1989). DNA index and cell cycle analysis of primary breast cancer and synchronous axillary lymph node metastases. Breast Cancer Res. Treat., 13, 17-22.

Giaretti, W. (1994). A model of DNA aneuploidization and evolution in colorectal cancer. Lab Investigation, 71, 904.

Goodson III, W.H., Ljung, B.M., Moore II, D.H., Mayall, B., Waldman, F.M., Chew, K., Benz, C.C., and Smith, H.S. (1993). Tumor labeling indices of primary breast cancers and their regional lymph node metastases. Cancer 71, 3914-3919.

Joensuu, H., Alanen, K., Falkmer, U.G., Klemi, P., Nordling, S. Remvikos, Y., and Toikkanen, S. (1992). Effect of DNA ploidy classification on prognosis in breast cancer. Int. J. Cancer 52, 701-706.

Killeen, J.L., and Namiki, H. (1991). DNA analysis of ductal carcinoma in situ of the breast. A comparison with histologic features. Cancer, 68, 2602-2607.

Lennington, W.J., Jensen, R.A., Dalton, L.W., and Page, D.L. (1994). Ductal carcinoma in situ of the breast. Heterogeneity of individual lesions. Cancer 73, 118-124.

Micale, M.A., Visscher, D.W., Gulino, S.E., and Wolman, S.R. (1994). Chromosomal aneuploidy in proliferative breast disease. Hum. Pathol., 25, 29-35.

Nielsen, K.V., Andersen, J.A., and Blichert-Toft, M. (1987). Chromosome changes of in situ carcinomas in the female breast. Eur. J. Surg. Onc. 13, 225-229.

Nielsen, K.V., Blichert-Toft, M., and Andersen, J. (1989). Chromosome analysis of in situ breast cancer. Acta. Oncologica 28, 919-922.

O'Connell, P.O., Pekkel, V., Fuqua, S., Osborne, C.K., and Allred, D.C. (1994). Molecular genetic studies of early breast cancer evolution. Breast Cancer Res. Treat., 32, 5-12.

Odagiri, E., Kanda, N., Jibiki, K., Demura, R., Aikawa, E., and Demura, H. (1994). Reduction of telomeric length and *c-erbB-2* gene amplification in human breast cancer, fibroadenoma, and gynecomastia. Cancer 73, 2978-2984.

Remvikos, Y., Gerbault-Seureau, M., Magdelénat, H., Prieur, M., and Dutrillaux, B. (1992). Proliferative activity of breast cancers increases in the course of genetic evolution as defined by cytogenetic analysis. Breast Cancer Res. Treat. 23, 43-49.

Remvikos, Y., Gerbault-Seurreau, M., Vielh, P., Zafrani, B., Magdelénat, H., and Dutrillaux, B. (1988). Relevance of DNA ploidy as a measure of genetic deviation: A comparison of flow cytometry and cytogenetics in 25 cases of human breast cancer. Cytometry, 9, 612-618.

Sennerstam, R. and Auer, G. (1993). Polyploidization by means of endoduplication in a human breast cancer cell line. Analyt. Quant. Cytol. 15, 303-310.

Schmitt, F.C., Leal, C., Lopes, C. (1995). p53 protein expression and nuclear DNA content in breast intraductal proliferations. J. Pathol., 176, 233-241.

Shay, J.W., Wright, W.E., and Werbin, H. (1993). Toward a molecular understanding of human breast cancer: A hypothesis. Breast Cancer Res. Treat. 25, 83-94.

Shaw, J.A., Walsh, T., Chappell, S.A., Carey, N., Johnson, K., and Walker, R.A. (1996). Microsatellite instability in early sporadic breast cancer. Br. J. Cancer 73, 1393-1397.

Siitonen, S.M., Kallioniemi, O., Helin, H.J., and Isola, J.J. (1993). Prognostic value of cells with more than 5c DNA content in node-negative breast cancer as determined by image cytometry from tissue sections. Hum. Pathol. 24, 1348-1353.

Silverberg, S.G. and Chitale, A.R. (1973). Assessment of significance of proportions of intraductal and infiltrating tumor growth in ductal carcinoma of the breast. Cancer, 32, 830-837.

Sneige, N., Sahin, A., Dinh, M., and El-Naggar, A. (1996). Interphase cytogenetics in mammographically detected breast lesions. Hum. Pathol., 27, 330-335.

Symmans, W.F., Liu, J., Knowles, D.M., and Inghirami, G. (1995). Breast Cancer Heterogeneity: Evaluation of clonality in primary and metastatic lesions. Hum. Pathol., 26:210-216.

Teixeira, M.R., Pandis, N., Bardi, G., Andersen, J.A., and Heim, S. (1996). Karyotypic comparisons of multiple tumorous and macroscopically normal surrounding tissue samples from patients with breast cancer. Cancer Res. 56, 855-859.

Younes, M., Lebovitz, R.M., Bommer, K.E., Cagle, P.T., Morton, D., Khan, S., and Laucirica, R. (1995). p53 accumulation in benign breast biopsy specimens. Hum. Pathol. 26, 155-158.

Visscher, D.W., Sarkar, F.H., and Crissman, J.D. (1991). Correlation of DNA ploidy with c-erbB-2 expression in preinvasive and invasive breast tumors. Analyt. Quant. Cytol. Histol. 13, 418-424.

Visscher, D.A., Shaheen, C., Drozdowicz, S., and Crissman, J.D. (1993). Image cytophotometric DNA histogram heterogeneity in adenocarcinoma of the breast. Analyt. Quant. Cytol. Histol., 15, 206-212.

Visscher, D., Dolan, P., Ottosen, S., and Crissman, J. (1995a). Geographic distribution of co-dominant DNA stemlines in breast carcinoma. Cytometry 21, 14-17.

Visscher, D.W., Sochacki, P., Ottosen, S., Wykes, S., and Crissman, J.D. (1995b). Assessment and significance of diploid-range epithelial populations in DNA aneuploid breast carcinomas using multi- parametric flow cytometry. Analyt. Cell. Pathol. 8, 267-277.

Visscher, D.W., Wallis, T., Ritchie, C.A. (1995c). Detection of chromosome aneuploidy in breast lesions with fluorescence in situ hybridization: Comparison of whole nuclei to thin tissue sections and correlation with flow cytometric DNA analysis. Cytometry, 21, 95-100.

Visscher, D.W., Wallis, T., and Crissman, J.D. (1996). Evaluation of chromosome aneuploidy in tissue sections of pre-invasive breast carcinomas using interphase cytogenetics. Cancer 77, 315-320.

Volpe, J.P.G. (1988). Genetic instability of cancer. Why a metastatic tumor is unstable and a benign tumor is stable. Cancer Genet. Cytogenet 34, 125-134.

THE HISTOPATHOLOGY OF TRANSGENES AND KNOCKOUTS IN THE MAMMARY GLAND

Robert D. Cardiff and Robert J. Munn

Advances in Oncobiology
Volume 2, pages 177-202.

ISBN: 0-7623-0388-3

INTRODUCTION

Transgenic biology is now in its second decade. The development of transgenic mice and of "knock-out" mice has created a dramatic revolution in experimental biology. Coupled with the promise of the Human Genome Project, the mouse has emerged as the major model for biological research. The laboratory mouse has been referred to as the **E. coli** of modern biology and the surrogate for human biology (Paigen, 1995). The mouse is relatively inexpensive to maintain, is relatively easy to manipulate and its genome is syntetic with the human. The mouse has become the mammal of choice for the analysis of interaction of specific genes with the whole animal. Various aspects of transgenic mouse mammary biology have been reviewed by ourselves and others (Cardiff and Muller, 1993; Dickson et al., 1991; Muller, 1991; Cardiff et al., 1992; Cardiff, 1988; Cardiff and Aguilar-Cordova, 1988; Strange and Cardiff, 1989; Cardiff, 1995; Cardiff and Munn, 1995; Webster and Muller, 1994; Groner 1992; Callahan 1991).

The investigator wishing to understand the pathophysiology associated with genetic manipulation in the transgenic or knock-out mouse needs to understand the normal variations that occur in the animal and the target organ. Genetic manipulation can result in unexpected alterations in growth and development that defy traditional pathologic interpretation. Variations in a normal can be misinterpreted as phenotypic changes associated with a transgene. Nonneoplastic inflammatory infiltrates have been interpreted as malignancies. Animals with knocked out or altered immune systems are particularly susceptible to opportunistic infections that can be misinterpreted as a "phenotype" by overeager investigators.

This chapter is designed to provide the beginning investigator with an overview of mammary histopathology. It will comment on the key findings in the normal growth and development of the mammary gland to establish baseline parameters for comparison with potential effects of transgenes or knock out of suppressor and other genes. Some of the more common experiences in alteration of growth and development of the mammary gland in genetically altered mice will be described. Finally, some of the dysplastic changes and progression to neoplasia will be described.

The current chapter primarily relates our experience with the mammary gland in transgenic and/or knock-out mice. It is intended as a useful guide for the investigator who is either targeting the mammary gland or has unexpected alterations in the mammary gland. Our experience is based on the examination of gross and/or microscopic samples from over 5000 genetically manipulated animals and controls at the Transgenic Histopathology Laboratory of the University of California, Davis, School of Medicine (http://www-mp.ucdavis.edu/tgmice/firststop.html//). Our archives contain mammary gland samples from over 2000 mice with over 100 transgenes and/or knock-outs including combinations sent to us by 137 investigators from 37 different laboratories. This collection permits a unique comparative study of the biology of the transgenic mammary glands from a number of different sources. The chapter will review normal mammary gland, hyperplasias, dysplasias, and tumors. We will also comment on neoplastic progression and compare the histopathology of mouse and human mammary glands.

HISTOPATHOLOGY OF THE WILD-TYPE MAMMARY GLAND

Growth and Development

The mammary glands form from a thickening of the epidermis called the milk line. The milk lines develop in all mammals and extend from the axilla to the groin. Later in development, milk buds form as discrete epidermal buds in most mammals. In the mouse, 10 nipples and 10 glands form, five on each side. The number 1 fat pad and gland is in the neck region adjacent to the salivary glands. Tumors of the salivary glands are frequently mistaken for mammary tumors and visa versa. The number 2 and 3 mammary glands are on the chest wall and overlap but are divided by the pectoralis muscle of the chest. The number 4 gland is on the abdomen and is the easiest to identify and dissect (Figure 1A). The number 5 gland is in the groin region.

Since the number 4 gland and fat pad is the most accessible, it is most frequently used for experimental transplantation (Figure 1B). The favored transplantation system requires that the mammary gland tissue is cleared from the fat pad within the first three weeks of the female's life. The mammary fat pads form as local areas of adipose tissue between the subcutaneous tissue and the body wall. They are ensheathed between two fascial planes. The primitive mammary gland of the mouse originates at the milk bud, forming the primitive nipple that extends into the mammary fat pad as a series of branched ducts. The growth rate is slow in the first three postpartum weeks, not reaching the lymph node of the number 4 fat pad (Figure 1B). By identifying the mammary vein and the lymph node (Figure 1B), the mammary fat pad between the lymph node and the nipple can be removed. Since this region in the three-week-old female contains the mammary gland, the entire area is removed and the mammary fat pad is free of mammary epithelium. The gland-free (or cleared) fat pad is then used for transplantation of other tissues.

After three weeks, the female gland is subject to the growth stimulus of puberty and grows at a more rapid rate to fill the mammary fat pad between nine and 12 weeks. The growth is almost exclusively at the terminal end bud that has a cap of undifferentiated "stem" cells (Daniel and Robinson, 1992; Humphreys et al., 1996).

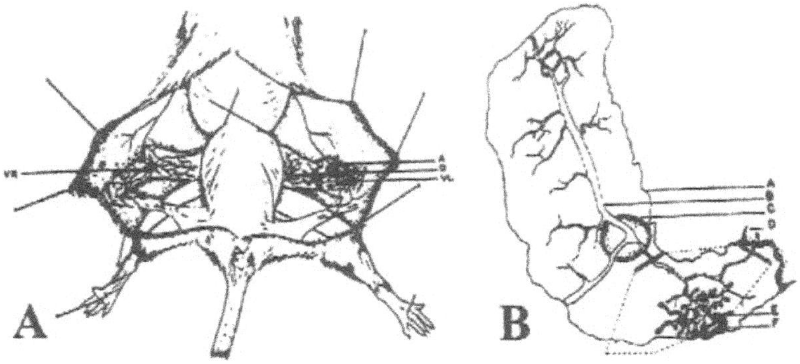

Figure 1. Drawings of the fourth mammary fat pad of the mouse. (**A**) The skin has been reflected and pinned back to expose the number four fat pads. (**B**) The right number four fat pad dissected free to show the development in a three week old female. A: the edge of the fat pad. B: the location of the mammary artery. C: the mammary lymph node. D: the portion of the fat pad to be removed to insure a gland-free fat pad for transplantation. E: the growing mammary bud. F: the nipple. VR: veins. VL: the location of the arteries. (Compliments of Larry J.T. Young and the CRGL Staff, University of California, Berkeley).

The mammary gland of the virginal female forms a series of ducts with a limited number of branches (Figure 2A). Furthermore, the ducts in wild-type mice without the Mouse Mammary Tumor Virus (MMTV) do not form side buds or tertiary branches until stimulated by hormones during pregnancy. With pregnancy, the gland begins forming side buds that either branch as tertiary ducts or form alveoli. The process results in the progressive acquisition of alveoli culminating in the filling of the entire fat pad with alveolar structures (Figure 2B). After the weaning of the pups, the maternal gland begins to regress at different rates. The early phases of regression are characterized by fragmentation of nuclei (programmed cell death or apoptosis) and infiltrates of macrophages (Schedin et al., 1996; Li et al., 1997. The fully regressed mammary gland retains many of the branches, resulting in short, pointed terminal ducts with a spiculated appearance resembling thorns that distinguish it from a nulliparous mammary gland. The areas around the terminal ducts often have brown and yellow pigments that stain for iron.

The mammary gland of the retired breeder female will often regress to a point that it is difficult to distinguish from the straight mammary tree of the virgin animal but, most of the time, it retains its peculiar spiculated appearance.

Hyperplasia

Numerous studies of the pathology of the wild-type mammary gland have been published. The classical comparative study came from Dr. Thelma Dunn, whose classification is still used today (Dunn, 1959). The greatest pathologic changes oc-

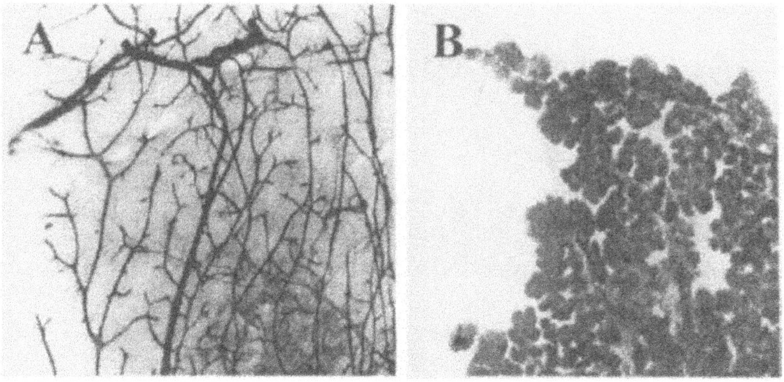

Figure 2. Images of whole mount preparations of portions of the mammary tree from FVB female mice. (**A**) An image from a virgin mature animal. Note the straight ducts with limited branching and lack of alveolar development. (**B**) An image from a late pregnant female. Note that the extensive lobulo-alveolar development fills the fat pad. (Images from samples submitted by Drs. Webster and Muller, McMaster University to the Archives of the Transgenic Histopathology Laboratory, University of California, Davis.)(Magnification = 5X).

cur in the mammary gland of MMTV-infected mice and mice treated with chemical carcinogens. Morphometric studies by Squartini and others have suggested that MMTV-infected, virginal mammary glands have more extensive lobulo-alveolar development than mammary glands from noninfected animals (Squartini et al., 1983; Pingitore and Squartini, 1982; Squartini et al., 1981). However, the growth rates appear to be similar. Wild feral mice were found that did not carry the mouse mammary tumor virus in their genome or as an infectious agent (Cohen et al., 1982; Cohen and Varmus, 1979; Gallahan et al., 1986). These mice rarely develop mammary tumors (Faulkin, 1984; Gardner, 1985). Furthermore, in our experience , they have limited acinar development (Faulkin, 1984).

Most biologists focus on focal lesions in the mammary gland because they clearly stand out from the background and are potentially premalignant (Cardiff, 1984; Morris and Cardiff, 1987) (Figure 3A). In addition, focal lesions can be identified in situ, isolated, and transplanted to test for biological potential. The donor transplant is generally placed either into the subcutaneous tissue or into the mammary fat pad that has been cleared of host mammary gland. Transplants of normal gland will not grow in the subcutaneous tissue but can be serially passed into the gland-cleared, mammary fat pad for five to 10 transplants (Daniel et al., 1966). Foci of malignant cells will develop into tumors in either subcutaneous or mammary fat tissues (Cardiff, 1984). Foci containing neoplastic but premalignant tissues (protoneoplasia) will not grow in subcutaneous tissue but can be transplanted indefinitely in the gland-cleared fat pad. Furthermore, tumor will emerge from protoneoplastic transplants but not from transplants of normal ducts. This trans-

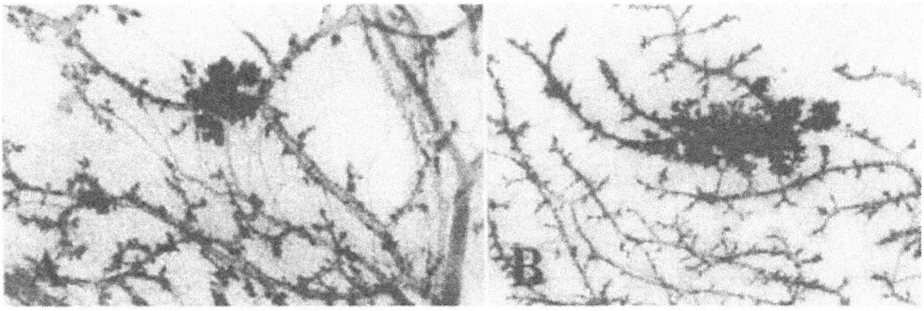

Figure 3. Images of whole mount preparations stained with hematoxylin showing typical hyperplastic alveolar nodules (HAN). Image **A** is from a Balb/cf.C3H female showing a typical mouse mammary tumor virus (MTV)-induced HAN (Images from samples submitted by Larry J.T. Young to the Archives of the Transgenic Histopathology Laboratory, University of California, Davis.). Image **B** shows a HAN developed in a transgenic mouse bearing the activated *src* transgene behind the MTV LTR promoter. (Images from samples submitted by Drs. Webster and Muller, McMaster University to the Archives of the Transgenic Histopathology Laboratory, University of California, Davis.) (Magnification = 7X).

plant system has created the operational definition of benign, premalignant, and malignant mammary tissue (Cardiff, 1984).

The wild-type mammary gland can contain a variety of nonmalignant focal abnormalities. The inflammatory nodule and the squamous nodule are two of the most frequent lesions. The inflammatory nodule is characterized by focal infiltrates of plasma cells and lymphocytes around residual acini. The squamous nodule has acini lined by squamous rather than columnar epithelium. These lesions can be found in noninfected and MMTV-infected mice. Both are focal and, when transplanted, rarely progress to tumors.

Two classical focal lesions appear in the mammary glands of MMTV-infected female mice: the Hyperplastic Alveolar Nodule (HAN) (Figure 3) and the hormone-dependent tumor or plaque. The HAN is found in most MMTV-infected females. It appears as a small nodule, sometimes outlined by yellowish-white pigment, that stands out from the background as small, whitish foci following withdrawal of experimental or pregnancy-associated hormone stimulation. Histologically, the HAN appears as a small focus of crowded acinar units without significant cytological dysplasia. It may or may not be surrounded by inflammation. Transplantation of the HAN results in immortal hyperplastic outgrowths with a high risk of malignant transformation.

The plaque is a focal mass that appears during pregnancy or experimental hormonal stimulation but regresses upon withdrawal of the stimuli. It will reappear with repeated stimulation and will eventually become a hormone-independent tumor (Foulds, 1956). Histologically, the plaque appears as a series of radiating, cell-filled ducts and surrounded by a dense connective tissue (Morris et al., 1990).

The term "dysplasia", used to describe transgenic lesions, is rarely applied to wild-type, nontransgenic mouse mammary lesions. This is primarily because the biological potential of the focal lesions has been characterized by transplantation (Cardiff, 1984; Morris and Cardiff, 1987). As a result, there has been little need to provide additional descriptive terms to describe the structure or predicted biological potential of a given lesion. Chemically induced dysplasias tend to be papillary, ductal lesions with a tendency toward squamous metaplasia (Medina, 1982).

Tumors

Dr. Dunn's description of the types of tumors found in the wild-type mouse is still accurate today and we cannot improve on her observations (Dunn, 1959). It is very important to note that around 90% of all MMTV-induced tumors are either microalveolar (Type A) , ductal (Type B), or a mixture (Type C) (Figure 4). It is important to note that these types of tumors are rarely seen in the transgenic mouse without virus infection. The keratoacanthoma or adenosquamous carcinoma is another type of tumor that is more commonly found in the animal exposed to chemical carcinogens. It is also the most common sporadic tumor found (in our experience) in the wild-type mammary gland in animals that have not been exposed to virus.

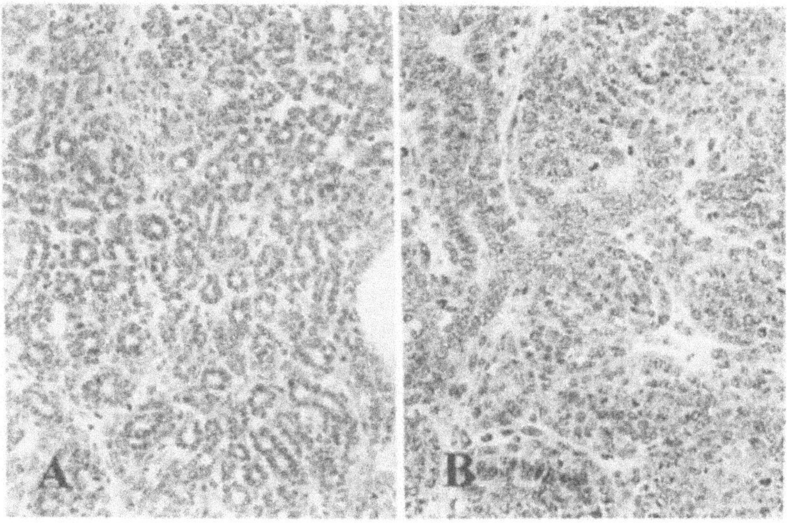

Figure 4. Images of MTV-associated mammary tumors showing patterns typical of those classified by Dr. Thelma Dunn as Type A mammary tumors (**A**) with the microacinar pattern and Type B mammary tumors (**B**) with the solid cords and nests (Images the Archives of the Transgenic Histopathology Laboratory, University of California, Davis.) (Magnification = 140X).

HISTOPATHOLOGY OF THE TRANSGENIC MAMMARY GLAND

Growth and Development

In general, the transgenes driven by the MMTV LTR promoter systems express relatively low levels of the transgenes. Most experiments with this and other promoters do not reveal significant developmental abnormalities. Our own review of the extensive collection of mammary whole mounts collected randomly from some of the early studies in Dr. Philip Leder's laboratory did not reveal significant disturbances of growth and development in transgenic mammary glands (Cardiff et al., 1992). However, subsequent analysis has shown a number of developmental abnormalities in mice with activated oncogenes (Robinson et al. 1996). For example, the activated *src* transgene results in a lactational defect (Webster et al., 1995). Increasing numbers of studies with knock-out mice have revealed early but very subtle developmental anomalies. For example, virgin *src* knock-out mice have dilated ducts with limited branching (Guy et al., 1994). *Cyclin D1* knock-outs are reported to have similar abnormalities (Fantl et al., 1995), while *p53* knock-outs have abnormalities of lobuloalveolar development (Li et al., 1994).

Abnormal mammary development now has been described in many transgenic strains (Muller et al., 1988; Muller et al., 1990; Ornitz et al., 1991; Matsui et al., 1990; Halter et al., 1992; Wang et al., 1994, Lin et al., 1992; Jhappan et al., 1992; Pierce et al., 1993). Targeted overexpression of transgenes and very powerful transgenes can result in developmental anomalies in the virgin mammary gland. For example, the Ornitz GAL-4 promoter system using the *int*-2 transgene resulted in dilated, short ducts with abnormal mammary buds whereas many of the *int*-2 transgenic animals promoted by the MMTV LTR had normal virginal mammary trees (Ornitz et al., 1991). The polyoma middle T (*PyV-MT*) transgene promoted by either the MMTV LTR or C(3) resulted in the very rapid development of focal mammary lesions in virgin mice (Guy et al., 1992a; Tehranian et al., 1996). In addition, we have observed altered development or inappropriate development of the mammary lobules in virgin mice bearing several different transgenes such as *neu* or *TGF*α (Muller et al., 1996). The TGFα results in cystic ducts and extensive dilated lobules. Some of the *neu* transgenes result in shortened ducts with dilated, cystic end buds.

The newcomer to this field must be alert to the fact that many transgenic strains have defects in mammary gland growth and development that lead to defective lactation (Sakai et al., 1994; Hennighausen et al., 1994; Pierce et al., 1993; Jhappan et al., 1992). This deficiency is sometimes acknowledged by a somewhat cryptic comment that the females did not nurse their young. However, most of these defects have gone unnoticed or have not been described in print. When the mammary gland is examined as a whole mount, it is often quite apparent that the transgene has resulted in incomplete development of the mammary tree. The mouse with activated *src* is one such example of defective lactation in which the acini are irregular and cystic and do not produce lipids (Webster et al., 1995).

At the other end of the process, some transgenic animals have a persistent hyperplasia following pregnancy (Wang et al., 1994). One of the more subtle examples we have seen is in the cyclin D1 animals, which had what would appear to be fully lactating mammary glands six to nine months following weaning of the pups. It appears that the mammary epithelium is unable to regress.

The point to be emphasized is that the genetically altered animals may have major developmental defects in the mammary gland that do not result in a tumor phenotype or a grossly obvious lesion. However, if these more subtle defects are to be discovered and understood, the mammary gland must be systematically studied using both whole mounts and microscopic slides. The development of the mammary gland should be followed through the first 12 to 16 weeks of pubertal development, through the pregnancy–regression cycle and in retired breeders. Although some genetic alterations will not result in cancer, they can reveal a great deal about the growth and development of the mammary gland.

Hyperplasia

A simple definition of hyperplasia is an increase in the number of cells. In the context of the mammary gland, lactation may be considered a form of physiological hyperplasia. In the context of the transgenic or knock-out animal, we have used the diagnosis hyperplasia to indicate increased numbers of cytologically normal cells without architectural or nuclear abnormalities.

Hyperplasias in transgenic mammary glands can be either focal or diffuse (Table 1). Generally, focal hyperplasias are of greater interest since they suggest the possibility of neoplastic transformation (Cardiff, 1984). The most dramatic development

Table 1. Classification of Transgenic Hyperplasia and Dysplasia

Tumor Type	*Prototype Transgenes*
• **Diffuse:**	
A. Ductal	*int*-2
B. Tubular	TGFα
C. Residual Lactation (nonregression)	cyclin D1
D. Acinar	*int*-1 (*wnt*-1)
A. Lobular	*int*-3 (*notch*)
• **Focal:**	
Hyperplastic Alveolar Nodule	*src*
Squamous Nodule	cyclin D1
Inflammatory Nodule	Casein Kinase II
Sclerosing adenosis	*src*
Cystic hyperplasia	TGFα
Periductal fibrosis	Stromelysin
Papillary atypia	cyclin D1

of diffuse hyperplasias of the mammary gland has been observed in the *c-neu* and the *int*-2 transgenic animals (Muller et al., 1988; Muller et al., 1990; Ornitz et al., 1991). *Int*-3 or *notch* and *int*-1 also lead to characteristic diffuse hyperplasias (Jhappan et al., 1992; Tsukamoto et al., 1988). It would appear that hormone-induced hyperplasia leads to diffuse hyperplasia in most, but not all, transgenic females (Cardiff and Muller, 1993). When the hormone stimulus is withdrawn, the mammary gland may regress to a normal postpartum state. However, the diffuse hyperplasia in the GAL-4–promoted *int-2* animal persists for the rest of the animal's life (Ornitz et al., 1991).

There has been a disappointing scarcity of residual hyperplastic foci in transgenic mice that might represent the characteristic preneoplastic hyperplastic alveolar nodule (HAN) found in the spontaneous mouse mammary tumor virus-induced lesions (Cardiff and Muller, 1993). However, studies are now being performed in a more systematic manner and are including retired breeders. Increasing numbers of strains have identifiable focal lesions that closely resemble the classic HAN. For example, the animal with the activated *src* has many HAN-type lesions (Webster et al., 1995) (Figure 3B). The *neu* transgene results in focal epithelial hyperplasias that have been described in several founders and their progeny (Guy et al., 1992b; Bouchard, et al., 1989) (Table 1). Furthermore, animals with the *src* transgene form lesions resembling sclerosing adenosis (Webster et al., 1995). Many transgenic strains in our archives form inflammatory and squamous nodules. The widespread incidence of these types of lesion may be the result of other injuries to the mammary gland such as obstruction of the duct rather than the direct result of the transgene expression. The TGFα mammary glands exhibit adenosis and cystic hyperplasia (Halter et al., 1992). *Myc* results in a variety of focal hyperplastic and dysplastic lesions in the regressing gland (Tulchin et al., 1995). A complete catalog of focal mammary hyperplasias has not yet been produced. Furthermore, the biological potential of such lesions has not been tested by transplantation.

Dysplasia

The terms dysplasia or atypia have rarely been used in classical mouse mammary tumor biology because each lesion could be readily characterized by transplantation. The lesions of the human mammary gland have been characterized by a large number of terms such as atypia or dysplasia. These terms are used in reference to lesions with cytological abnormalities. These abnormalities typically involve changes in the shape and size of the nuclei (pleomorphism), increased staining density of nuclei (hyperchromasia), and increased size or number of nucleoli. When lesions showing dysplastic cytological changes appear in the mammary glands of genetically altered mice that cannot be transplanted, we have adopted the human nomenclature, preferring to use dysplasia. This designation could be quite misleading and will be tentative until formal experimental proof is provided.

Although the classical preneoplastic HANs have not been consistently observed in the transgenic mammary tree, focal atypical lesions have been. These lesions take many forms. They are best identified in the whole mounts as focal lesions that stand out from the general background or in histological sections as microscopic foci containing atypical cells. The feature unique to the transgenic mouse is the appearance of these lesions in the virgin animal (Cardiff, 1995).

Focal atypias have been described in transgenic animals bearing the *cyclin-D1* gene (Wang et al., 1994). These tend to be in the form of atypical papillary regions in the major ducts. We have also seen focal atypias in transgenics with activated *neu*, *src*, *myc*, *PyV-MT* and *TGFα*. The atypical lesions observed in the TGFα transgenics are typically tubular arrays while the cyclin-D1 atypias are frequently squamous (Wang et al., 1994; Halter et al., 1992; Muller et al., 1996); the *myc*-related lesions are acinar and frequently associated with extensive fibrosis (Tulchin et al., 1995).

Serial transplantation experiments with these lesions have not been done and will be required to assess their biological potential. Our own experiments have demonstrated that the mammary hyperplasias can be transplanted into the mammary gland but not into the subcutaneous fat (Ornitz et al., 1992). It is conceivable that some of these focal mammary atypias have already undergone malignant transformation but this requires experimental proof.

The focal nature of the atypias is important because it implies that the transgene itself is not sufficient for neoplastic or malignant transformation. Other molecular events are required for complete pre- or protoneoplastic transformation (Cardiff, 1984; Cardiff et al., 1996; Cardiff, 1995). The focal nature of the lesions suggests that the secondary molecular events result in a clonal proliferation. While these are intriguing and plausible hypotheses, we must again point out that they have not been tested by traditional and rigorous transplantation into the gland-cleared fat pad. Thus, the designation of dysplasia is morphological, based on the severity of nuclear hyperchromatism and pleomorphism.

It is important to recognize that the morphology of many of the transgene-induced hyperplasias and dysplasias is characteristic of the transgene. The nulliparous GAL-4/*int*-2 transgenic is the prototype, developing bulbous, distorted dysfunctional mammary trees (Ornitz et al., 1992). The nulliparous polyoma virus middle T (*PyV-MT*) animals also have very distorted, branching mammary trees with dysplastic lesions as early as five weeks of age (Guy et al., 1992a). Stromelysin expression is associated with extensive periductal fibrosis (Sympson et al., 1994). TGFα is associated with cystic hyperplasia (Halter et al., 1992). *Cyclin D1* is associated with persistent lactation and squamous nodules (Wang et al., 1994). *Src* results in sclerosing adenosis (Webster et al., 1995). The same type of lesion may occur in a variety of types of transgenic animals. The most common lesion is the squamous nodule that has been observed in mammary glands of many different transgenic animals. However, each transgene seems to have a predominance of a single type of lesion, suggesting that the transgene influences the morphogenesis of the lesion.

Primary Mammary Tumors

Transgenic mouse mammary tumors are quite remarkable and quite unique (Munn et al., 1995). Most tumors do not resemble the spontaneous mammary tumors described by Dunn (see above) (Figure 4) (Dunn, 1959). Only animals that express the transgenes most commonly found in MMTV-infected mice develop tumors that can be classified by the Dunn nomenclature. Therefore, mice with the *int*-1 transgene generally have the alveolar type A tumors that are found in MMTV-infected mice (Cardiff and Munn, 1995; Lin et al., 1992; Kwan et al. 1992). Animals with the *int*-2 transgene form hyperplasias that resemble in detail the hormone-dependent plaques found in mouse strains GR/A and DD (Cardiff and Munn, 1995). Interestingly, the GR/a hormone-dependent tumors frequently have insertion activation of *int*-2 (Morris et al., 1990). They also form type B and papillary tumors (Cardiff and Munn, 1995).

All other transgenic and knock-out animals have tumors that do not resemble those described by Dunn. In fact, all other genetically manipulated animals have relatively unique tumor phenotypes. This suggested to us that the genotype determined the mammary tumor phenotype (Cardiff and Munn, 1995; Cardiff et al., 1991). In a study of over 700 tumors, the *myc* transgene was found to result in adenocarcinomas with characteristic large dark-staining cells. The *ras* transgene was associated with papillary transitional carcinomas with small cells. The *neu* transgene was associated with nodular tumors having intermediate cells. Subsequent studies have confirmed and reinforced this impression.

Additional tumor phenotypes have now been found. TGFα tumors tend to be tubular adenocarcinomas with an inflammatory stroma (Muller et al., 1996; Cardiff et al., 1995). Polyoma virus middle T antigen tumors tend to be papillary and are associated with extensive fibrosis (Guy et al., 1992a). Other examples of transgene-specific tumor phenotypes may be given and are summarized elsewhere and in Table 2 (Cardiff, 1995).

Table 2. Classification of Transgenic Tumors

Tumor Type	*Prototype Transgenes*
• Papillary transitional cell carcinoma (small cell)	• *ras*
• Adenocarcinoma	
a. papillary	• PyV-MT
b. tubular	• TGFα
c. acinar (large cell)	• *myc*
d. alveolar	• *int-1*
e. large glands	• *met-1*
• Nodular carcinoma (intermediate cell)	• *neu*
• Scirrhous carcinoma	• *src*
• Adenosquamous	• cyclin D1
• Undifferentiated/NOS	

The tumor phenotypes found in *myc, ras,* and *neu* have, in our experience, the most characteristic phenotypes (Figure 5). They have not been found in mice with other transgenes. We have now observed these three tumor phenotypes associated with the same transgene in animals from different laboratories and different constructs. Therefore, the tumor phenotype can be consistently associated with these transgenes. *Myc* is the most dominant transgene. Any combination of *myc* with another transgene results in tumors with large blue staining cells. *Myc* and *neu* , or *myc* and *ras* result in tumors with the *myc* phenotype (Cardiff et al., 1991). This has proven true with an increasing number of transgene and knock-out combinations with *myc*. They all produce *myc*-type tumors.

We have had extensive experience with *neu*-related animals. The solid nodular intermediate cell phenotype is consistent in animals from different founder strains and different promoter systems. The *neu* phenotype appears to be the dominant phenotype in most bigenic crosses, excluding the *neu* crossed with *myc*. Furthermore, various types of mutationally activated or inactivated *neu* result in tumors with the *neu* phenotype (Guy et al., 1992b; Siegel et al., 1994; Deckard-Janatpour et al., 1997; Guy et al., 1996; Muller et al., 1996; Krane and Leder, 1996). This is of particular interest since the *neu* transgene is the murine homologue to *c-erb*-B2 or *Her*-2 associated with the comedocarcinomas in humans (Barnes et al., 1992; All-

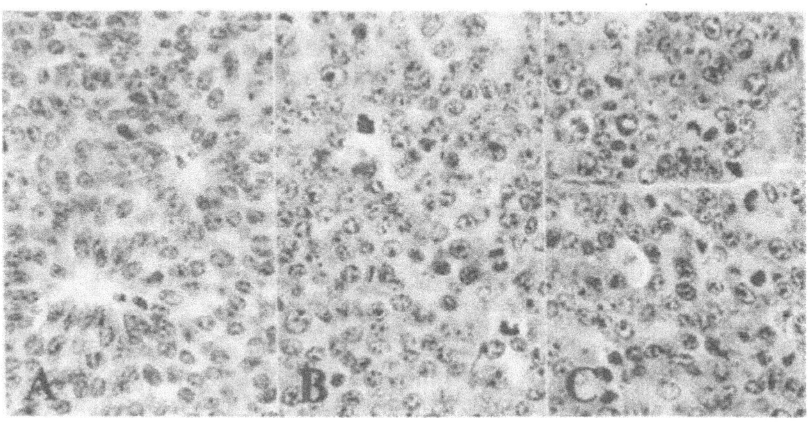

Figure 5. High magnification images showing the typical cytological features of mammary tumors induced by (**A**) the *ras* transgene, (**B**) the *neu* transgene, and (**C**) the *myc* transgene, (Images the Archives of the Transgenic Histopathology Laboratory, University of California, Davis.). The typical *ras* tumor has cells with smaller, more uniform nuclei and red cytoplasm that cluster around blood vessels in a papillary pattern. The *neu* tumors are nodular with more pleomorphic, larger nuclei with a brisk mitotic rate and pale pink cytoplasm. The *myc* tumors have the largest cells with large pleomorphic nuclei with clumped chromatin, prominent nucleoli, and dark basophilic cytoplasm. They frequently form irregular glands (Magnification = 400X).

red et al., 1992; Lodato et al., 1990; Bartkova, et al. 1990). As will be discussed later, the mouse and human lesions are almost identical.

As is the case with the transgenic dysplasias, most of the transgenic tumors do not have as distinctive characteristics as the *myc*, *ras*, and *neu* tumors. Many different transgenes result in adenocarcinomas that are difficult to distinguish from each other. On the other hand, the pattern of tumor is very consistent within the transgenic strain. For example, almost all of the *tpr/met* tumors are adenocarcinomas, *int*-2 and PyV-MT tumors are typically papillary and *src* tumors are typically scirrhous carcinomas. We have observed some subtle cytological differences in the tumors. However, we cannot be confident, given the differences in the fixation and the fixation schedules, that these differences are biologically significant. More detailed observation with rigorous morphometric techniques under controlled conditions will be required to sort these tumors.

Suppressor genes have been assessed primarily using knock-out technology (Donehower et al., 1992; Jacks et al., 1994). Increasing numbers of backcrosses with transgenic animals are now being assessed. We have a large number of *p53* null crossed with a variety of oncogene-bearing transgenic mice. The *p53* knock-outs develop leukemia and sarcomas more rapidly than mammary tumors. Therefore, our experience with mammary tumors from *p53* null animals is limited. However, the mammary tumors in *p53* null are generally adenocarcinomas with relatively large, pink staining cells with large nuclei and a delicate chromatin. They stand out as a different type of cytological phenotype.

The knock-out animals crossed with transgenic mice develop tumors that are identical to the transgene. Null crosses with *myc* result in large cell tumors (Krane and Leder, 1996). While crosses between *p53* null and *neu*-bearing mice result in tumors with the *neu* phenotype. This pattern is true with other null animals such as *p21* and *grb*-2.

However, the investigator must be aware of unique biological patterns sometimes developing with combinations of transgenes and combinations of transgenes and suppressor null mice. Most experiments document the acceleration of tumorigenesis with combinations of transgenes or of transgenes and suppressor gene knock-outs (Jacks et al., 1994). However, some combinations of transgenes or transgenic and null animals lead to tumor suppression. For example, the combination of *neu* and *ras* bigenics resulted in a suppression of the tumor incidence (Cardiff et al., 1991). The combination of *src* null and *PyV-MT* resulted in an almost complete ablation of the tumor phenotype (Guy et al., 1994).

Nonmammary Tumors

Another word of caution should be inserted here. Not all lumps in the mammary fat pad are mammary tumors. For example, tumors in the neck region sometimes originate in the salivary gland. The salivary tumors are typically acinar or squamous carcinomas. The acinar tumors have characteristic small acinar structures containing five to eight cells with basal nuclei and brighter red apical cytoplasm surrounding tiny lumina. In well-preserved, well-stained sections, the apical cytoplasm may have a granular appearance. These can usually be distinguished from tumors originating in the mammary gland.

Other tumors appearing in the mammary fat pad are more obvious. For example, many transgenic animals will have high rates of leukemia. The leukemia results in enlarged lymph nodes or actual infiltrates of the mammary gland. For example, whole mounts of stromelysin-bearing mice with leukemia have multiple nodules along the mammary tree. Histological examination will demonstrate that the nodules are periductal and composed of malignant lymphocytes. Spindle cell sarcomas and hemangiosarcomas can frequently be observed in the mammary glands of *p53* null mice. Samples of ovaries and testes have been submitted as examples of tumors in the number five fat pads. Occasionally, abscesses and granulomas have been submitted as mammary neoplasms. Other examples can be cited that prove that not all lumps in the fat pad are tumors of the mammary epithelium.

Metastatic Tumors

There has been considerable debate about the frequency of metastases from "spontaneous" mammary tumors in wild-type mice. Many investigators have regarded the tumors induced by virus or carcinogens as benign with rare metastases (Webster and Muller, 1994). However, the primary mammary tumors are often huge and the focus of the investigator's attention. Careful observation for metastases has rarely been carried out. When the search for metastases has involved extraction of tumor cells from the lung or molecular probes to identify malignant cells, the metastatic rate from "spontaneous" tumors was as high as 80% (Drohan et al., 1980).

In contrast to humans, the primary tumor is rarely removed from the mouse. This creates a vastly different clinical situation in which the primary tumor is allowed to grow until the animal is destroyed. We found that many metastases were not detected until the primary tumor was removed and the metastasis had an opportunity to grow (Drohan et al., 1980).

Although the exact rate of metastasis has not been measured in many transgenic strains, metastases have been readily found in association with the PyV-MT (Figure 6) and some *neu* mammary tumors (Guy et al., 1992 a, b). Invasion of the tumor vasculature is common. The most metastatic tumor in our experience is the PyV-MT tumor, which results in 100% metastases in the first six weeks of life (Guy et al., 1992a). Our own observations have been that the metastases are almost exclusively found in the lung (Cheung et al., 1996). Experimental tail vein or intraperitoneal injection of tumor cells results in pulmonary metastases, and metastases are rarely observed in other organs. The local mammary lymph nodes are the next most frequent site of metastases.

The metastatic pulmonary lesions closely resemble the lesions found in the mammary gland. They are sometimes even secretory and may even contain clear lipid vacuoles. They are generally found in the blood vessels but may invade sur-

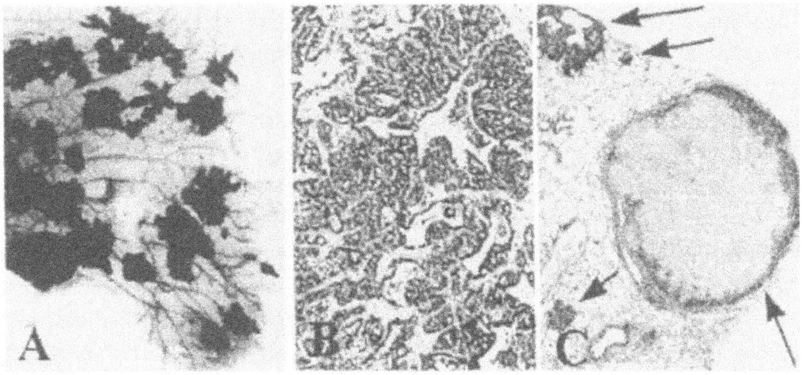

Figure 6. Images of (**A**) a whole mount of the mammary gland and of (**B**) the histology of the mammary gland and (**C**) the lung from a five-week-old mouse with the PyV-MT transgene. (**A**) Even at five weeks of age, the mammary gland has multifocal lesions (Magnification = 5X). (**B**) The tumors at this early age tend to be papillary with a dense stroma (Magnification = 50X). The limited lung field has at least four foci of metastatic tumor cells (arrows). The largest tumor mass has a light-staining, necrotic center (**C**) (Magnification = 16X. (Images from samples submitted by Drs. Guy and Muller, McMaster University to the Archives of the Transgenic Histopathology Laboratory, University of California, Davis.).

rounding tissue. These observations are very important because, in our experience, many transgenic animals that have mammary tumors may also develop pulmonary adenomas and adenocarcinomas. The pulmonary lesions are generally papillary and are covered by a uniform epithelium with relatively pale cells. Furthermore, the pulmonary tumors are in association with the bronchi and not the blood vessels. Awareness of the pulmonary lesions prevents confusing them with mammary tumor metastases.

Neoplastic Progression

The above descriptions of mouse mammary tumors reflect a morphological progression of events during which focal lesions emerge as subsets from the previous background. Detailed discussions of neoplastic progression in the mammary gland can be found elsewhere (Cardiff 1984; Cardiff et al., 1986; Foulds, 1956; Foulds, 1969). Most tumors arise in a background of proliferative hyperplasia or dysplasia. Animals with dysplastic lesions develop tumors. Although this chapter focuses on the histopathology, it is important to note that the progressive changes found in some transgenic models are associated with changes in the molecular biology of the epithelium (Webster and Muller, 1994: Cardiff and Muller, 1993; Cardiff, 1995; Callahan and Campbell, 1989).

We have yet to observe convincing evidence of a morphological progression that corresponds with biological progression of malignant tumors. Our most extensive comparative studies were based on serial transplants of wild-type and mutant PyV-MT tumors (Cheung et al., 1997). In this model system, the degree of morphological differentiation of the tumors could be maintained in serial transplant and the better-differentiated tumor type had a higher biological potential. The only structural feature of these low and high metastatic lines that correlated with their biological potential was the structure and density of the microcirculation. However, this sample is too limited to warrant a general conclusion about morphological progression.

Evidence that additional transgene mutational events contribute to tumorigenesis has been documented molecularly in the c-*neu* transgenic system. Transgenic animals with the c-*neu* transgene develop tumors (Guy et al., 1992b). The c-*neu* is overexpressed in tumors as opposed to non-neoplastic mammary gland (Deckard-Janatpour et al., 1997). Does overexpression alone result in transformation? Muller's group has found that a majority of the tumors which arise in the *neu* transgenic animals have an additional mutation resulting in the expression of a protein with a deletion in the extracellular domain (Siegel et al., 1994). Since the deleted region was nonrandom, the region is clearly critical in the progression of the neoplasm. In addition, c-*src* is overexpressed in these tumors (Muthuswamy et al., 1994) and PEA is overexpressed in metastatic tumors (Trimble et al., 1993). In some cases tumorigenesis is associated with overexpression of *ras* (Mangues et al., 1992). Thus, mutation and elevated expression are clearly involved in neoplastic progression. It is interesting to note that Muller and his colleagues have found that the metastatic PyV-MT tumors have acquired additional molecular traits that suggest that they are the result of further neoplastic progression (Webster and Muller, 1994).

THE COMPARATIVE PATHOLOGY OF HUMAN AND MOUSE MAMMARY GLANDS

Introduction

Wellings and his colleagues have pointed out the similarities between mouse and human mammary development and pathological lesions (Wellings et al., 1975; Jensen et al., 1976). Their pioneering studies have only recently been accepted by the surgical pathology community (Page, 1989; Dupont and Page, 1985). We have the perspective of having worked in surgical pathology with Dr. Wellings in an academic institution for 30 years as well as being involved in "transgenic pathology". Using this as a basis, we will point out the morphological similarities that we have observed between the lesions found in the mouse and the human. Because there is an increasing array of lesions that are identical, we are confident that the mouse represents a very important model for human breast disease. In our opinion, there is one medicine.

It should be recognized that the bulk of lesions and tumors found in the wild-type mouse do not resemble the common lesions of the human breast. Moreover, the murine breast is significantly different in scale and structure. The experienced pathologist should have no difficulty distinguishing between the two species. The mouse mammary fat pad is small relative to the size of the human breast. The human mammary ducts are sheathed in a dense connective tissue. The normal human has well-developed terminal ducts and lobulo-alveolar units lying in a loose connective tissue. The normal wild-type mouse gland does not exhibit persistent lobules between pregnancies. The mouse will undergo relatively more cycles of lactation and regression than human. These and other factors influence our interpretation of the histopathology of the normal and abnormal mammary gland in the two species.

Hyperplasia and Dysplasia

Wellings and colleagues identified 36 types of focal lesions in the human mammary gland (Wellings et al., 1975). The most significant lesions consisted of epithelial lesions originating in the terminal lobulo-alveolar units (Wellings et al., 1975). They found a continuum of progressive atypical cytological changes in the epithelium, an observation that could be correlated with the history of neoplasia in the given breast or the ipsilateral breast (Jensen et al., 1976). They postulated that these lesions were the preneoplastic equivalent of the mouse HAN (Wellings et al., 1975; Cardiff et al., 1977). Dr. Wellings and his colleague, Dr. Jensen, tried to prove the biological potential of these lesions by a number of indirect techniques, but were never able to secure unequivocal evidence.

The mouse HAN can be transplanted into the mammary fat pad cleared of gland. The HAN grows into the empty fat pad to produce a hyperplastic outgrowth (Cardiff, 1984). The hyperplastic tissue is immortal and can be serially transplanted proving its altered biological potential. In contrast, transplants of normal ducts grow into normal ductal tissue and can be transplanted for a limited number of generations (Daniel et al., 1966). The hyperplastic outgrowth will develop focal proliferations that emerge as tumors (Cardiff, 1984). Transplants of normal duct do not form tumors. This type of experiment provides formal proof that the classical HAN has malignant potential, a proof lacking in human mammary biology.

As more and more types of transgenic mammary glands have become available, several investigators have commented on the similarities between nonmalignant dysplasias of the mouse and the human gland (Halter et al., 1992; Bouchard et al., 1989). Common dysplastic lesions of both species include various cysts (Halter et al., 1992), epithelial proliferations of ducts and lobules (Cardiff, 1995), ductal carcinoma-in-situ (Bouchard et al., 1989), and sclerosing adenosis (Webster et al., 1995). Although rare, we have seen several classical fibroadenomas in the mouse gland. Perhaps the transgenic mouse mammary lesions that most closely resemble the complex patterns of human fibrocystic disease occur in retired breeders that have the *myc* transgene. Although these animals are rare, their mammary glands are most remarkable.

Tumors

Since the "spontaneous" mouse mammary tumor does not morphologically or biologically resemble most human breast cancers, early investigators did not always recognize the utility of the system. However, the emergence of transgenic biology has provided remarkable evidence that mouse tumors produced by the same genes as those implicated in human breast cancer result in tumors that closely resemble the human cancer.

As investigators with experience in both human and murine biology, we present a series of images from our archives (see Figures 7-10). These images illustrate the remarkable similarity between the lesions in humans and those in transgenic mice. Slides or photographs of limited samples from human and mouse tumors are indistinguishable. For example, the "sclerosing adenosis" found in humans has a counterpart found in transgenic mice with the activated *src* transgene (Figure 7). Duct ectasias can be found in both species. The nodular, ductal tumors associated with the *neu* family of oncogenes in transgenic mice are morphologically indistinguishable from comedocarcinomas from the human mammary gland (Figure 8). Papillary carcinomas of humans resemble those of the PyV-MT mouse (Figure 9). Tubular tumors of human and mouse are similar. Acinar tumors of humans are rare but resemble those from the *int*-1 transgene. The most common breast cancer of the human breast is the scirrhous carcinoma, which has a virtually identical counterpart

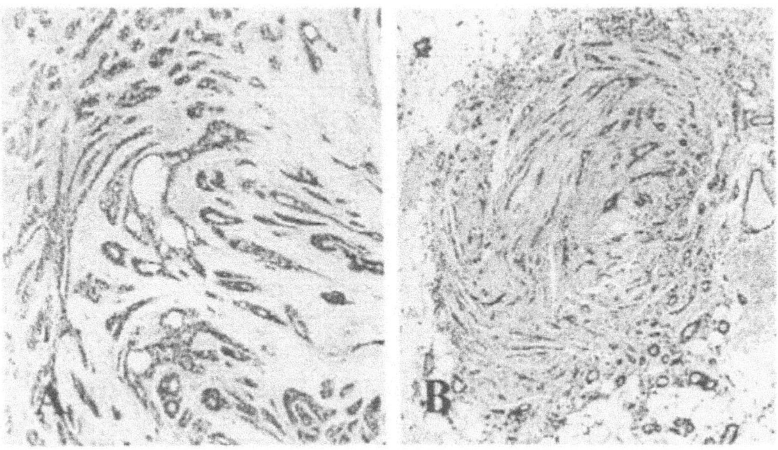

Figure 7. **(A)** I mages of mammary tissue disrupted by the process known as sclerosing adenosis as found in the human breast (Image from samples submitted by Dr. Wellings to the Archives of the Transgenic Histopathology Laboratory, University of California, Davis) and **(B)** in a mouse with the activated *src* transgene (Image from samples submitted by Drs. Webster and Muller, McMaster University to the Archives of the Transgenic Histopathology Laboratory, University of California, Davis.). Both are characterized by foci of distorted epithelial cords in a dense connective tissue (Magnification = 50X).

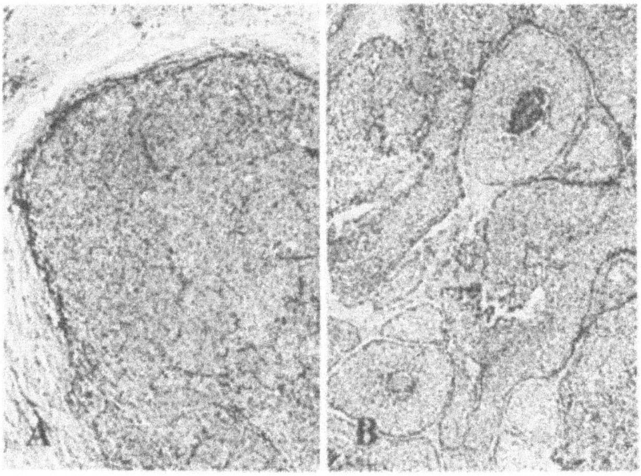

Figure 8. Images of ductal carcinoma in situ (DCIS) from (**A**) a human breast (Image from the Archives of the Transgenic Histopathology Laboratory, University of California, Davis) and (**B**) from a transgenic mouse bearing the *neu* transgene driven by the MTV LTR promoter (Image from samples submitted by Dr. Coffey, Vanderbilt University and Dr. Muller, McMaster University to the Archives of the Transgenic Histopathology Laboratory, University of California, Davis) (Magnification = 65X).

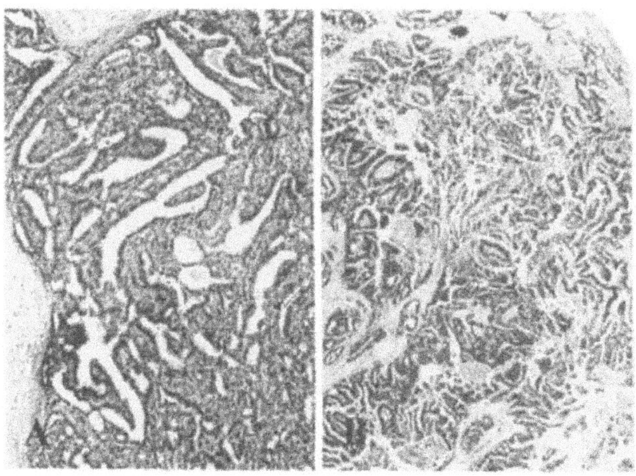

Figure 9. (**A**) Images of papillary carcinomas from a human breast (Image from the Archives of the Transgenic Histopathology Laboratory, University of California, Davis) and (**B**) from a transgenic mouse bearing the *Casein Kinase II* transgene driven by the MTV LTR promoter (Image from samples submitted by Dr. Seldin, Boston University to the Archives of the Transgenic Histopathology Laboratory, University of California, Davis) (Magnification = 50X).

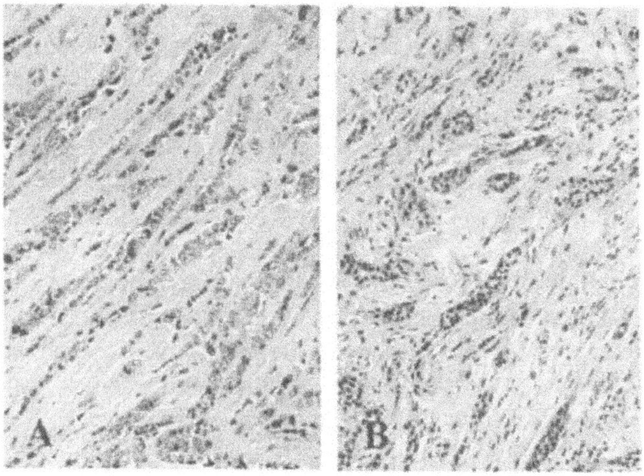

Figure 10. (**A**) Images of ductal carcinoma with a scirrhous pattern from a human breast (Image from the Archives of the Transgenic Histopathology Laboratory, University of California, Davis) and (**B**) from a transgenic mouse bearing the *src* transgene driven by the MTV LTR promoter (Image from samples submitted by Drs. Webster and Muller, McMaster University to the Archives of the Transgenic Histopathology Laboratory, University of California, Davis) (Magnification = 150X).

in the mouse with an activated *src* transgene (Figure 10). These and other examples provide some insight into the important and consistent similarities between human and mouse breast cancer.

In retrospect, the pathological analysis of the mouse has been limited by the inbreeding of the laboratory mouse. Dr. Otto Muhlbock, the grand old man of Dutch mammary biology, once remarked during a speech that the mouse was so inbred that our experience with an entire mouse strain was equivalent to seeing a single human. Muhlbock cautioned us to be circumspect with our interpretation of the mouse data. In retrospect, the vast majority of mouse mammary tumors are associated with the activation of only three genes, *int*-1, *int*-2, and *int*-3 (Callahan, 1996). None of these genes is found activated in a significant number of human breast cancers (Callahan, 1996).

The opportunity to study the effects of specific genes in the mammary gland has broadened our horizons. The insertion of genes associated with human breast disease has resulted in the development of benign and malignant lesions that closely resemble the human counterpart.

SUMMARY AND CONCLUSIONS

An understanding of the histopathology of transgenic mouse mammary gland requires a comparison of glands from many different founder strains. The insertion and expression of a transgene or the knock-out of a gene may result in abnormal growth and development of the gland which can only be appreciated by detailed examination of the entire growth, development, pregnancy, lactation, and regression cycle. Mammary tumors from transgenic mice differ dramatically from "spontaneous," virus-induced tumors. Their histological and cytological phenotype is very dependent on the transgene expressed in the mammary gland. The lesions in the mouse show progressive changes in morphology and biological behavior that suggest neoplastic progression. These changes can be correlated with molecular lesions. The histology of the lesions found in the transgenic mammary gland closely resembles that found in the human breast. Given this remarkable resemblance, the biological and molecular changes found in the progressive mouse lesions should be reflected in the human breast. The rules of biological parsimony dictate that the transgenic mouse is an accurate model of human breast disease.

ACKNOWLEDGMENTS

The authors are indebted to Ms. Judy Walls who has provided the outstanding histotechnology. We appreciate the many valuable insights and editorial suggestions provided from the graduate student perspective by Rachel Kravitz, Arash Tehranian, and Ruth Lang. We are also indebted to the following collaborators who have so generously shared their materials with the University of California, Davis Transgenic Histopathology Laboratory and who have shared their experimental insights with us: Drs. Wm.. Muller, P. Leder, A. Leder, F. Lee, E. Sinn, N. Tulchin, R. Coffey, E. Schmidt, M. Bissell, C. Sympson, R. Nandi, R. Guzman, H. Varmus, J. Liang, D. Morris, T. Vogt, D. Seldin, R. Skoda, M. Shen, D. Ornitz, R. Tepper, L. Young, T. Wang, L. Zuckerberg, E. Lees, C. Guy, M. Webster, M. Schaller, T. Parsons, S. Muthuswamy, A. Elson, D. Deng, A. Luster, A. Arnold, P. Barry, K. Ravid, A. Cheung, P. Siegel, K. Siminovitch, P. Soriano, Z. Werb and many others.

REFERENCES

Allred, D.C., Clark, G.M., Molina, R., Tandon, A.K., Schnitt, S.J., Gilchrist, K.W., Osborne, C.K., Tormey, D.C., and McGuire, W.L. (1992). Overexpression of *HER-2/neu* and its relationship with other prognostic factors change during the progression of in situ to invasive breast cancer. Hum. Pathol. 23, 974-799.

Barnes, D.M., Bartkova, J., Camplejohn, R.S., Gullick, W.J., Smith, P.J., and Millis, R.R. (1992). Overexpression of the c-erbB-2 oncoprotein: why does this occur more frequently in ductal carcinoma in situ than in invasive mammary carcinoma and is this of prognostic significance? Eur. J. Cancer 28, 644-648.

Bartkova, J., Barnes, D.M., Millis, R.R., and Gullick, W.J. (1990). Immunohistochemical demonstration of c-erbB-2 protein in mammary ductal carcinoma in situ. Hum. Pathol. 21, 1164-1167.

Bouchard, L., Lamarre, L., Tremblay, P.J., and Jolicoeur, P. (1989). Stochastic appearance of mammary tumors in transgenic mice carrying the MMTV/*c-neu* oncogene. Cell 57, 931-936.

Callahan, R. (1996). MMTV-induced mutations in mouse mammary tumors: their potential relevance to human breast cancer. Breast Cancer Res. Treat. 39 (1), 33-44.

Callahan, R. and Campbell, G. (1989). Mutations in human breast cancer: an overview. J. Natl. Cancer Inst. 81 (23), 1780-1786.

Callahan, R. (1991). Oncogenes and breast cancer progression. Basic Life Sci. 57, 143-153.

Cardiff, R.D. and Muller, W.J. (1993). Transgenic mouse models of mammary tumorigenesis. Cancer Surveys 16, 97-137.

Cardiff, R.D., Ornitz, D., Lee, F., Moreadith, R., Sinn, E., Muller, W., and Leder, P. (1992). Mammary Morphogenesis and Oncogenes. In: Breast Cancer: Progress in Biology, Clinical Management and Prevention (Dogliotti, L., Sapino, A., and Bussolati, G., Eds.). pp. 41-55. Kluwer Acad. Publ., Boston.

Cardiff R.D., Sinn, E., Muller, W., and Leder, P. (1991). Transgenic oncogene mice: Tumor phenotype predicts genotype. Am. J. Path. 139, 495-501.

Cardiff, R.D and Munn, R.J. (1995). Comparative pathology of mammary tumorigenesis in transgenic mice. Cancer Letters 90, 13-19.

Cardiff, R.D. (1984). Protoneoplasia: The molecular biology of mouse mammary hyperplasia. Adv. Cancer Res. 42, 167-190.

Cardiff, R.D. and Aguilar-Cordova, E. (1988). Protoneoplasia Revisited: The molecular biology of mouse mammary hyperplasia. Anticancer Res. 8 (5), 925-933.

Cardiff, R.D. (1988). Cellular and molecular aspects of neoplastic progression in the mammary gland. Eur J. Can and Clin. Oncol. 24, 15-20.

Cardiff, R.D. (1995). Understanding transgenes in mammary tumorigenesis: Five Rules. J. Mam. Gland Bio. Neo. 1, 61-73.

Cardiff, R.D., Morris, D.W., Young, L.J.T., and Strange, R. (1986). MuMTV genotype, protoneoplasia,and tumor progression. In: Breast Cancer: Origins, Detection and Treats. (Rich M., Hager J. and Taylor-Papdimitriou, J., Eds.). pp. 156-166, M. Nijhoff Publishing, Boston, Dordrecht, and Lancaster.

Cardiff, R.D., Wellings, S.R., Faulkin, L.J. (1977) . Biology of breast preneoplasia. Cancer 39, 2734-2746.

Cheung, A., Chen, P., Young, L., Chao, C., Ndoye, A., Kong, A., and Cardiff, R. (1996). Angiogenesis and metastasis in breast tumors. pp 23-26. Proceedings of the Sixth World Congress for Microcirculation.

Cheung, A.T.W., Chao, C.Y, Young, L.T.J., Chen , P.C.Y., Ndoye, A., Chanberlain, S.R., Muller, W.J., and Cardiff, R.D. (1997). Angiogenesis, microvessel density and metastasis in a metastatic mammary tumor model. Int. J. Oncol. 11, 69-77.

Cohen, J.C., Traina, V.L., Breznik, T., and Gardner, M. (1982). Development of a mouse mammary tumor virus-negative mouse strain: A new system for the study of mammary carcinogenesis. J. Virol. 44, 882-885.

Cohen, J.C., and Varmus, H.E. (1979). Endogenous mammary tumour virus DNA varies among wild mice and segregates during inbreeding. Nature 278, 418-423.

Cunha, G.R. (1994). Role of mesenchymal-epithelial interactions in normal and abnormal development of the mammary gland and prostate. Cancer 74, 1030-1044.

Daniel, C.W., De Ome, K.B., Young, J.T., Blair, P.B., and Faulkin, L.J., Jr. (1968). The in vivo life span of normal and preneoplastic mouse mammary glands: A serial transplantation study. Proc. Natl. Acad. Sci. U. S. A. 61 (1), 53-60.

Daniel, C.W. and Robinson, S.D. (1992). Regulation of mammary growth and function by TGF-β. Mol. Reprod. Dev. 32, 145-151.

Deckard-Janatpour, K., Muller, W.J., Chodosh, L., Coffey, R.J., and Cardiff, R.D.. (1997). Differential expression of the *neu* transgene in murine mammary tissues. Int. J. Oncol. 11, 235-244.

Dickson, R.B., Gottardis, M., and Merlino, G.T. (1991). Molecular insights into breast cancer from transgenic mouse models. Bioessays 13, 591-596.

Donehower, L.A., Harvey, M., Slagle, B.L., McArthur, M.J., Montgomery, C.A., Jr., Butel, J.S., and Bradley, A. (1992). Mice deficient for *p53* are developmentally normal but susceptible to spontaneous tumours. Nature 356, 215-221.

Drohan, W., Cardiff, R.D., Lund, J.K., and Schlom, J. (1980). The use of nucleic acid hybridization to detect pulmonary metastases in mice bearing spontaneous mammary tumors. Can. Res. 40, 2316-2322.

Dunn, T.B. (1959). Morphology of mammary tumors in mice. In: Physiopathology of Cancer (Homburger, F., Ed.). pp. 38-83. A. J. Phiebig, New York.

Dupont, W.D. and Page, D.L. (1985). Risk factors for breast cancer in women with proliferative breast disease. New Eng. J. Med. 312, 146-151.

Fantl, V., Stamp, G., Andrews, A., Rosewell, I., and Dickson, C. (1995). Mice lacking cyclin D1 are small and show defects in eye and mammary gland development. Genes Dev. 19, 2364-2372.

Faulkin, L.J., Mitchell, D.J., Young, L.J.T., Morris, D.W., Malone, R.W., Cardiff, R.D., and Gardner, M.B. (1984). Hyperplastic and neoplastic changes in the mammary glands of feral mice free of endogenous mouse mammary tumor virus provirus. J. Natl. Cancer Inst. 73, 971-982.

Foulds L. (1956). The histologic analysis of mammary tumors of mice. II. The histology of responsiveness and progression. The origins of tumors. J. Natl. Cancer Inst. 17, 713-753.

Foulds, L. (1969) Neoplastic development. Academic Press, London, New York.

Gallahan, D., Escot, C., Hogg, E., and Callahan, R. (1986). Mammary tumorigenesis in feral species of the genus Mus. Curr. Top. Microbiol. Immunol. 127, 354-361.

Gardner, M.B., Malone, R.W., Morris, D.W., Faulkin, L.J., Mitchell, D.J., Young, L.J.T., Strange, R., and Cardiff, R.D. (1985). Mammary tumors in feral mice lacking MuMTV DNA. J. Exp. Path. 2, 93-98.

Groner, B. (1992). Oncogene expression in mammary epithelial cells [see comments] J. Cell. Biochem. 49, 128-136.

Guy C.T, Cardiff, R.D., and Muller, W.J. (1992a). Induction of metastatic mammary tumors by expression of polyomavirus middle T oncogene: A transgenic mouse model for metastatic disease. Mol. Cell. Biol. 12, 954-961.

Guy C.T., Webster, M.A., Schaller, M., Parsons, T.J., Cardiff, R.D., and Muller, W.J. (1992b). Expression of the *c-neu* proto-oncogene in the mammary epithelium of transgenic mice induces metastatic disease. Proc. Natl. Acad. Sci. U.S.A. 89, 10578-10582.

Guy, C.T., Muthuswamy, S.K., Soriano, P., Cardiff R.D., and Muller, W.J. (1994). Activation of c-*src* tyrosine kinase is required for the induction of mammary tumors in transgenic mice. Genes Dev. 8, 23-32.

Guy, C.T., Cardiff, R.D., and Muller, W.J. (1996). Activated *neu* induces rapid tumor progression. J. Biol. Chem. 271, 7673-7678.

Halter S.A., Dempsey, P., Matsui, Y., Stokes, M.K., Graves-Deal, R., Hogan, B.L., and Coffey, R.J. (1992). Distinctive Patterns of hyperplasia in transgenic mice with mouse mammary tumor virus transforming growth factor-α. Characterization of mammary gland and skin proliferations. Am. J. Path. 149, 1131-1146.

Hennighausen L., McKnight,R., Burdon, T., Baik, M., Wall, R.J., and Smith, G.H. (1994). Whey acidic protein extrinsically expressed from the mouse mammary tumor virus long terminal repeat results in hyperplasia of the coagulation gland epithelium and impaired mammary development. Cell Growth Differ. 5, 607-613.

Humphreys, R.C., Krajewska, M., Krnacik, S., Jaeger, R., Weiher, H., Krajewski, S., Reed, J.C., and Rosen, J.M. (1996). Apoptosis in the terminal endbud of the murine mammary gland: A mechanism of ductal morphogenesis. Development 12, 4013-4022.

Jacks T., Remington, L., Williams, B.O., Schmitt, E.M., Halachi, S., Bronson, R.T., and Weinberg, R.A. (1994). Tumor spectrum analysis in *p53* mutant mice. Curr. Biol. 4, 1-7.

Jensen, H.M., Rice, J.R., and Wellings, S.R. (1976). Preneoplastic lesions in the human breast. Science 191, 295-297.

Jensen, H.M. and Wellings, S.R. (1976). Preneoplastic lesions of the human mammary gland transplanted into the nude athymic mouse. Cancer Res. 36, 2605-2610.

Jhappan, C., Gallahan, D., Stahle, C., Chu, E., Smith, G.H., Merlino, G., and Callahan, R. (1992). Expression of an activated *Notch*-related *int-3* transgene interferes with cell differentiation and induces neoplastic transformation in mammary and salivary glands. Genes Dev. 6, 345-355.

Krane, I.M. and Leder, P. (1996). NDF/heregulin induces persistence of terminal end buds and adenocarcinomas in the mammary glands of transgenic mice. Oncogene. 128, 1781-1788.

Kwan H., Pecenka, V., Tsukamoto, A., Parslow, T.G., Guzman, R., Lin, T.P., Muller, W.J., Lee, F.S., Leder, P., and Varmus, H.E.. (1992). Transgenes expressing the *Wnt-1* and *int-2* proto-oncogenes cooperate during mammary carcinogenesis in doubly transgenic mice. Mol. and Cell Biol. 12, 147-154.

Li, Li B., Greenberg, N., Stephens, L.C., Meyn, R., Medina, D., and Rosen, J.M. (1994). Preferential overexpression of a 172Arg→Leu mutant *p53* in the mammary gland of transgenic mice results in altered lobuloalveolar development. Cell Growth Differ. 7, 711-721.

Li, Li M., Liu, X., Robinson, G., Bar-Peled, U., Wagner, K.U., Young, W.S., Hennighausen, L., and Furth, P.A. (1997). Mammary-derived signals activate programmed cell death during the first stage of mammary gland involution. Proc. Natl. Acad. Sci. USA 94 (7), 3425-3430.

Lin, T.P., Guzman, R.C., Osborn, R.C., Thordarson, G., and Nandi, S. (1992). Role of endocrine, autocrine, and paracrine interactions in the development of mammary hyperplasia in *Wnt-1* transgenic mice. Cancer Res. 52, 4413-4419.

Lodato, R.F., Maguire, H.C. Jr., Greene, M.I., Weiner, D,B., and LiVolsi, V.A. (1990). Immunohistochemical evaluation of c-erbB-2 oncogene expression in ductal carcinoma in situ and atypical ductal hyperplasia of the breast. Mod. Pathol. 3, 449-454.

Mangues, R., Seidman, I., Gordon, J.W., and Pellicer, A. (1992). Overexpression of the N-*ras* proto-oncogene, not somatic mutational activation, associated with malignant tumors in transgenic mice. Oncogene. 7, 2073-2076.

Matsui, Y., Halter, S.A., Holt, J.T., Hogan, B.L. and, Coffey, R.J. (1990). Development of mammary hyperplasia and neoplasia in MMTV-TGF α transgenic mice. Cell 61, 1147-1155.

Medina, D. (1982). Mammary tumors. In The mouse in Biomedical Research (Foster, S.L., Small, J.D., and Fox, J.G., Eds.). pp 373-396. Academic Press, New York.

Morris, D.W. and Cardiff, R.D.. (1987). The multistep model of mouse mammary tumor development. Adv. Viral Oncol. 7, 123-140.

Morris, D.W., Barry, P.A., Bradshaw, H.D., and Cardiff, R.D. (1990). Insertion mutagenesis of the *int-1* and *int-2* loci by mouse mammary tumor virus in premalignant and malignant neoplasms from the GR mouse strain. J. Virol. 64, 1794-1802.

Morris, D.W. and Cardiff, R.D. (1987). The multistep model of mouse mammary tumor development. Adv. in Viral. Oncol. 7, 123-140.

Muller, W.J., Lee, F.S., Dickson, C., Peters, G., Pattengale, P., and Leder, P. (1990) The *int-2* gene product acts as an epithelial growth factor in transgenic mice. EMBO J. 9, 907-913.

Muller, W.J. (1991). Expression of activated oncogenes in the murine mammary gland: Transgenic models for human breast cancer. Cancer Metastasis Rev. 10, 217-227.

Muller, W.J., Sinn, E., Pattengale, P.K., Wallace, R., and Leder, P. (1988). Single step induction of mammary adenocarcinoma in transgenic mice bearing the activated *c-neu* oncogene. Cell 54, 105-115.

Muller, W.J., Arteaga, C.L., Muthuswamy, S.K., Siegel, P.M., Webster, M.A., Cardiff, R.D., Meise, K.S., Li, F., Halter, S.A., and Coffey, R. (1996). Synergistic interaction of the *Neu* protooncogene and TGFα in the mammary epithelium of transgenic mice. Mol. Cell. Biol. 16, 5726-5736.

Munn, R.J., Webster, M., Muller, W.J., and Cardiff, R.D. (1995). Histopathology of transgenic mouse mammary tumorigenesis (A Short Atlas). Seminars in Cancer Biology 6, 153-158.

Muthuswamy, S.K., Siegel, P.M., Dankort, D.L., Webster, M.A., and Muller, W.J. (1994). Mammary tumors expressing the *neu* proto-oncogene possess elevated c-*Src* tyrosine kinase activity. Mol. Cell. Biol. 14, 735-743.Ornitz, D.J., Moreadith, R.W., and Leder, P. (1991). Binary system for regulating transgene expression in Mice: Targeting *int-2* expression with yeast GAL4/UAS control elements. Proc. Natl. Acad. Sci. USA 88, 698-702.

Ornitz, D.M., Cardiff, R.D., Kuo, A., and Leder, P. (1992). *Int-2*, an autocrine and/or ultra-short-range effector in transgenic mammary tissue transplants. J. Natl. Cancer Inst. 84, 887-892.

Page, D.L. (1989). Pathology of preinvasive and early breast cancer. Curr. Opin. Oncol. 1, 277-283.

Paigen, K. (1995). A miracle enough: The power of mice. Nature Medicine 1, 215-220.

Pierce, Jr., D.F., Johnson, M.D., Matsui, Y., Robinson, S.D., Gold, L.I., Purchio, A.F., Daniel, C.W., Hogan, B.L., and Moses, H.L.. (1993). Inhibition of mammary duct development but not alveolar outgrowth during pregnancy in transgenic mice expressing active TGF-β 1. Genes Dev. 7, 2308-2317.

Pingitore, R. and Squartini, F. (1982). Relationships between morphology and viral etiology in mammary tumors of BALB/cfC3H and BALB/cfRIII mice. Tumori 68, 199-203.

Robinson, G.W., Smith, G.H., Gallahan, D., Zimmer, A., Furth, P.A., and Hennighausen, L. (1996). Understanding mammary gland development through the imbalanced expression of growth regulators. Dev. Dyn. 206 (2), 159-168.

Sakai, S., Mizuno, M., Harigaya, T., Yamamoto, K., Mori, T., Coffey, R.J., and Nagasawa, H. (1994). Cause of failure of lactation in mouse mammary tumor virus/ human transforming growth factor α transgenic mice. Proc. Soc. Exp. Biol. Med. 205, 236-242.

Schedin, P.J., Thackray, L.B., Malone, P., Fontaine, S.C., Friis, R.R., and Strange, R. (1996). Programmed cell death and mammary neoplasia. Cancer Treat. Res. 83, 3-22.

Siegel, P.M., Dankort, D.L., Hardy, W.R., and Muller, W.J. (1994). Novel activating mutations in the *neu* proto-oncogene involved in induction of mammary tumors. Mol. Cell. Biol. 14, 7068-7077.

Squartini, F., Bistocchi, M., and Buongiorno, L. (1981). Development, morphology, and progression of mammary tumors during and after fertile life in BALB/cfRIII mice. J. Natl. Cancer Inst. 66, 311-319.

Squartini, F., Basolo, F., and Bistocchi, M. (1983). Lobuloalveolar differentiation and tumorigenesis: two separate activities of mouse mammary tumor virus. Cancer Res. 43, 5879-5882.

Strange, R. and Cardiff, R.D. (1989). Transgenic animals and transgenic mammary glands. In: Breast Cancer Progress in Biology, Clinical Management and Prevention. (Rich, M.A., Hager, J.C., Keydar, I., Eds.). pp 1-14. Kluwer Academic Publishers, Boston, Dordecht, London.

Sympson, C.J., Talhouk, R.S., Alexander, C.M., Chin, J.R., Clift, S.M., Bissell, M.J., and Werb, Z. (1994). Targeted expression of stromelysin-1 in mammary gland provides evidence for a role of proteinases in branching morphogenesis and the requirement for an intact basement membrane for tissue-specific gene expression. J. Cell. Biol. 125, 681-693.

Tehranian, A., Morris, D.W., Min, B.H., Bird, D.J., Cardiff, R.D., and Barry, P.A. (1996). Neoplastic transformation of prostatic and urogenital epithelium by the Polyoma virus *Middle T* gene. Am. J. Path. 149, 1177-1191.

Trimble, M.S., Xin, J.H., Guy, C.T., Muller, W.J., and Hassell, J.A. (1993). PEA3 is overexpressed in mouse metastatic mammary adenocarcinomas. Oncogene 8, 3037-3042.

Tsukamoto, A.S., Grosschedl, R., Guzman, R.C., Parslow, T., Varmus, H.E. (1988). Expression of the *int-1* gene in transgenic mice is associated with mammary gland hyperplasia and adenocarcinomas in male and female mice. Cell 55, 619-625.

Tulchin, N., Lee, F.S., Ornstein, L., Strauchen, J. and Cardiff, R.D. (1995). c-*myc* protein distribution: Mammary adenocarcinomas of MTV/*MYC* transgenic mice. Int. J. Oncol. 7, 5-9.

Wang, T.C., Cardiff, R.D., Zuckerberg, L., Lees, E., Arnold, A., and Schmidt, E.V. (1994). Mammary hyperplasia and carcinoma in MMTV-cyclin D1 transgenic mice. Nature 369, 669-671.

Webster, M.A and Muller, W.J. (1994). Mammary tumorigenesis and metastasis in transgenic mice. Sem. Cancer Biol. 5, 69-76.

Webster, M.A., Cardiff, R.D., and Muller, W.J. (1995). Induction of mammary epithelia hyperplasias and mammary tumors in transgenic mice expressing an MMTV/activated c-*src* fusion gene. Proc. Natl. Acad. Sci. U. S. A. 92, 7849-7853.

Wellings, S.R., Jensen, H.M., and Marcum, R.G. (1975). An atlas of subgross pathology of the human breast with special reference to possible precancerous lesions. J. Natl. Cancer Inst. 55, 231-273.

IMMUNE FACILITATION OF BREAST CANCER

T.H.M. Stewart and Gloria H. Heppner

Advances in Oncobiology
Volume 2, pages 203-228.

ISBN: 0-7623-0388-3

INTRODUCTION

This chapter is a collaboration between a clinician (T.S.) and a tumor biologist (G.H.) who have been involved in the study of breast cancer immunology since the 1960's and who, despite having taken quite different pathways, have arrived at similar, albeit controversial, conclusions about the role of the immune response in the development of breast cancer. Our work has been done during a time when the main body of opinion is that the immune response to cancer, including breast cancer, is a defense reaction and that the development of cancer is evidence either of a failure in the immune system or of the ability of the cancer to circumvent or escape from its effect. It follows from this that cancer ought to be prevented or cured by measures that restore immune effectiveness. In this age of gene and molecular therapy, and with increasing understanding of the molecular and cellular mechanisms that underlie cancer, the goal of utilizing the immune response for prevention or therapy may at last be achievable. To do so, however, it will be necessary to reevaluate the role of the immune response in the *natural history* of cancer. It is also necessary to focus specifically on the biology of each type of cancer and not to assume that the lessons learned from say, melanoma, are applicable to, for example, breast cancer.

This chapter will discuss the possible involvement of the immune response in breast cancer development. On the clinical side, we consider the literature from pathologists and surgeons who sought evidence for immune recognition in breast cancer tissue and its relationship to prognosis. We also review studies that sought to link specific or nonspecific immune reactivity with breast cancer prognosis. We consider studies on breast cancer incidence or outcome in situations of immune suppression. From the experimental literature, principally in the mouse, we review evidence for immune system influence in breast cancer development. As with the clinical literature, our focus is on the natural immunobiology of breast cancer, not on the perturbations of the immune system in advanced disease, nor, in the case of the mouse, on the response to transplantable tumors or lines. As will be seen, our experiences as well as those of others suggest that in *breast cancer*, the immune response may facilitate disease development, rather than suppress it. We discuss potential mechanisms to explain this unsettling possibility.

GENERAL CONSIDERATIONS

Breast cancer is a complex disease. Its etiology is multifactorial, its clinical manifestations are diverse, and its outcome is variable. The natural history of breast cancer is prolonged; from epidemiological studies we know that circumstances early in life, in preadolescence or early adulthood, can influence whether or not cancer develops many decades later (Henderson, et al, 1981). This prolonged natural history is carried out against a background of profound physiological change, particularly in regard to reproductive hormone status as well as during the gradual unfolding of the processes of maturation and aging. The breast itself is a complex and dynamic organ. For most of life, the epithelial components are relatively inconspicuous, but, when called upon, they are capable of marked hyperplasia, structural differentiation and function followed in due time by rapid involution and quiescence. The breast is also an immunological organ. It is the source of immunoglobulins and T-cells passed from mother to nursing child (Eglinton et al., 1994) and the breast must therefore be hospitable to cells of the immune system.

IMMUNE RECOGNITION IN BREAST CANCER PATIENTS

Immune Competence and Breast Cancer Prognosis

Both of us have recently co-authored reviews on the topic of the immune system in breast cancer, and the reader is referred to them for more detailed documentation of the sources of information discussed here (Stewart and Tsai, 1993; Wei and Heppner, 1996). Overall, it appears that breast cancer patients are not grossly abnormal immunologically, at least not until the time of terminal disease. Thus, T and B cell counts and functional assays are well within the admittedly broad range of normal. This conclusion is based on a large number of studies, spanning several decades over which our ability to measure immune competence has become more refined without, however, substantially altering the picture (For a recent study, see Head et al., 1993). Most importantly, there appears to be no strong correlation among various parameters of immune competence and accepted clinical indices of breast cancer prognosis, although it is usually possible to tease out some association if the patient numbers are large enough and the statistician persistent.

Inflammatory Infiltrates in Breast Cancer

A potential indicator of the involvement of the immune response in breast cancer development is the extent of inflammatory cell infiltration into the tumor mass. Unfortunately, the literature is not consistent on this point. A 1993 review (Stewart and Tsai) involved the dividing of 43 articles into three groups: Nine articles reported good prognosis when there was an intense infiltrate; six of these were descriptions of medullary cancer. Two articles concluded that there was no relationship, of any sort, and 23 others showed a worsening of prognosis when the stromal infiltrate was intense. Eleven of these latter papers concern inflammatory breast cancer. Of the remaining 12 publications, six are particularly impressive: Champion et al. (1972), Fisher et al. (1974), Meyer and Hixon (1979), Fisher et al. (1983), Parkes et al. (1988), and Rosen and Groshen (1990). Overall, one may speculate that although the immune system may downregulate growth of medullary cancer, it appears to stimulate other types of breast cancers. A particularly telling study was provided by

Kurtz and colleagues (1990). They studied 18 factors in 496 stage I–II ductal cancers treated by conservative surgery and radiotherapy. Multivariate analysis showed that in younger women, a major lymphocytic stromal reaction was the factor most strongly correlated with recurrence. The authors concluded that the intensity of the cellular reaction may reflect a cancer-host response that favors, rather than impedes, cancer growth.

In a retrospective study, Pupa and colleagues (1992) reported that overexpression of the c-erbB2 oncogene indicates a bad prognosis in node-positive breast cancer patients (length of follow-up was 19 years). Oncogene overexpression was strongly associated with the presence of lymphoplasmacytic infiltration. These results are similar to those of a study of 106 primary breast cancers by Tang and colleagues (1990), who found a very strong association between amplification of the oncogenes *c-erbB2*, *int-2,* or *c-myc* and dense lymphocyte infiltration of the tumor. This association was highly significant in those tumors that presented high-level amplification of *c-erb B2* and *int-2*. The authors concluded that cytokine production may be associated with paracrine immunological phenomena, which are themselves associated with poor prognosis.

A very recent publication (Leek et al., 1996) focused specifically on the relationship of infiltrating macrophages, degree of tumor angiogenesis (vascularity), and survival of breast cancer patients. There was a strong positive correlation between the first two parameters and between both and reduced survival. The authors suggested that macrophages promote angiogenesis by secreting a variety of endothelial cell cytokines or extracellular matrix–degrading enzymes. In turn, angiogenesis promotes metastatic spread, establishment of systemic disease, and ultimately, poor survival.

As indicated, interpretation of studies on the significance of inflammatory infiltrates in breast cancer specimens is complicated by the heterogeneity of diseases lumped under the term "breast cancer." Many of the studies may be skewed by having significant numbers of medullary carcinomas in their mix, a disease with a good prognosis, or inflammatory carcinoma, a type with an overall very bad prognosis.

Furthermore, as mentioned above, the subclinical period of breast cancer development appears to be quite prolonged so that the significance of inflammatory cell infiltration within a clinical cancer may reflect earlier events lost in time. In this context, the early work of Black (1972), who focused on morphological and functional indicators of immune reactivity in patients with "precancerous" lesions or carcinoma-in-situ (CIS), may be relevant. Black used a variety of approaches to show that immune-associated parameters may signify a good prognosis. His work remains unique in its emphasis on immune events in early disease progression, an area that needs revisiting in light of more recent findings.

Another problem in interpreting histological reports of inflammatory cell infiltration is that it is difficult to know the functional status of the infiltrating cells. A number of investigators have attempted to isolate lymphocytes from primary breast cancers and to test their functionality in in vitro assays (Vose and Moore, 1979; Eremin et al., 1981; Whiteside et al., 1986). In general, T cells tend to predominate in isolated infiltrates, with CD8 outnumbering CD4 cells. NK cells are relatively infrequent. Functionally, the cells tend to be deficient or even suppressive in lymphocyte proliferation or cytotoxicity assays (Vose and Moore, 1979; Eremin et al., 1981). However, a detailed clonal analysis by Whiteside and colleagues (1986) of T cells isolated from breast cancers revealed the presence of potential cytolytic function, demonstrable in expanded cell cultures in vitro. The relevance of these observations to cell function in situ, however, is not clear.

Specific and Nonspecific Immune Reactivity in Breast Cancer Patients

Hope that the immune response can be used to interfere with cancer growth and development has undergone dramatic shifts since we began work in this area. For T.S., the initial excitement came from the area of autoimmune disease. In the late 1950s and early 60s evidence was rapidly accumulating that autoimmune disease was a reality. This was seen vividly in the thyroid clinic: Grave's disease was found to be the result of stimulation of thyroid function by antibodies (McKenzie and Zakerija, 1986) and Hashimoto's thyroiditis resulted in destruction of the thyroid by cellular immune mechanisms (reviewed by Volpe, 1986). An occasional patient would show features of each disease, with a period of stimulation of function in an otherwise dying gland. As a young physician, T.S. became aware of the possibility that the immune system could either stimulate or depress function.

Animal models were developed to study the mechanisms involved in the initiation of autoimmune diseases. In several models, a specific disease could be induced by inoculating the animal with the appropriate organ antigen, autologous or allogeneic in origin, homogenized in Freund's complete adjuvant (FCA). In 1959 and again in 1962, Waksman reviewed some 600 published reports. An important generalization was made: The severity of the induced disease correlated very well with the intensity of the delayed hypersensitivity reaction (DHR) to the organ antigen in the animal in which the disease was produced. Severe disease was accompanied by an intense DHR, mild disease with a weak DHR. For example, Brent and colleagues (1958) showed that the tempo of graft rejection was reflected in the DHR expressed by recipient animals when challenged by injection of soluble histocompatibility antigens of the donor animal. Where the barrier was strong, rejection was rapid and the DHR was intense. Where the histocompatibility barrier was weak, there was a sluggish, prolonged rejection and weak DHR to the antigen. In autoimmune disease, the severity and tempo of the immune reaction also correlated well with the DHR to the appropriate antigen.

A further correlation could be made. In rapid graft rejection or severe autoimmune disease, the "round cell" stromal infiltrate of the organ would be marked. In contrast, in sluggish, prolonged rejection, or mild disease, the infiltrate was much less pronounced.

In 1964, a paper was published showing that 27% of 50 patients with cancer gave a positive DHR when skin tested with acellular extracts of their own tumors (Hughes and Lytton, 1964). This experiment was done following a suggestion by Medawar, who thought it would be interesting to ascertain if there was evidence of an autoimmune reaction to cancer in humans. On returning to Ottawa from a postgraduate year in the unit of William Beierwaltes at the University of Michigan, T.S. decided to repeat this experiment. From 1966 to 1969, 144 patients were tested with extracts of their own tumors (Stewart, 1969). The earlier results were confirmed: 26% of T.S.'s patients had such a reaction. Furthermore, it was discovered that the strongest reaction was induced by cell membranes and that there was a low incidence of a positive reaction when no lymphocytic or round cell infiltrate of the tumor stroma was seen. Only 6% of such patients were positive compared to a much greater frequency of positive reactions when the infiltrate was marked, 44%, or very marked, 100%.

These were exciting findings. They suggested that there was an autoimmune reaction to autologous tumor in at least some patients with a solid tumor. Would it be of benefit to patients to strengthen this reaction, using the method that successfully induced autoimmune disease in animals, namely, vaccination with membrane antigen homogenized in FCA? The literature of the time supported this possibility. In postgestational choriocarcinoma, a tumor of fetal tissue proliferating in the maternal host, significant survival advantage was observed for those women whose tumors showed a marked infiltrate in the tumor stroma as compared with those with a mild infiltrate (Elston, 1969). In testicular seminoma, 56% of 179 patients who survived more than 10 years had tumors with a definite lymphoid stroma, whereas only 29% survived when such an infiltrate was absent (Martin et al., 1965). In colon cancer, a highly significant five-year survival advantage was reported for 78 patients who had a local round cell reaction to their tumor compared to 70 in whom this was absent (Murray et al., 1967). Spratt and Spjut (1967) studied 1,137 patients with colon cancer, followed them for 10 years, and also showed that an absence of such an infiltrate carried a very ominous prognosis. In nonsmall carcinoma of the lung, favorable prognosis was associated with the degree of lymphocytic infiltration of the tumor (Di Paola et al., 1977), particularly in the case of patients with squamous cell carcinoma (Yesner, 1992). In medullary carcinoma of the breast, a marked survival advantage was seen in 104 cases out of a total breast cancer cohort of 1,411 cancers, followed for 20 years by Bloom and colleagues (1970): 74% of the medullary breast cancer patients were still alive compared to 14% of cases with similar-stage, nonmedullary breast cancer. A striking feature of medullary cancer, which makes up only 5% of breast cancers in the western world, is the rich stromal infiltrate by lymphocytes. However, improperly treated medullary cancers had a median survival of 2.2 years indicating that such tumors are potentially highly malignant. The authors speculated that "their biological potential is countered by an effective host resistance" (Bloom et al., 1970).

Thus it was evident that in some types of cancer, a host immune reaction to the tumor gave survival advantage. It was therefore logical to ask whether one could achieve therapeutic advantage by increasing the intensity and frequency of such favorable reactions in the adjuvant setting following "curative" surgery. However, although the prospect of using cell membrane tumor antigens homogenized with FCA to vaccinate patients intradermally held promise, it also had potential dangers. If normal tissue antigens were present, there was danger of inducing severe autoimmune disease as had been shown in men in whom severe acute orchitis was induced following autovaccination with an homogenate of normal testicular tissue in FCA prior to completing orchictomy as treatment for prostate cancer (Mancini, 1965).

So, before starting a potentially dangerous experiment, T.S. thought it prudent to go back and examine the survival experience of what was the largest cohort of tumor patients that had been skin- tested earlier with extracts of their own tumors, namely, women with breast cancer. There were 56 such women of whom 12 had shown a DHR to extracts of their tumors (Stewart, 1969, a, b). Although the longest follow-up was only six years, T.S. was confident that he would see a survival advantage in these 12 women. The results of this study were published in 1971 (Stewart and Orizaga, 1971) and, shockingly, were exactly the opposite of that prediction. There were three major findings. Firstly, of 52 patients tested and followed for at least 2.5 years, 40 had a negative DHR to their own tumor; of these 31 (77.5%) were alive and 9 (22.5%) had died of the cancer. Of 12 patients showing a positive DHR to their tumors, 5 (41.5%) were alive and 7 (58.5%) were dead. The difference between the two groups was statistically significant. Secondly, a significant association was found between a positive DHR and the degree of nuclear differentiation; DHR was seen only in two of 21 patients with moderately to highly differentiated tumors but in 10 of 35 patients with anaplastic tumors. Thirdly, antigens from one patient gave a positive reaction in another patient with breast cancer, suggesting common antigens. Two conclusions were suggested: that the immune system might facilitate tumor growth and that a viral influence might be present in the origin of breast cancer (see below).

In 1971, T.S. abandoned any further study of breast cancer. For the next 20 years, he collaborated with Ariel Hollinshead in the study of immune reactivity of nonsmall-cell lung cancer patients (Hollinshead, et al, 1974). In 1973, they started a trial of specific active immunotherapy in patients having had curative surgery. Over the years their initial positive findings of improved survival in those patients who had received adjuvant vaccination (Stewart, et al, 1977) were independently confirmed by Takita and colleagues (1985) in a study of squamous-cell lung cancer patients and eventually in a large Canadian–U.S. multicenter trial. However, this last trial gave a negative overall result, a cause of much tribulation. In the paper describing this study, three pages were used to describe the massive protocol violations that were recognized (Stewart et al., 1984).

By 1991, T.S. and Hollinshead had accumulated six cases in which nonregional metastases regrew many years following curative resection of lung cancer, with no new primary tumor, an exquisitely rare observation in lung cancer (Stewart et al., 1990, 1991). On the strength of these observations, the first international workshop on tumor dormancy was held in Ottawa in October 1991 (Stewart and Wheelock, 1992). In preparation for this workshop, two meta-analyses of lung cancer trials were performed. The first examined 10 randomized trials of alkylating agents given as adjuvant therapy for nonsmall-cell lung cancer (Stewart and Raman, 1992). A significant worsening of survival was seen in patients receiving such therapy as compared with controls. [This finding has been confirmed in a much larger meta-analysis published in 1995, with a combined hazard ratio of 1.15 ($P=0.005$) (Lung Cancer Collaborative Group, 1995)]. The second meta-analysis showed a highly significant survival advantage in patients able to mount a ≥2 cm DHR to tumor antigen at one year postsurgery compared to nonimmunized, randomized controls who did not show this reaction (Raman and Stewart, 1992).

Taken together, T.S. concluded that boosting autoimmune reactivity to lung cancer is beneficial in the adjuvant setting, whereas the use of strongly immunosuppressive alkylating agents worsens the prognosis, perhaps by disarming such helpful reactivity. This was in contrast, however, to the growing experience in the adjuvant treatment of breast cancer, where such immunosuppressive drugs as melphalan or, more convincingly, CMF (Cyclophosphamide, methotrexate, 5 fluorouracil) are beneficial (Early Breast Cancer Trialists Collaborative Group, 1992). Similarly, the use of cortisone in advanced lung cancer was found to be detrimental in a randomized trial (Wolf et al., 1960), causing more rapid death due to lung cancer growth, whereas, in advanced breast cancer, cortisol can be beneficial (Stewart, et al, 1984) and nonspecific immunostimulation can be detrimental (see below).

IMMUNE FACILITATION OF HUMAN BREAST CANCER

Cell-Mediated Reactivity and Prognosis

Thus, for T.S., the question of possible immune facilitation of breast cancer was again raised along with the suspicion that the role of the immune system in breast cancer may be unique. A first approach to following up these possibilities was to review the experience of others in regard to DHR and related assays of cell-mediated immunity in breast cancer patients. This review has been published (Stewart and Tsai, 1993) and only the highlights are reiterated here:

- Nonspecific stimulation of breast cancer patients using the method of Bacillus Calmette Guerin inoculation or scarification as an adjuvant therapy significantly worsened prognosis by 20% (Early Breast Cancer Trialists Collaborative Group, 1992).
- In a study of 134 patients with breast cancer and 63 patients with benign breast lesions, low lymphocyte reactivity in vitro to various mitogens and an-

tigens correlated with low overall risk compared to that of patients with high or intermediate risk disease (Wanebo et al., 1976).

- In a series of 77 patients with inflammatory breast cancer treated with radiotherapy and chemotherapy, half of whom were randomized to receive BCG, patients who were tuberculin negative on skin testing survived significantly better than those who were tuberculin positive (Pouillart, et al, 1981).
- Tunisian women, who have a high frequency of inflammatory breast cancer, had a DHR to soluble breast cancer antigen with a frequency three times greater than that in women with other forms of breast cancer (Mourali et al., 1978). Tunisian adults were shown to have increased immune reactivity when compared to a comparable cohort of adult Americans (Levine et al., 1981). As mentioned before, inflammatory breast carcinoma is extremely aggressive.
- The thymidine labeling index (TLI) was studied in primary invasive breast carcinomas of 133 patients (Meyer and Hixon, 1979). Operable patients with TLIs above the median had a significantly higher rate of occurrence than those with indices below. There was a significant linear increase of log TLI with increasing degrees of inflammatory cell reaction at the margin of the breast cancer.
- Cannon and colleagues (1981) studied the lymphoproliferative responses of peripheral blood mononuclear cells from 95 stage I and II breast cancer patients and 35 lung cancer patients in the early postoperative period following mastectomy or full or partial lobectomy. The test used was one-way mixed leukocyte culture (MLC) against a pool of mitomycin-C–treated lymphocytes from allogeneic donors. Depressed lymphoproliferative responses were associated with significantly longer disease-free intervals in breast cancer patients and significantly shorter disease-free intervals in lung cancer patients.

This literature, although clearly not definitive nor easily interpreted, certainly suggested that if the immune system has a role in breast cancer development at all, it is not as an unequivocal host defense response .

Breast Cancer in Immunosuppressed Patients

Reinforcing the suspicion of immune facilitation of tumor growth in a substantial subset of women with breast cancer is the fact that all nonsurgical treatments that have given positive results in breast cancer have one thing in common—they are immunosuppressive to a greater or lesser degree. Again this topic has been reviewed in detail (Stewart et al., 1994) and only salient points will be emphasized here:

- Median response rates of 30% lasting 3–14 months following treatment with corticosteroids have been reported in patients with metastatic breast cancer.

Additive effects are seen when corticosteroids are combined with endocrine therapy. The immunosuppressive effects of corticosteroids are quite clear cut: a clear dose response exists for suppression of the production of lymphokines including IL-1, IL-2, IL-6, and TNF (Stewart et al., 1994).

- Adjuvant chemotherapy with CMF is immunosuppressive and causes prolonged impairment of certain aspects of B, T, and NK cell function. Zielinski and colleagues (1990) reviewed their studies showing a depression of antibody production following vaccination, prolonged impairment of mitogen-induced, soluble IL-2 receptor production, and a decrease in proliferation of peripheral blood mononuclear cells following phytohemagglutinin stimulation. The depressed activity was seen up to three years following cessation of CMF therapy.
- Escalating doses of chemotherapy followed by allogeneic or autologous bone marrow transplantation cause marked and prolonged immunosuppression lasting up to two years (Welte et al., 1984; Olsen et al., 1988). Defective production of IL-2 is seen up to 18 months following high-dose cyclophosphamide and whole body irradiation.
- Locoregional radiotherapy following mastectomy for breast cancer has been found to confer a systemic benefit for survival (Stewart, 1994). One explanation for improved survival is that there is a significant T-cell lymphopenia and decreases in T-cell responses that can persist for as long as 11 years in women so treated for breast cancer.
- Robinson and colleagues (1993) found that, in women treated for bilateral breast cancer and without evidence of disease, NK cell activity was higher than in normal controls, but that this activity declined on long-term tamoxifen therapy. Tamoxifen also reduced CD4 cells and the CD4/CD8 ratio in such patients.

If successful treatment regimens for breast cancer are immunosuppressive, one might ask whether women who are chronically immunosuppressed show a low population incidence of de novo breast cancers. Such a prediction was made April 1994, in Brugge at the Lancet conference, *The Challenge of Breast Cancer* (Evans, 1994). To investigate this hypothesis the incidence of breast cancer was assessed in just such a population—female transplant recipients. Results based on data provided since 1983 to the Collaborative Transplant Study in Heidelberg were published in 1995 (Stewart et al., 1995). Indeed, overall incidence of breast cancer was found to be significantly lower among 25,914 transplant recipients than would be expected from background rates. During a follow-up period of 1–11 years, 86 cases were observed compared to 113.8 expected (X_1^2=6.75, p =0.009). Incidence was particularly low in the first year following transplant with a relative risk of 0.49 (95% Cl 0.64 to 1.03). A subset of 13,003 women received a combination of cyclosporin, azathioprine, and steroids (CSA), and the remainder, only one or two of these drugs. There were 30 cases of breast cancer in the 13,003 patients against 53.8

expected, and 56 cases in 12, 911 patients treated otherwise against 60 expected, (RR 0.58, 95% Cl 0.36 to 0.93, p = 0.011). Furthermore the CSA group had a very low rate in the years following the first year, with 24 observed cases rather than the 39.8 expected, giving an SIR of 0.60. The low breast cancer rate applied only to kidney recipients, particularly the 8,166 North American women in whom the incidence was halved. (The number of heart recipients was small, 2,185, and the results are less reliable.) *All other major cancers* had higher than expected incidence, in some cases substantially, with marginal increases in others.

IMMUNOBIOLOGY OF MOUSE MAMMARY TUMORS

In vivo Evidence of Immune Recognition

About the time that T.S. was beginning to think about the possibility of treating breast cancer patients with preparations of their cancer, G.H. was being introduced to the new era of experimental tumor immunology by her graduate school advisor, David Weiss. There, in the laboratory, hopes for using the immune response to prevent or treat cancers were also running high. Pioneering work by Prehn and Main (1957) and the Klein's laboratory (1960), among others, had demonstrated potent immune responses specifically able to inhibit cancer growth in syngeneic animals that had been previously exposed to the cancer cells. These studies carried out with a variety of nonmammary tumor models had also revealed some apparent rules; for example, cancers induced by chemical carcinogens express immunogens unique to each cancer, whereas virus-induced cancers are immunologically cross-reactive (Habel, 1961).

In light of these promising results, Weiss had decided to give up his career on the immunology of tuberculosis and shift to cancer immunology, specifically breast cancer immunology. He joined the University of California (Berkeley) Cancer Research Genetics Laboratory, which was a mecca for workers in mouse mammary biology, and developed a team of students and post-docs to investigate aspects of the immune response to mammary tumors.

Over the next few years, a number of important observations on the immunobiology of mouse mammary cancers were made by the Weiss group (Weiss et al., 1966). Firstly, it was possible to immunize mice, by a number of ways, to transplants of either autografts or isografts of mammary tumors. However, the specificity of the response depended upon the type of host being tested. The tumors used were ones associated with the mouse Mammary Tumor Virus (MMTV). In hosts that had been neonatally infected (i.e., through their mother's milk) with MMTV, the immune response only extended to the tumor used for immunization. In syngeneic but MMTV-free hosts, immune responsiveness included other tumors of the same origin as the immunizing tumor. The interpretation of these results was that neonatal infection resulted in a state of effective "tolerance" to MMTV and associated antigens, so that only noninfected mice could respond to the common, viral-associated antigens expressed by tumor cells. The fact that infected mice could, however, mount an immune response to the particular tumor used for immunization meant that there were other, non-MMTV–associated immunogens in the system. The direction of the immune response was also interesting. Although immunized mice could respond to a challenge by inhibiting tumor growth, the opposite was also seen, that is, stimulation of tumor growth in previously sensitized hosts.

Much of this work, as well as the vast majority of studies in experimental tumor immunology, was done with transplantable tumors. However, Weiss and associates (1964, 1966) also studied the immune reactions to autochthonous mammary tumors, that is, to tumors as they developed, in their original host. The overall results of these studies were sobering: About 1/3 of the cancers produced no detectable response in the host of origin, whereas about 1/3 grew better than when transplanted into naive, syngeneic mice. Only 1/3 appeared to induce a protective response. Although these studies were logistically difficult and statistically problematic, it was clear that even in inbred strains of mice the immunological relationship between mammary cancers and their own hosts were heterogeneous and not necessarily protective.

G.H.'s work in Weiss' laboratory took a different approach, although with similarly sobering results. A number of investigators had shown that immunosuppression of mice by neonatal thymectomy greatly increased susceptibility to the growth of various kinds of nonmammary tumors (Law, 1966; Miller et al., 1963). G.H. decided to follow their lead to see whether neonatal thymectomy would alter the spontaneous development of mammary tumors in MTV-infected mice. Much to her surprise, the development of tumors as well as of their morphological precursors, hyperplastic alveolar nodules (HANs), was inhibited or delayed by neonatal thymectomy (Heppner et al., 1968), a finding also reported by others (Martinez, 1964; Squartini et al, 1970; Belyaev and Gruntenko, 1972). Other means of immunosuppression, for example, by anti-lymphocyte serum, had the same effects (Lappe and Blair, 1970). The mechanisms behind these unexpected findings were complex and included hormonal as well as immunological and viral factors. However, the "bottom line" result did not suggest that the immune response was protective.

In vitro Assays of Mammary Tumor Immunity

In the late 1960s the overall excitement in the field of tumor immunology had led to attempts to develop in vitro assays of anti-tumor reactivity in patients. A variety of in vitro assays were described in which lymphocytes from patients or animal hosts were tested for their ability to kill or to inhibit the proliferation of cancer cells. As a post-doctoral student, G.H. went to work in the laboratory of Karl-Eric and Ingegerd Hellstrom in Seattle and applied the then current "colony- inhibition assay" to the mouse mammary tumor system. Using lymphocytes from mice that had been transplanted with syngeneic mammary tumors, she was able to reproduce the observations from the in vivo transplantation experiments in Weiss' laboratory (Heppner and Pierce, 1969). Again depending upon the MMTV status of the host, immunity (i.e., inhibition of tumor colony formation) was demonstrable either only to the immunizing tumor or extended to other tumors of similar origin. With this assay, inhibition was fairly easy to demonstrate with lymphocytes that had been isolated from their host. However, the addition of serum from the same host often abrogated the inhibitory activity of their lymphocytes, a circumstance that was attributed to the presence of "blocking antibody" or antigen–antibody complexes (Heppner, 1969). These observations were not unique to mammary tumors, but were reproduced in a variety of other tumor types, both animal and human (Heppner, 1972).

"Immune Stimulation" of Mammary Tumor Cell Proliferation

The generally optimistic view of tumor immunology during the 1960s and 1970s was not without challenge. Prehn and Lappe (1971) suggested that a weak immune response, of the sort most likely to be seen against naturally occurring (nontransplanted) tumors that gradually develop without the involvement of "strong" carcinogens and become established without alerting their host in a threatening way, might be stimulatory rather than inhibitory (see Prehn, 1994 for a recent expansion of this concept). This suggestion was not well-received by the rapidly growing number of tumor immunologists; Indeed, it was mostly treated with ridicule. By this time, however, G.H., in collaboration with Dan Medina, had decided to test another mouse mammary tumor model involving tumors that arose spontaneously in mice implanted with a series of HAN lines that Medina (1973) had established and characterized. With these transplantable HAN lines, particularly the lines known as D2 and C4, it was possible to focus on the immune events during the critical time of progression from the preneoplastic to neoplastic phenotype. A first effort was to see whether the colony inhibition or newer microcytotoxicity assays for lymphocyte reactivity would reveal an immune response to D2 or C4 HANs and tumors. However, as reported in 1973 (Medina and Heppner) and more extensively in 1976 (Heppner et al.), stimulation, rather than inhibition, of HAN or tumor cell growth was often the result of in vitro exposure to host lymphocytes. Stimulation was not a random, or "feeder," effect, but was instead specific to particular combinations of lymphocytes and tumors.

In Fidler et al, (1974) stimulation of tumor growth in vitro by low ratios of sensitized lymphocytes to tumor cells was reported. Of the 13 dogs used in his studies, two had breast cancer—a miniature poodle and a beagle. The addition of autologous serum to the lymphocyte–tumor cultures potentiated the stimulation of growth. These authors concluded, as had Medina and Heppner (1973), that their results supported the work and hypothesis of Prehn.

More recently, Ögmundsdottir and colleagues (1995) were able to establish primary cultures of human breast cancers and unaffected tissues in the serum-free, hormonally defined, highly supplemented growth medium, CDM3, of Petersen and van Deurs (1987). Co-cultures of fresh samples of breast carcinoma and autologous peripheral lymphocytes were successful in 20 cases. Cancer cell growth significantly above control levels was seen with lymphocytes in 11 such cultures. Optimal growth was also seen in five of 17 co-cultures of uninvolved breast tissue and lymphocytes. Growth stimulation in response to lymphocytes was significantly associated with the expression of MHC class I by the tumor cells, suggesting that it depended upon immune recognition. Growth stimulation was not seen with a series of breast cell lines—only with primary cultures.

Immune Facilitation of Mouse Mammary Tumor Progression

As mentioned, the HAN model of Medina offered a good opportunity to investigate whether the immune response affects preneoplastic progression to mammary cancer. The particular HAN model chosen for further work was the preneoplastic C4 HAN line. The HAN lines are transplanted into "cleared" (i.e., epithelium-free) mammary fat pads of syngeneic, strain BALB/c mice; the HAN implants grow out to fill the pads; eventually, after periods of many months, focal tumors arise within the HAN tissue. Thus, tumor development is spontaneous and within the mammary site, although it occurs in transplanted HAN tissue. With this model, a variety of approaches were tried to detect an immune response to mammary tumor development.

The results with the C4 model were clear: The host response appeared to stimulate the development of the cancers by a mechanism that involved NK cells. Activated NK cells were thus found to be present in higher numbers in HANs (as compared with normal mammary epithelium) (Wei and Heppner, 1987), and suppression of NK activity was accompanied by a lengthening of the latency period and reduction in the frequency of tumor formation (Wei et al., 1989). In contrast, NK stimulation correlated with shortening of the latency period and increased tumor formation (Wei et al., 1989; Tsai et al., 1992). The idea that NK activity may be a stimulatory factor in mammary tumor growth is also supported by results in another animal model, the androgen-responsive Shionogi's mouse mammary tumor, in which Rowse and associates (1995) have demonstrated a positive correlation between tumor size and NK activity as a function of stress.

As stated above, NK activity has been reported to be low in human breast cancers, and breast cell infiltrates have been found capable of suppressing NK activity (Vose and Moore, 1979; Eremin et al., 1981). Similarly, although the numbers of NK cells were elevated in both C4 HAN and tumors, in comparison to normal glands, unlike in the HAN, tumor NK activity was depressed, probably due to the presence of suppressor cells. These C4 results resemble those of Pross and coworkers (1984), who found that a subset of women characterized as being of high-risk for developing breast cancer due to a mammographic pattern defined as "benign breast syndrome" had a history of abnormally high, peripheral NK activity. Pross and associates speculated that prolactin (PRL), reported to be at high levels in patients with proliferative breast disease (Cole et al., 1977; Franks et al., 1974), might contribute to NK stimulation. Interestingly, inhibition of pituitary PRL production by bromocriptine simultaneously decreases both progression of C4 HAN to tumors and the HAN-infiltrating NK activity (Tsai et al, 1992). Prolactin also stimulates NK activity in vitro (Tsai and Heppner, 1994). Taken together, these results suggest a possible influence at the level of the interface of the immune and endocrine systems during the early progression of breast cancer.

One reason that the C4 system was chosen for studies of immune activity during neoplastic progression was the belief that there was no viral agent (i.e., MMTV) present. However, this belief has since been shown to be false. Wei and associates (1991) have described a new, milk-transmitted virus, MMTV (C4), which is produced by C4 HAN. This virus encodes a superantigen that causes thymic deletion of T cells with the Vβ2 segment in their T cell receptor. MMTV (C4) has been shown to be responsible for at least some of the NK activation in this system (Gill et al., 1994).

The involvement of MMTV (C4) in the NK-mediated stimulation of preneoplastic progression, the evidence that PRL also contributes to this stimulation, and the work of Pross and colleagues (1984) indicating the possibility that PRL-driven NK cells are a feature in a subset of women at high risk of developing breast cancer raises the old question of a possible MMTV-like agent in human breast cancer. This is a topic that has had a rocky history. However, there are data that show 38–40% of human breast cancers contain gene sequences homologous to the MMTV env gene that are absent from other tumors and tissues (Wang et al., 1995). Indeed, Pross and co-workers (1984) suggested that the increased NK activity they observed could be due to viral stimulation of interferon production. Of particular relevance is the observation made by Cannon and colleauges (1982) that women with hyperplastic benign breast disease had a higher frequency of positive MMTV antigen leucocyte migration assays (a test of cell-mediated immunity) than did women without. Such hyperplastic lesions were judged homologous to the HAN in mice. Once again this calls to mind the work of Black (1972) who, using similar techniques, demonstrated enhanced migration to cryostat sections of noninvasive, in situ lesions with much less reactivity to invasive cancers, a study that was reproduced in detail by Wei and colleagues (1979) in the C4 HAN and tumor system. The parallel effects of immunosuppression on mice bearing preneoplastic lesions and on women immunosuppressed for organ transplantation also suggest that a viral cause of breast cancer in a subset of women should be considered. If true, future strategies could be developed to reduce the incidence of this pandemic cancer.

MECHANISMS OF IMMUNE FACILITATION OF BREAST CANCER DEVELOPMENT

The idea that immune responses may facilitate cancer development is initially counter- intuitive, given our basic assumption that immunity is a "host-defense" reaction. The immune system does protect us from external threats, ones that are sufficiently foreign to be perceived as not-self. There are many examples from parasitology, however, in which the foreign invader avoids immune destruction by becoming "like" the host. It should not be surprising that cancer cells, which differ from their hosts in only the most minute ways, can also escape immune destruction. Nevertheless, escaping the immune response is not the same as using it to promote growth. What special mechanisms might underlie immune facilitation?

To begin with, it is important to consider the difference between a cause–effect relationship, versus an association of events that are themselves quite independent. It has become recognized in recent years that the immune system is not an "island" with its own, unique mechanisms of modulation and control, but rather is connected with the other organ systems of the body via a variety of shared growth factors and hormones. Many, if not all, of the same factors that control the differentiation or proliferation of lymphoid cells also control other cell types (Paneri, 1993; Weigent and Blalock, 1995). For example, in the C4 HAN system described earlier, PRL appears to act as a mitogen for infiltrating lymphocytes (Tsai and Heppner, 1994) and possesses its well-known stimulatory effects on mammary epithelial cells (Welsch annd Nagasawa, 1977). Thus, although suppression of pituitary PRL suppresses HAN-infiltrating NK activity and simultaneously interferes with C4 HAN progression, this correlation is not in itself evidence that NK activity facilitates HAN progression. Quite simply, both cell types may be sensitive to the same control factor and yet be independent of each other.

There is evidence, however, that tumor-associated inflammatory cells are themselves the source of many factors that can act, in a paracrine fashion, as direct growth factors for cancer cells, and can act to influence cancer growth indirectly by, for example, enhancing angiogenesis. Again, in reference to the C4 HAN example, activated lymphocytes have been reported to produce a molecule with PRL activity (Montgomery et al., 1990), so that C4 HAN epithelial cells may be stimulated by PRL produced in situ by infiltrating lymphocytes, as well as by that produced by the pituitary. Tumor-associated macrophages produce a variety of endothelial cell cytokines and proteases that contribute to angiogenesis (Leek et al., 1996) thereby indirectly aiding tumor progression. They are also a source of active oxyradicals that could act as endogenous mutagens to increase the genetic instability of tumor cells and enhance the probability of progression (Mahoney et al., 1989; Yamashina et al., 1986).

Human breast tumor-infiltrating lymphocytes (TILs) have been extensively studied for their capacity to produce cytokines (Rubbert et al., 1991; Schwartzentruber et al., 1992; Vitolo et al., 1992). Two recent reviews (Mantovani et al, 1992; Michiel and Oppenheim, 1992) show clearly that cancer-infiltrating mononuclear cells produce cytokines that can either downregulate *or* stimulate cancer growth. This is an important conclusion that must be emphasized as it is not universally appreciated. Rubbert and colleagues (1991) found that TILs could produce tumor necrosis factor (TNF) when exposed to autologous tumor cells in vitro.

Mitogen-stimulated TILs also produced TNF, as well as IL2 and IFNα. Vitolo and colleagues (1992) studied the TILs expression of cytokine mRNAs. In the stroma of ductal breast cancers that contained intracellular or intraductal mucous, up to 30% of lymphoid cells expressed IL-2, TNFα, IFNα, and IL-2R mRNA. Schwartzentruber and colleagues (1992) showed that TILs stimulated by autologous breast cancers secreted TNFα, GM-CSF, and IFNα. Peoples and colleagues (1995) have shown that TILs produce heparin-binding epidermal growth factor–like growth factor (HB-EGF) and basic fibroblast growth factor (bFGF) in vitro as well as in vivo. HB-EGF and bFGF derived from TILs directly stimulated breast cancer cell proliferation in vitro and also stimulated vascular smooth muscle cells, whereas bFGF displayed angiogenic properties. Another potent angiogenic factor, vascular endothelial growth factor (VEGF), is produced by TIL in situ at bioactive concentrations in human prostate and bladder cancers (Freeman et al., 1995). Recent data (Lu and Brodie, 1996) have shown that both hormone-dependent and -independent human breast cancer cell lines are stimulated to grow by VEGF. Thus, a wealth of cytokines are produced by infiltrating inflammatory cells. The next question is which of these are most likely to be responsible for immune facilitation of breast cancer development?

TNF is a leading candidate for an inflammatory cell–produced cytokine that stimulates breast cancer development. Using a model of rat mammary epithelial cells growing in serum-free medium in a reconstituted basement membrane, Ip and associates (1992) found that TNFα stimulates epithelial cell proliferation at optimal growth conditions and markedly stimulates morphogenesis in suboptimal or EGF-deficient medium. Tsai and colleauges (1992) have found TNFα to be a positive growth factor for normal mouse mammary cells as well as for preneoplastic C4 HAN and C4 tumor cells. She has also demonstrated TNFα mRNA expression in both HAN-infiltrating lymphocytes (HILs) and HAN epithelial cells. In the milieu of HAN tissues, HILs are in close contact with HAN cells and may be constitutively secreting TNFα in response to autologous HAN stimulation. C4 HAN epithelial cells also express mRNA for both the p55-60 receptor and the p75-80 TNFα receptors. Most importantly, in vivo treatment of C4 HAN–bearers with TNFα decreases the latency period and enhances the frequency of HAN progression to tumor (Heppner and Miller, 1996). These results indicate that stimulation of epithelial cell proliferation by HIL-produced TNFα is one mechanism responsible for the immune facilitation of neoplastic progression in the HAN model. However, growth stimulation may not be the only mechanism by which TNFα affects progression; TNFα may promote tumor progression indirectly by its ability to stimulate angiogenesis (Vukanovic and Isaacs, 1995), to enhance epithelial cell motility (Rosen et al., 1991), or to induce increased expression of adhesion molecules (Spriggs et al., 1987), among other possibilities.

Other candidate inflammatory cell–produced, breast cancer–stimulating cytokines include IL-6 and colony-stimulating factor (CSF). Although IL-6 can inhibit the proliferation of several duct carcinomas of breast in vitro, it also enhances motility of breast cancer cells and causes increased cell–cell separation with decreased adherens-type function formation (Sehgal and Tamm, 1991). Furthermore, human mammary epithelial cells transfected with the *int-2* gene are stimulated to grow by IL-6 (Basolo et al., 1993). CSF-1 has been suggested as a key mediator of breast cancer invasion and metastasis (Scholl et al., 1993). Autocrine growth stimulation due to the combined expression of CSF-1 and its receptor in breast cancer cells is possible.

In addition to cytokines, infiltrating lymphocytes may be the source of other factors of specific relevance to breast cancer. Prolactin has already been mentioned. A number of investigators are exploring the possibility that lymphocytes either produce enzymes involved in estrogen synthesis (Berstein et al, 1993), or that inflammatory cell–produced cytokines affect the activity of such enzymes (Reed et al., 1995). Indeed, Singh and colleagues (1997) have recently shown that exposure of fibroblasts from normal breast tissue to conditioned medium from cultures of lipopolysaccharide- stimulated monocytes and lymphocytes obtained from an immunosuppressed kidney transplant recipient receiving cyclosporin A therapy resulted in a significant reduction in aromatase activity stimulation. This was in comparison with the marked stimulation seen using conditioned medium derived from monocytes and lymphocytes taken from a woman with breast cancer. The authors postulate that reduction in cytokine-induced estrogen synthesis in breast tissue of immunosuppressed patients may play an important role in lessening the incidence of breast cancer in such patients (Stewart et al., 1995). So far, there appear to be no studies of estrogen synthesis modulation by TIL, although Berstein and colleagues (1993) have reported that peripheral lymphocytes from breast cancer patients have a higher capacity to convert androstendione to estrogen than do peripheral lymphocytes of age- matched control women.

Antibodies are also a possible mechanism of breast cancer stimulation. Parkes and colleagues (1988) analyzed material taken from 34 patients who had been operated on for primary breast cancer in 1976 and followed for 11 years. Plasma cells were found in cancers from 84% of women who had relapsed and died (19 patients) whereas they were detected in only one tumor of 15 women who had survived (6%), a case of medullary carcinoma, free of disease. The authors concluded that the presence of plasma cells in infiltrating duct carcinoma and mixed infiltrating duct and lobular carcinoma is associated with a poor prognosis.

Using the Prausnitz-Kustner reaction, Grace and Dao (1958) found that a patient's antibodies may be responsible for inflammation in inflammatory breast cancer. HER-2/neu protein is amplified and overexpressed in inflammatory breast cancer and noninflammatory breast cancer with positive axillary nodes (Guerin et al., 1989). Antibody responses have been seen in breast cancer patients having this oncoprotein, in 647 cases that also had a lympho-plasma cell infiltrate (Pupa et al, 1993), and in 11 of 20 premenopausal breast cancer patients (Disis et al., 1994). One patient also showed a significant proliferative T cell response to the HER-2/neu protein and peptides. Antibodies to growth factors or growth factor receptors may facilitate cancer cell growth by contributing to the stimulus for signal transduction through that receptor. For example, in a recent study of a newly isolated human breast cancer cell line, it was shown that a monoclonal antibody that binds to the extracellular domain of erb B-2 is a potent growth factor in vitro for these cells (Ethier et al., 1996). Complimenting this finding is an earlier study by Stancovski and colleagues (1991) showing that monoclonal antibodies to the extracellular portion of the erbB2 protein of human breast cancer cell lines could either inhibit or strongly stimulate the growth of these cancers in athymic mice. They caution that antibody therapy could accelerate tumor growth.

CONCLUDING COMMENTS

The hope that the immune response may be harnessed to control cancer development and growth continues to burn brightly and, indeed, has received new fuel in the form of recent advances in gene and molecular therapy. It is likely that the treatment of many types of cancers will be impacted favorably by these advances and we surely wish that breast cancer will be one of these cancers. For this to be so, however, it is first necessary to confront the data on the role of immunity in the natural history of breast cancer. Unfortunately, these data are not encouraging. To be sure, immune responses of various types can be demonstrated and, in some types of laboratory settings, shown to be growth inhibitory. However, in human breast cancer as well as in experimental models of mammary cancer development (as opposed to growth of transplantable lines), it often seems that the inflammatory/immune response is associated with facilitation of cancer growth and progression. Potential mechanisms of facilitation include inflammatory cell-produced cytokines, modulators of steroid hormone metabolism, and antibodies, working either directly on the cancer cells themselves or indirectly through effects on angiogenesis or stromal cell behavior. The implications of these observations include the possibility that immunosuppression may actually be therapeutic for breast cancer. Optimistically, one may argue that any type of immune response to cancer may be better than no response at all in the sense that it indicates that the host does recognize the presence of "an invader" and, therefore, that there may be a basis for engineering that recognition to achieve a clinically favorable outcome. To do so, it will be necessary to select those immunogens and conditions that are able either to overcome or to deflect the immune-faciliatory responses that appear to play a significant role in the natural history of breast cancer development.

As stated at the outset of this chapter, our views of the role of the immune response in breast cancer development are a convergence of two, quite different pathways of investigation: clinical observation of human populations and laboratory studies of mouse models. The mouse models have all involved the presence of an overt viral agent. This would not seem to be the case for human breast cancer. However, mouse tumor–associated viruses are, in reality, endogenous genes packaged for export. One of the most fascinating recent findings of MMTV biology is the recognition that the murine *mls* genes, the genes that specify T cell–receptor, Vβ-chain specificity, and the *Mtv* provirus genes are, in fact, the same (Wei et al., 1993). The MMTV (*C4*) virus is the exogenous form of a gene that leads to deletion of the Vβ2 specificity from the T cell repertoire and, as reviewed above, appears to contribute to the enhanced NK activity observed in the C4 model. Thus, it may be that the involvement of MTV-like genes (if not MTV viruses) in human breast cancer immunobiology is not as far-fetched a concept as it might have appeared even a few years ago. If so, there are some clear opportunities for clinical exploitation.

ACKNOWLEDGMENTS

Our acknowledgments to all our colleagues and co-workers whose work has contributed to our views includes Drs. Bonnie Miller, Fred Miller, S-C Jane Tsai, and Wei-Zen Wei. We thank Dr. Jacob Karsh for editing the manuscript and Liz Lacasse, Mackenzie Duke, and Elaine Weber for their care in typing it. We also acknowledge the support of USPHS Grant CA61217 and of the Concern and Concern II Foundation.

REFERENCES

Basolo, F., Calvo, S., Fiore, L., Conaldi, P.G., Falcon, and Toniolo A. (1993). Growth-stimulating activity of interleukin-6 on human mammary epithelial cells transfected with the *int-2* gene. Cancer Res. 53, 2957-2960.

Belyaev, D.K. and Gruntenko, E.V. (1972). The influence of the thymus on the development of transplanted mammary tumors in mice. Int. J. Cancer 9, 1-7.

Berstein, L.M., Larionov, A.A., Krjukova, O.G., and Semiglazov, V.F. (1993). Conversion of androstendione in blood lymphocytes of breast carcinoma patients. J. Endrocrinol. 139, 17.

Berstein, L.M., Santner, S.J., Brodie, A.M., Koos, R.D., Naftolin, F., and Santen, R.J. (1993). Psuedoaromatase in circulating lymphocytes. J. Steroid Biochem. Molec.Biol. 44, 647-649.

Black, M.M. (1972) Cellular and biologic manifestations of immunogenicity in precancerous mastopathy. Natl. Cancer Inst. Monographs 35, 73-82.

Bloom, H.J.G., Richardson, W.W., and Field, J.R. (1970). Host resistance and survival in carcinoma of the breast: A study of 104 cases of medullary carcinoma in a series of 1,411 cases of breast cancer followed for 20 years. Brit. Med. J. 3, 181-188.

Brent, L., Brown and J., Medawar, P.B. (1958). Skin transplantation immunity in relation to hypersensitivity. Lancet 34, 561-563.

Cannon, G.B., Barsky, S.H., Alford, T.C., Jerome, L.F., Tinley, V., McCoy, J.L., and Dean, J.H. (1982). Cell-mediated immunity to mouse mammary tumor virus antigens by patients with hyperplastic benign breast disease. J. Natl. Cancer Inst. 68, 935-943.

Cannon, G.B., Dean, J.H., and Herberman, R.B. (1981). Lymphoproliferative responses to alloantigen by postoperative patients with stage I lung and breast cancer and their application to prognosis. In: Biological Relevance of Immune Suppression. (Dean, J.H. and Padarathsingh, M., Eds.), pp. 98-118, Van Nostrand Reinhold, New York.

Champion, H.R., Wallace I.W.J., and Prescott, R.J. (1972). Histology in breast cancer prognosis. Brit. J. of Cancer 26, 129-138.

Cole, E.N., Sellwood, R.A., England, P.C. and Griffiths, K. (1977). Serum prolactin concentrations in benign disease throughout the menstrual cycle. Europ. J. Cancer 13, 597-603.

DiPaola, M., Bertolotti, A., Colizza, S., and Coli, M. (1977). Histology of bronchial carcinoma and regional lymph nodes as putative immune response of the host to the tumor. J. of Thoracic and Cardiovas. Surg. 73, 531-537.

Disis, M.L., Calenoff, E., McLaughlin, G., Murphy, M.E., Chen, W., Groner, B., Teschke, M., Lydon, L., McGlynn, E., Livingston, R.B., Moe, R., and Cheever, M.A. (1994). Existent T-cell and antibody immunity to HER-2/neu protein in patients with breast cancer. Cancer Res. 54, 16-20.

Early Breast Cancer Trialists Collaborative Group (1992). Systemic treatment of early breast cancer by hormonal, cytotoxic, or immune therapy. Lancet 339 (8784), 1-15.

Eglinton, B.A., Roberton, DM, and Cummings A.G. (1994). Phenotype of T-cells, their soluble receptor levels, and cytokine profile of human breast milk. Immunol. and Cell Biol. 72, 306-313.

Elslton, C.W. (1969). Cellular reaction to choriocarcinoma. J. Path. 97, 261-268.

Eremin, O., Coombs, R.R.A., and Ashby, J. (1981). Lymphocytes infiltrating human breast cancers lack K-cell activity and show low levels of NK-cell activity. Brit. J. Cancer 4, 166-176.

Ethier, S.P., Kokeny, K.E., Ridings, J.W., and Dilts, C.A. (1996). ErbB family receptor expression and growth regulation in a newly isolated human breast cancer cell line. Cancer Res. 56, 899- 907.

Evans, I. (1994). The challenge of breast cancer (Lancet conference). Lancet 343, 1085-1086.

Fidler, I.J., Brodey, R.S., and Beck-Nielsen, S. (1974). In vitro immune stimulation—inhibition to spontaneous canine tumors of various histologic types. J. Immunol. 112, 1051-1060.

Fisher, B., Saffer, E.A. and Fisher, E.R. (1974). Studies concerning the regional lymph nodes in cancers VII. Thymidine uptake by cells from nodes of breast cancer patients relative to axillary location and histopathologic discrimination. Cancer 33, 271-279.

Fisher, E.R., Kotwal, N. Hermann, C., and Fisher, B. (1983) Types of tumor lymphoid response and sinus histiocytosis. Arch. of Path. and Lab. Medicine 107, 222-227.

Franks, S., Ralphs, D.N.L., Seagroatt, V., and Jacobs, H.S. (1974). Prolactin concentrations in patients with breast cancer. Brit. Med. J. 2, 320-321.

Freeman, M.R., Schneck, F.X., Gagnon, M.L., Corless, C., Soker, S., Niknejad, K., Peoples, G.E. and Klagsbrun, M. (1995). Peripheral blood T lymphocytes and lymphocytes infiltrating human cancer express vascular endothelial growth factor: A potential role for T cells in angiogenesis. Cancer Res. 55, 4140-4145.

Gill, R., Wang, H., Bluethmann, H., Inglesias, A., and Wei-W-Z. (1994). Activation of natural killer cells by mouse mammary tumor virus C4 in BALB/c mice. Cancer Res. 54, 1529-1535.

Grace, J.T. and Dao, T.L. (1958). Etiology of inflammatory reactions in breast cancer. Surgical Forum 9, 611-614.

Guerin, M., Gabillot, M., Mathieu, M.C., Travalgi, J.P., Spielman, M., Andrieu, N., and Riou, G. (1989). Structure and expression of *C-erb B-2* and EGF receptor genes in inflammatory and non-inflammatory breast cancer: Prognostic significance. Int. J. Cancer 43, 201-208.

Habel, K. (1961). Resistance of polyoma virus immune animals to transplanted polyoma tumors. Proc. Soc. Exptl. Biol. Med. 106, 722-725.

Head, J.F., Elliott, R.L. and McCoy, J.L. (1993). Evaluation of lymphocyte immunity in breast cancer patients. Breast Can. Res. and Trtmt. 26, 77-88.

Henderson. B.E., Pike, M.C., and Gray, G.E. (1981). The epidemiology of breast cancer. In: Breast Cancer (B Hoogstraten, RW McDivitt, Eds.), pp 1-25, CRC Press, Boca Raton.

Heppner, G.H. (1969). Studies on serum-mediated inhibition of cellular immunity to spontaneous mouse mammary tumors. Int. J. Cancer 4, 608-615.

Heppner, G.H. (1972). Blocking antibodies and enhancement. Series Hematologica 4, 41-66.

Heppner, G.H., Kopp, J.S., and Medina, D. (1976). Microcytotoxicity assay of immune responses to non-MTV-induced, preneoplastic, and neoplastic mammary lesions in strain BALB/c mice. Cancer Res.38, 753-758.

Heppner, G.H., Miller, B.E. (1996). Enhanced tumor progression following TNFα or interferon treatment of mice bearing preneoplastic lesions. Proc. Am. Assoc. Cancer Res. 37, 157.

Heppner, G.H. and Pierce, G.E. (1969). In vitro demonstration of tumor specific antigens in spontaneous mammary tumors of mice. Int J. Cancer 4, 212-218.

Heppner, G.H., Wood, P.C., and Weiss, D.W. (1968). Studies on the role of thymus in viral tumorigenesis and effect of thymectomy on induction of hyperplastic alveolar nodules and mammary tumors in Balb/cf C_3H mice. Israel J. Med. Sci. 4, 1195-1203.

Hollinshead, A.C., Stewart, T.H.M., and Herberman, R.B. (1974). Delayed hypersensitivity reactions to soluble membrane antigens of human malignant lung cells. J. Natl. Cancer Inst. 51, 1603-1607.

Hughes, L.E. and Lytton, B. (1964). Antigenic properties of human tumors: Delayed cutaneous hypersensitivity reactions. Brit. Med. J. 1, 209-212.

Ip, M.M., Shoemaker, S.F., and Darcy, K.M. (1992). Regulation of rat mammary epithelial cell proliferation and differentiation by tumor necrosis factor-α. Endocrinol. 130, 2833-2840.

Klein, G., Sjögren, H.O., Klein, E., and Hellström, K.E. (1960). Demonstration of resistance against methylcholanthrene-induced sarcomas in the primary autochthonous host. Cancer Res. 20, 1561-1572.

Kurtz, J.M., Jacquemier, J., Almaric, R., Brandone, H., Ayme, Y., Hans, D., Bressac, C., and Spitalier, J.M. (1990). Why are local recurrences after breast- conserving therapy more frequent in younger patients? J. Clin. Oncology 8, 591-598.

Lappe, M.A. and Blair, P.B. (1970). Interference with mammary tumorigenesis by antilymphocyte serum. Proc. Amer. Assoc. Cancer Res. 11, 47.

Law, L.W. (1966). Studies on thymic function with emphasis on the role of the thymus in oncogenesis. Cancer Res. 26, 551-574.

Leek, R.D., Lewis, C.E., Whitehouse, R., Greenall, M., Clarke, J., and Harris, A.L. (1996). Association of macrophage infiltration with angiogenesis and prognosis in invasive breast carcinoma. Cancer Res. 56, 4625-4629.

Levine, P.H., Mourali, N., Tabanne, F., Loon, J., Terasaki, P., Tsang, P., and Bekesi, J.G. (1981). Studies on the role of cellular immunity and genetics in the etiology of rapidly progressing breast cancer in Tunisia. Int. J. Cancer 27, 611-615.

Lu Q. and Brodie, A. (1996). Stimulation of the growth of MCF-7 and MDA-MB-468 breast cancer cells by vascular endothelial growth factor. Proc. Am. Assoc. Cancer Res. 37, 220.

Mahoney, K.H., Fulton, A.M., and Heppner, G.H. (1983). Tumor-associated macrophages of mouse mammary tumors. II. Differential distribution of macrophages from metastatic and non-metastatic tumors. J. Immunol. 131, 2079-2085.

Mancini, R.E., Anchada, J.A., Saraceni, A.E., Bachman, J.C., Lavieri, J.C., Remirousky, M.J. (1965). Immunological and testicular response in man sensitized with human testicular homogenate. J. Clin. Endocrinol. Metab. 25, 859-875.

Mantovani, A., Botazzi, B., Colotta, F., Sozzani, S., and Ruco, L. (1992). The origin and function of tumor-associated macrophages. Immunology Today 13, 265-270.

Martin, L.J.S., Woodruff, M.W., Webster, J.H., and Picken, J.W. (1965). Testicular seminoma: A review of 179 patients treated over a 50 year period. Arch. Surg. 90, 306-312.

Martinez, C. (1964). Effect of early thymectomy on development of mammary tumors in mice. Nature 203, 1188.

McKenzie, J.M. and Zakarija, M. (1986). Assays of thyroid-stimulating antibodies (TSAb) of Grave's disease. In: Werners's The Thyroid, a fundamental and clinical text (Ingbar, S.H., Braverman, L.E., Eds.), 5th ed., pp. 559-575, J.P. Lippincott Company, Philadelphia.

Medina, D. (1973). Preneoplastic lesions in mouse mammary tumorigenesis. Methods in Cancer Res. 7, 3-53.

Medina, D. and Heppner, G.H. (1973). Cell-mediated immunostimulation induced by mammary tumor virus-free BALB/c mammary tumors. Nature 242, 329-330.

Meyer, J.S. and Hixon, B. (1979). Advanced stage and early relapse of breast carcinomas associated with high thymidine-labeling indices. Cancer Res. 39, 4042-4047.

Michiel, D.F. and Oppenheim, J.J., (1992). Cytokines as positive and negative regulators of tumor promotion and progression. Cancer Bio. 3, 3-15.

Miller, J.F.A.P., Grant, G.H., and Roe, F.J.C. (1963). Effect of thymectomy on the induction of skin tumors by 3, 4 -benzopyrene. Nature 199, 920-922, 1963.

Montgomery, D.W., Le Fevre, J.A., Ulrich, E.D., Adamson, C.R., and Zuko, C.F. (1990). Identification of prolactin-like proteins synthesized by normal murine lymphocytes. Endocrinol. 127, 2601-2605.

Mourali N., Levine, P.H., Tabanne, F., Belhassen, S., Baki, J., Bennaceur, M., and Herberman, R. (1978). Rapidly progressing breast cancer (pousée évolutive) in Tunisia: Studies on delayed hypersensitivity. Int. J. Cancer 22, 1-3.

Murray, D., Hreno, A., Dutton, J., and Hampson, L.G. (1967). Prognosis in colon cancer, a pathologic reassessment. Cancer 20, 1976-1985.

Non-small-cell Lung Cancer Collaborative Group. (1995). Chemotherapy in non-small-cell lung cancer: A meta-analysis using updated data on individual patients from 52 randomized clinical trials. Brit. Med. J. 311, 899-909.

Ögmundsdottir, H.M., Petersdottir, I., Gudmunsdottir, I. (1995). Interactions between the immune system and breast cancer. Acta Oncologica 34, 647-650.

Olsen, G.A., Gockerman, J.P., Basr, R.C., Borowitz, M., and Peters, W.P. (1988). Altered immunologic reconstitution after standard-dose chemotherapy or high-dose chemotherapy with autologous bone marrow support. Transplantation 46, 57-60.

Panerai, A.E. (1993). Lymphocytes as a source of hormones and peptides. J. Endocrinol. Invest. 16, 549-557.

Parkes, H., Collis, C., Baildam, A., Ralphs, D., Lyons, B., Howell, A., and Craig, R. (1988). In situ hybridization and S_1 mapping show that the presence of plasma cells is associated with poor prognosis in breast cancer. Brit. J. of Cancer 58, 715-722.

Peoples, G.E., Blotnik, S., Takahashi, K., Freeman, M.R., Klagsbrun, M., and Eberlen, T.J. (1995). T lymphocytes that infiltrate tumors and atherosclerotic plaques produce heparin-binding epidermal growth factor-like growth factor and basic fibroblast growth factor: a potential pathologic role. Proc. Natl. Acad. Sci. USA 92, 6547-6551.

Petersen, O.W. and van Deurs, B. (1987). Preservation of defined phenotypic traits in short-term cultured human breast carcinoma derived epithelial cells. Cancer Res. 47, 850-866.

Pouillart, P., Palangie, T., Jouve, M., Garcia-Giralt, E., Vilcoq, J.R., Bataini, J.P., Calle, R., Fenton, J., Mathieu, S., Rousseau, J., and Asselain, B. (1981). Cancer inflammatoire du sein traité par une association de chimio-therapie et d'irradiation, resultat d'un essai randomise etudiant le rôle d'une immunotherapie par le BCq, Bulletin du Cancer 68, 171-186.

Prehn, R.T. (1994). Stimulatory effects of immune reactions upon the growth of untransplanted tumors. Cancer Res. 54, 908-914.

Prehn, R.T. and Lappe, M.A. (1971). An immunostimulation theory of tumor development. Transplant Rev. 7, 26-54.

Prehn, R. T. and Main, J.M. (1957). Immunity to methylcholanthrene-induced sarcomas. J. Natl. Cancer Inst. 18, 769-778.

Pross, H.F., Sterns, E., and MacGillis, D.R. (1984). Natural killer cell activity in women at "high risk" for breast cancer, with or without benign breast syndrome. Int. J. Cancer 34, 303-308.

Pupa, S.M., Menard, S., Andreola, S., Rilke, F., Cascinelli, N., and Colnaghi, M.I. (1992). Prognostic significance for C-erb B-2 oncoprotein overexpression in breast carcinoma patients and its immunological role. Proc. Amer. Assoc. of Cancer Res. 31, 312.

Pupa, S.M., Menard, S., Andreola, S., and Colnaghi, M.I. (1993). Antibody response against the c-erb B-2 oncoprotein in breast carcinoma patients. Cancer Res. 53, 5864-5866.

Raman, S., Stewart, T.H.M. (1992). Meta-analysis of clinical trials of specific immunotherapy of non-small-cell lung cancer. In: Cellular Immune Mechanisms and Tumor Dormancy (Stewart, T.H.M. and Wheelock, E.F., Eds.), pp. 333-347, Boca Raton, CRC Press.

Reed, M.J., Purobit, A., Duncan, L.J., Singh, A., Roberts, C.J., Williams, G. J., and Potter, B.V.L. (1995). The role of cytokines and sulphatase inhibitors in regulating estrogen synthesis in breast tumors. J. Steroid Biochem. Molec. Biol. 53, 413-420.

Robinson, E., Rubin, D., Mekori, T., Segal, R., and Pollack, S. (1993). In vivo modulation of natural killer cell activity by tamoxifen in patients with bilateral primary breast cancer. Cancer Immunol. Immunother. 37, 209-212.

Rosen, E.M., Goldberg, I.D., Liu, D., Setter, E., Donovan, M.A., Bhargava, M., Reiss, M., and Kacinksi, B.M. (1991) Tumor necrosis factor stimulates epithelial tumor cell motility. Cancer Res., 51, 5315-5321.

Rosen, P.P. and Groshen, S. (1990). Factors influencing survival and prognosis in early breast carcinoma: assessment of 644 patients with median follow-up of 18 years. Surg. Clin. Of North America 70, 937-962.

Rowse, G., Weinberg, J. and Emerman, J. (1995). Role of natural killer cells in psychosocial, stressor-induced changes in mouse mammary tumor growth. Cancer Res. 55, 617-622.

Rubbert, A., Manger, B. Lang, N., Kalden, J.R., and Platzer, E. (1991). Functional characterization of tumor-infiltrating lymphocytes, lymph-node lymphocytes and peripheral blood lymphocytes from patients with breast cancer. Int. J. Cancer 49, 25-31.

Scholl, S.M., Crocker, P., Tang, R., Pouillart P., and Pollard, J.W. (1993). Is colony-stimulating factor-1 a key mediator of breast cancer invasion and metastasis. Molecular Carcinogenesis 7, 207-211.

Schwartzentruber, D.J., Solomon, D., Rosenberg, S.A., and Topalian, S.L. (1992). Characterization of lymphocytes infiltrating human breast cancer: Specific immune reactivity detected by measuring cytokine secretion. J. of Immunotherapy 12, 1-12.

Seghal, P.B. and Tamm, I. (1991). Interleukin-6 enhances motility of breast carcinoma cells. In: Cell Motility Factors (Goldberg, I.D., Ed.), pp. 178-193, Birkhäuser Verlag Basel, Switzerland.

Singh, A., Purohit, A., Duncan, L.J., Mokbel, K., Ghilchik, M.W., and Reed, M.J. (1997). Control of aromatase activity in breast tumors: The role of the immune system. J. Steroid Biochem. Molec. Biol. 61, 185-192.

Spratt, J.S. and Spjut, H.S. (1967). Prevalence and prognosis of individual and pathologic variables associated with colorectal carcinoma. Cancer 20, 1976-1985.

Spriggs, D., Inamura, K., Rodriguez, C., Horiguchi, J., and Kufe, D.W.(1987). Induction of tumor necrosis factor expression and resistance in a human breast tumor cell line. Proc. Natl. Acad. Sci. USA 84, 6563-6566.

Squartini, F., Olivi, M., and Bolis, G.B. (1970). Mouse strain and breeding stimulation as factors influencing the effect of thymectomy on mammary tumorigenesis. Cancer Res. 30, 2069-2072.

Stancovski, L., Hurwitz, E., Leitner, O., Ullrich, A., Yarden, Y., and Sela, M. (1991). Mechanistic aspects of the opposing effects of monoclonal antibodies to the ERBB-2 receptor on tumor growth. Proc. Natl. Acad. Sci., USA 88, 8691-8695.

Stewart, T.H.M. (1994). Post-mastectomy radiotherapy: More than a local matter. Lancet 344,351.

Stewart, T.H.M. (1969a). The presence of delayed hypersensitivity reactions in patients toward cellular extracts of their own tumors. 1. The role of tissue antigen, non-specific reactions of nuclear material, and bacterial antigen as a cause for this phenomenon. Cancer 23, 1368-1379.

Stewart, T.H.M. (1969b). The presence of delayed hypersensitivity reactions in patients toward cellular extracts of their own tumors. 2. A correlation between the histologic picture of lymphocytic infiltrate of the tumor stroma, the presence of such a reaction, and a discussion of the significance of this phenomenon. Cancer 23, 1380-87.

Stewart, T.H.M., Hollinshead, A.C., Harris, J.E., Ramen, S., Belanger, R., Crepeau, A., Crook, A.F., Hinte, W.E., Hooper, D., Klaasen, D.J., Rapp, E.F., and Sachs, H.J. (1977). Specific active immunochemotherapy in lung cancer: A survival study. Can. J. Surg. 20, 370-377.

Stewart, T.H.M., Hollinshead, A., and Raman, S. (1991). Tumor dormancy: Initiation, maintenance and termination in animals and humans. Can J. Surg. 34, 321-325.

Stewart, T.H.M. and Orizaga M.(1971). The presence of delayed hypersensitivity reactions in patients toward cellular extracts of their malignant tumors. 3. The frequency, duration, and cross reactivity of this phenomenon in patients with breast cancer and its correlation with survival. Cancer 28, 1472-1478.

Stewart, T.H.M. and Raman, S. (1992). Adjuvant treatment of non-small-cell lung cancer: a meta-analysis. In: Cellular Immune Mechanisms and Tumor Dormancy (Stewart, T.H.M. and Wheelock, E.F., Eds.), pp. 213-238, Boca Raton, CRC Press.

Stewart, T.H.M., Raman, S., Eidus, L., Sachs, H.J., Crepeau, A., Belanger, R., Stewart, D.J., Nair, B.D., Maroun, J.A., Verma, S., Yesner, R., and Hollinshead, A.C. (1990). Two patients with non-regional metastases of adenocarcinoma of the lung 11 and 14 years following surgery. Lung Cancer 6, 28-40.

Stewart, T.H.M., Retsky, M.W., Tsai, S.C.J., and Verma S. (1994). Dose response in the treatment of breast cancer. Lancet 343, 402-404.

Stewart, T.H.M., Shelley, W.E., Willan, A.R., and Hollinshead, A. (1986). An evaluation of the role of tumor-specific antigens. In: Annual Clinical Conference on Cancer; Lung Cancer: Current status and prospects for the future. The University of Texas System Cancer Center 28, 351-374.

Stewart, T.H.M. and Tsai, S-C., J. (1993). The possible role of stromal cell stimulation in worsening the prognosis of a subset of patients with breast cancer. Clin. Exp. Metastasis 11, 295-305.

Stewart, T.H.M., Tsai, S.C.J.,Grayson, H., Henderson, R., and Opelz, G. (1995). Incidence of de-novo breast cancer in women chronically immunosuppressed after organ transplantation. Lancet 346, 796-798.

Stewart, T.H.M. and Wheelock, E.F. (1992). Cellular Immune Mechanisms and Tumor Dormancy. CRC Press, Boca Raton.

Takita, H., Hollinshead, A., Hart, T. Jr., Bhayana, J., Adler, R., Rao, U., Moskowitz, R., and Ramundo, M. (1985). Adjuvant specific immunotherapy of respectable squamous cell lung carcinoma. Analysis at the eighth year. Cancer Immunol Immunother. 20, 231-235.

Tang, R., Kacinski, B., Validire, P., Beuron, F., Sastre, X., Benoit, P., de la Rochefordière, Mosseri, V., Pouillart, P., and Scholl, S. (1990). Oncogene amplification correlates with dense lymphocytic infiltration in human breast cancers: A role for hematopoietic growth factor release by tumor cells? J. of Cell Biochem. 44, 189-198.

Tsai, S-J. and Heppner, G.H. (1994). Immunendocrine mechanisms in mammary tumor progression: Direct prolactin modulation of peripheral and preneoplastic hyperplastic-alveolar-nodule infiltrating lymphocytes. Cancer Immunol. Immunother. 39, 291-298.

Tsai, S-J., Loeffler, D.A., and Heppner, G.H. (1992). Associated effects of bromocriptine on neoplastic progression of mouse mammary preneoplastic hyperplastic alveolar nodule (HAN) line C4 and on HAN-infiltrating and splenic lymphocyte function. Cancer Res.52, 2209-2215.

Tsai, S-J., Miller, F., Wei, W-Z, and Heppner, G.H. (1992). Tumor necrosis factor (TNF) modulates growth of normal and neoplastic murine mammary cells. Proc. Am. Assoc. Cancer Res. 33, 269.

Vitolo, D. Zerbe, T., Kanbour, A., Dahl, C., Heberman, R.B., and Whiteside, T.L. (1992). Expression of mRNA for cytokines in tumor-infiltrating mononuclear cells in ovarian adenocarcinoma and invasive breast cancer. Int. J. Cancer 51, 573-580.

Volpé, R. (1986). Pathogenesis of autoimmune thyroid disease. In: Werners's The Thyroid, a fundamental and clinical text (Ingbar, S.H. and Braverman, L.E., Eds.), 5th ed., pp.747-767, J.P. Lippincott Company, Philadelphia.

Vose, B.M. and Moore, M. (1979). Suppressor cell activity of lymphocytes infiltrating human lung and breast tumors. Int J. Cancer 24, 579-585.

Vukanovic, J. and Isaacs, J.T. (1995). Linomide inhibits angiogenesis, growth, metastasis, and macrophage infiltration within rat prostatic cancers. Cancer Res. 55, 1499-1504.

Waksman, B.H. (1959). Experimental allergic encephalomyelitis and the "autoallergic" diseases. Int. Arch. All., Suppl. 14, 1-87.

Waksman, B.H. (1962). Autoimmunization and the lesions of autoimmunity. Medicine (Balt) 14, 93-141.

Wanebo, H.J., Rosen, P.P., Thaler, T., Urban, J.A., and Oettgen, H.F. (1976). Immunobiology of operable breast cancer: An assessment of biologic risk by immunoparameters. Ann. of Surgery 184, 258-267.

Wang, Y., Holland, J.F., Bleiweiss, I.J., Melana, S., Liu, X., Pelisson, I., Cantarella, A., Stellrecht, K., Mani, S., and Pogo, B.G.T. (1995). Detection of mammary tumor virus ENV genelike sequences in human breast cancer. Cancer Res. 55, 5173-5179.

Wei, W-Z, Ficsor-Jacobs, R., Tsai, S.J., and Pauley, R. (1991). Elimination of Vβ2-bearing T cells in BALB/c mice implanted with syngeneic preneoplastic and neoplastic mammary lesions. Cancer Res. 51, 3331-3333.

Wei, W-Z., Fulton, A., Winkelhake, J., and Heppner, G. (1989). Correlation of natural killer activity with tumorigenesis of a preneoplastic mouse mammary lesion. Cancer Res. 49, 2709-2715.

Wei, W-Z., Gill, R.F., and Wang, H. (1993). Mouse mammary tumor virus associated antigens and superantigens: immuno-molecular correlates of neoplastic progression. Sem. In Cancer Biol. 4, 205-213.

Wei, W-Z. and Heppner, G.H. (1987). Natural killer activity of lymphocytic infiltrates in mouse mammary lesions. Br. J. Cancer 55, 589-594.

Wei, W-Z. and Heppner, G.H. (1996) Breast Cancer Immunology. In: Mammary Tumor Cell Cycle, Differentiation and Metastasis. (R. Dickson and M. Lippman, Eds.) pp. 395-410, Kluwer Acad. Pub., Boston.

Wei, W-Z., Miller, F.R., Blazar, B.A., Medina, D., and Heppner, G. (1979). Opposing effects of cryostat sections of preneoplastic and neoplastic mouse mammary lesions on in-vitro migration of peritoneal exudate cells. J. Immunol 122, 2059-2067.

Weigent, D.A. and Blalock, J.E. (1995). Associations between the neuroendocrine and immune systems. J. Leukocyte Biol. 58, 137-150.

Weiss, D.W., Faulkin, L.J., and De Ome, K.B. (1964). Acquisition of heightened resistance and susceptibility to spontaneous mouse mammary carcinomas in the original host. Cancer Res. 24, 732-741.

Weiss, D.W., Lavrin, D.H., Dezfulian, M., Vaage, J., and Blair, P.B. (1966). Studies on the immunology of spontaneous mammary carcinomas of mice. In: Viruses Inducing Cancer (W.J. Burdette, Ed.), pp. 138-168, Univ.Utah Press, Salt Lake City.

Welsch, C.W. and Nagasawa, H. (1977). Prolactin and murine mammary tumorigenesis: A review. Cancer Res. 37, 951-963.

Welte, K., Ciobañu, N., Moore, M.A.S., Gulati, S., O'Reilly, R.J., and Mertelsman, R. (1984). Defective interleukin 2 production in patients after bone marrow transplantation and in vitro restoration of defective T lymphocyte proliferation by highly purified interleukin 2. Blood 64, 380-385.

Whiteside, T.L., Miescher, S., Hurlimann, J., Moretta, L., and von Fliedner, V. (1986). Clonal analysis and in situ characterization of lymphocytes infiltrating human breast carcinomas. Can Immunol. Immunother. 23, 169-178.

Wolf, J., Spear, P., Yesner, R., and Patno, M.E. (1960). The sex hormones and cortisone in the treatment of inoperable bronchogenic carcinoma. In: Biological Activities of Steroids in Relation to Cancer (Pincus, G. and Vollmer, E.P., Eds.), pp. 413-426, Academic Press, New York.

Yamashina, K., Miller, B.E., and Heppner, G.H. (1986). Macrophage-mediated induction of drug-resistant variants in a mouse mammary tumor cell line. Cancer Res. 46, 2396-2401.

Yesner, R. (1992). A pathologist's view of tumor dormancy. In: Cellular Immune Mechanisms and Tumor Dormancy (Stewart, T.H.M. and Wheelock, E.F., Eds.), pp. 17-25, CRC Press, Boca Raton.

Zielinski, C.C., Müller, C., Kubista, E., Staffen, A., and Eibl, M.M. (1990). Effects of adjuvant chemotherapy on specific and nonspecific immune mechanisms. Acta. Med. Austr. 17, 11-14.

A HISTORY OF CANCER OF THE MALE BREAST

Peter J. Dawson

In this chapter, early work up to the end of the nineteenth century will be considered chronologically. Our understanding in the twentieth century of the clinical and pathologic features of the disease, the hormonal influences, and genetic factors will be discussed under these headings.

EARLY HISTORY

Diseases of the male breast have been known since antiquity, but male breast cancer was not recognized until the Middle Ages. Because the early history of male

Advances in Oncobiology
Volume 2, pages 229-243.

ISBN: 0-7623-0388-3

breast cancer does not appear to have been collected before, I will review it in some detail.

Gynecomastia has been recognized since ancient times and is represented in both early and primitive art. While it was known to Aristotle (Melville, 1933), the term gynecomastia was first coined by Galen in the second century B.C. (Erdheim, 1930), who used it to describe increased amounts of fat in the breast. There are several anecdotal cases of men with gynecomastia who have suckled children (Chase, 1964) but none of these stand up to close scrutiny (Karsner, 1946). Although, on occasion, carcinoma in situ may be seen in ducts in association with gynecomastia, there is no good evidence linking sporadic gynecomastia with an increased risk of breast cancer. A possible exception is in patients with Klinefelter's syndrome who frequently have longstanding gynecomastia and are at increased risk for the development of breast cancer (v. infra).

The earliest reference to a tumor in the male breast is to be found in Case 45 of the *Edwin Smith Surgical Papyrus* which dates from 1600 BC (Figure 1). The hieroglyphics have been translated (Breasted, 1930).

> If thou examinest a man having bulging tumors on his breast, (and) thou findest that [swellings] have spread over his breast; if thou puttest thy hand upon his breast upon these tumors (and) if thou findest them very cool, there being no fever at all therein when thy hand touches him; they have no granulation, they form no fluid, they do not generate secretions of fluid, and they are bulging to thy hand, thou shouldest say concerning him: One having bulging tumors. An ailment with which I will contend.

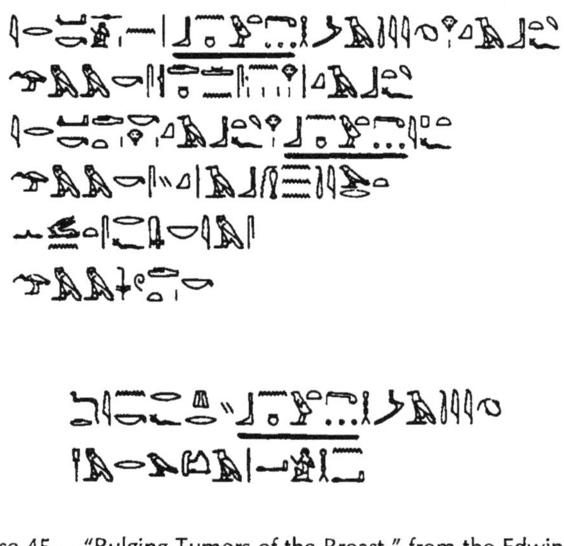

Figure 1. Case 45— "Bulging Tumors of the Breast," from the Edwin Smith Surgical Papyrus. A copy of the original hieroglyphics, which are translated in the text. The hieroglyphics for tumor are underlined. (Modified from Azar, A.H. *Pathology of Human Neoplasms,* (1988). p. 2. reprinted by permission of Lippincott-Raven Publishers.)

While this case has been interpreted as a tumor, Sandison and Wells, (1967) feel that it is better regarded as superficial presentation of caseous tuberculous osteitis. In their opinion there is no convincing evidence of breast cancer in ancient Egyptian literature.

The first description of male breast cancer in the Western world is attributed to John of Arderne (1307–1390), who came from an established family in the English Midlands (Power, 1910). He is thought to have been educated at Montpelier, and served as a surgeon in the British Army in the Hundred Years' War. He returned to England to practice first in Newark and later in London where he made his reputation by his operation for fistula-in-ano, a common and, at that time, incurable condition. In his description of male breast cancer, he recounts the story of "a priest of Colstone" to whom "there fell a sore of the right pap within the skin" (Graham, 1939). Over a 2-year period, the cancer extended and the skin became livid. After trying various medicines, the priest consulted a barber (surgeon) who promised to cure the condition. But before undergoing treatment, he consulted John of Arderne. When asked "if that he were curable or if that he might suffer any cutting or corrosive or any such medicine," Master John of Arderne's advice was "that he should in no manner put no corrosive nor none other violent medicines, nor let no cuttings come therenigh." He further warned the priest that any of these treatments "would bring him to death without any recovery." It seems likely that this is not only the first clinical description of breast cancer in a man, but also of Paget's disease of the nipple.

Francesco Arceo or Arceus (1493–1571), who practiced in Spain when medicine there was at its zenith, wrote (1588; Figure 2):

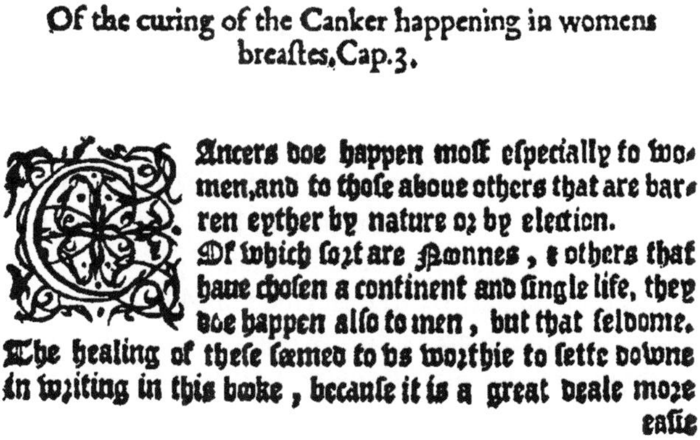
Of the curing of the Canker happening in womens breastes,Cap.3.

Cancers doe happen most especially to women,and to those aboue others that are barren eyther by nature or by election. Of which sort are Nonnes, & others that haue chosen a continent and single life, they doe happen also to men, but that seldome. The healing of these seemed to be worthie to sette downe in writing in this booke, because it is a great deale more easie

Figure 2. The beginning of Francesco Arceo's description of breast cancer from "A most excellent and COMPENDIOUS METHOD of curing woundes in the head, and in other partes of the body, with other precepts of the same Arte, practiced and written by that famous man FRANCISCVS ARCEVS, Doctor in Phisicke & Chirugery: and translated into English by John Read, Chirurgion," London, 1588.

Of the Curing of Canker Happening in Womens Breasts

> Cancers doe happen most especially to women, and to those above others that are barren either by nature or by election. Of which last are Nunnes, and others that have chosen a continent and single life, they doe happen also to men, but that seldome ...

He distinguished ulcerating from nonulcerating tumors and recognized that the former can only be palliated. For the latter, he advocated surgery. He recalls: "a certaine man, a priest, we did so preserve above XX yeeres with this kinde of cure, before he ended his life, he was thoroughlie cured. And when he deceased he was more than foure score yeeres of age."

Ambroise Paré (1510–1590), a leading French surgeon of the sixteenth century was also well aware of breast cancer, he wrote (1634):

> Every cancer is held almost incurable ...a Cancer is not easily staied until it hath eaten even to the innermost of the part which it possesses. It invades women more frequently than men, and those parts which are laxe rare, fungous, and glandulous are therefore opportune to receive a deflection of a grosse humor, such are the breasts...When it possesses the breasts, it often causes inflammation to the armholes, and sends the swelling even to the glandules thereof.

It is not clear that his reference to cancer being more common in women than men, refers specifically to breast cancer; nor is it clear that he actually was aware that the latter could occur in men. However, he did understand that it might spread to axillary lymph nodes. Paré also described a case of a woman who feigned breast cancer and was whipped for her wickedness.

Wilhelm Fabricius, also known as Fabricius Hildanus (1560–1636), was born in Germany at Hilden near Düsseldorf but practiced in Switzerland and is considered to be the founder of German surgery. He described, in a certain Monsieur Polier of Lausanne, an ulcerated tumor of the left breast the size of a hen's egg (Hildanus, 1606). Treatment with emollients caused inflammation and pain, and so was discontinued. The patient lived a long-time.

The Danish anatomist Thomas Bartholin (1616–1680) is often credited with an early description of breast cancer in a man (1671–2). However, the patient was called Christina and appears to have been a woman.

A German surgeon, named Wolff described (1742) a nodule in a 50-year-old man that appeared in the breast following a "pinch" that grew to the size of a fist over a three year period. The patient refused surgery and went to another physician who first treated it with caustics and then incised it. One week later the patient developed paralysis of the lower jaw, presumably tetanus, and died.

Lorenzo Heister (1683–1758) was born in Frankfurt and studied medicine at the University of Giessen. He was professor of Anatomy and Surgery at the University of Helmstaedt from 1720 to 1758. His textbook, A *General System of Surgery* was the first modern and comprehensive surgical text (Heister, 1750). In it he wrote:

> A Cancer, as well as a Scirrhus, will arise in almost any part of the Body, but most frequently in the Breasts of Women, nay sometimes of Men; a very memorable Instance of which you will find recorded by Bidloo.

Govert Bidloo (1649–1713) was professor of Anatomy at Leyden. He published in his *Opera Omnia Anatomico-Chirurgica* (1715) an account of a breast tumor that developed in a 22-year-old trumpeter following a hunting accident. The patient was cured by mastectomy. However, the lesion was almost certainly inflammatory.

Giovanni Batista Morgagni (1682–1771), Professor of Anatomy at the University of Padua for more than 50 years, published his magnum opus *De Sedibus et Causis Morborum* in 1761. In it, he mentions that he had seen a tumor in the breast of a goldsmith, but it is also not clear that it was carcinoma (Morgagni, 1769).

A novel approach to breast cancer was taken by Domenico Rigoni-Stern (1810–1855) who was educated at Padua and became deputy professor of clinical medicine there. He analyzed cancer incidence data over a long period of time in the expectation of increasing our understanding of the disease, and was thus one of the first cancer epidemiologists. He communicated to the surgical section of the IV Congress of Italian Scientists the results of his epidemiologic studies in the city and suburbs of Verona during the period of 1760 to 1839 (Rigoni-Stern, 1842). He observed that the highest frequency of breast cancer was between the 6th and 7th decades and that it was five times more frequent in nuns than among other women and that it was more common in the left than the right breast. He also recorded four cases of breast cancer in men, all of them priests. It will be recalled that both John of Arderne and Francesco Arceus also described breast cancer in priests. (It is not clear whether this is of etiological significance or simply reflects the higher probability of an educated man seeking medical attention.) Rigoni-Stern speculated that the high incidence of breast cancer among members of religious orders might be due to excessive use of fish or garlic or long periods of abstinence; or possibly to the effects of pressure from corsets or prolonged kneeling.

Velpeau (1795–1867), who was professor of surgery in Paris, published *Traité des Maladies du Sein et de la Region Mammaire* (1854). In the chapter on diseases of the male breast, he mentioned that he had observed nine or 10 cases of breast cancer in men and described several of them in detail. He concluded that the disease in men too closely resembled that in women to justify separate detailed treatment in his book.

There appears to have been continuing interest in the subject in France. Horteloup (1872) gave a comprehensive account of male breast disease in his thesis published in Paris in 1872, and Chenet (1876) addressed the same subject in his thesis. Poirier (1883) published the first systematic survey of cancer of the male breast, the first paragraph of which was quoted by Wainwright (1927).

> The condition more often begins in the second half of life, 45 to 65 years. It begins ordinarily as a small limited induration, under or near the nipple. It remains in this

state sometimes for a considerable period without causing appreciable pain and without affecting the general health. Then suddenly, after six months to five years, the tumor increases in volume and soon is the size of a nut or an egg. The skin becomes adherent and dimpled. The nipple retracts and becomes buried. The tumor becomes hard and nodular. Ordinarily about this period one or two glands appear in the axilla, sometimes they are connected to the tumor by a nodular cord. Quite often one now sees small nodules or tubercles appearing around the tumor, they have the characteristics of the principal mass. The patient now begins to complain of lancinating pains. Then only the skin, distended and adherent to the tumor, reddens and perforates, there is ulceration; the ulcer is excavated, the borders hard, the base is firm, it increases slowly and it may be the seat of hemorrhages. Such is the picture which presents ordinarily. The general health is good for a long time, progress is always slow. This clinical type corresponds to the anatomical form which we have described under the name of scirrhus carcinoma. But other cases with a different histological structure progress more rapidly. They are much more rare.

The earliest descriptions of the histopathology of male breast cancer were made by Poirier (1883) and by Courtade (1885). Imbert (1891) reported two cases of male breast cancer with microscopic descriptions of the tumors that were surprisingly modern and complete.

On low power, we observe a considerable dilation of external ducts; these are lined by a layer of epithelium much thicker than normal. Rare ducts have an intact basement membrane that is infiltrated and destroyed by an epithelial proliferation, focally or along its entire circumference..... The cells that invade the basement membrane around the acini are of cylindrical type; the most advanced are rounded and have a large nucleus. In the excretory ducts, the cells are cylindrical with multiple nuclei, but the lumina of these ducts are filled with necrotic cellular debris.... The sections from the nipple perpendicular to the lactiferous ducts allows (sic) us to notice that these ducts are large, tortuous and lined by several layers of cylindrical cells. In certain foci, these cells form from intracanalicular masses; it is at this level that the basement membrane of the duct is destroyed and invaded by epithelial cells that form irregular prolongations that are poorly limited but, however, following the form of circles or cylinders depending on the cut.

Working in Germany, Schuchardt compiled (1885) a remarkable series of 406 cases of breast disease in men, of which 348 were cancers. The majority of these came from Western Europe; cases from Russia, Japan, and North America were also included. In this series, patients ranged in age from under 20 to 84 years. Generally, the disease was far advanced when medical advice was sought; for example, 61 of 70 cases had ulcerated tumors. In 25 cases, the tumor was attributed to preexisting trauma or inflammation and in one case to suspenders. Survival ranged from three months to 18 years.

Williams established the now generally accepted relative incidence of breast cancer in women and men; he reported that, among 2422 cases of breast cancer from three London teaching hospitals, 25 were in men (incidence = 1%; Williams, 1889). He noted the same age range as Schuchardt, probably because many of the cases were common to the two series. His youngest patient was 20 years, the oldest 82 years. The fact that his mean age was 50 years (about 8 years less than that now generally accepted) may be a reflection of the shorter life span experienced then. Blodgett reported (1897) a histologically confirmed case in a youth of 14 years 8 months and Lunn described (1896–7) breast cancer in a 91-year-old.

Although Arderne's case involved the pap (nipple), it was not until 1880, 6 years after Paget's description that Forest described this condition in a man (Forrest, 1880). The patient, aged 72 years, gave a history of a "leakage on one side of the nipple," of a secretion described by the patient "as resembling women's milk." Subsequently, nipple retraction developed and an underlying carcinoma became apparent.

By the end of the nineteenth century, breast cancer in men had become a well recognized, albeit rare, entity. The clinical features were generally known, although the disease had been studied pathologically in only a few cases. Most cases were too far advanced for treatment to be effective and the prognosis was generally bad. Aside from the observations of Rigoni-Stern, little was known about etiology or pathogenesis.

MODERN CLINICOPATHOLOGIC STUDIES

Wainwright published the first modern description of breast cancer in men (1927). This was composed of 154 personally studied cases and an additional 264 from the literature. Two features distinguish this report from those of the nineteenth century, namely, a uniformity of clinical detail and histopathological confirmation of the diagnosis. Clinically, he emphasized that ulceration, pain, and nipple discharge were more frequent and more dangerous signals in men than in women and that the disease occurred at an older age in men. His paper includes a detailed account of the normal and pathologic anatomy including the microscopic findings, although his classification is archaic by modern standards.

A number of authors have shown that breast cancer is histopathologically indistinguishable in men relative to that seen in women (Norris et al., 1969; Visfeldt et al., 1973). The majority of cancers in men are infiltrating ductal carcinomas (NOS); however, medullary and mucoid types occur. The incidence of papillary carcinoma in men is much higher than the 1% usually given for women—8% and 13%, respectively, for the two series cited above. Lobular carcinoma is thought not to occur in men, although Sanchez and colleagues (1986) report one such case in a man with Klinefelter's syndrome. This is not totally unexpected as prolonged estrogen stimulation can result in acinar development (Sandison, 1962; Schwartz et al., 1963). Paget's disease of the nipple appears to be rarer in men than women.

Studies comparing tumor-related antigens in male and female breast cancer have generally failed to show differences between the sexes, and the oncogene c-*erb*B-2 appears to be overexpressed in a similar percentage of cases in both sexes (Lundy, et al., 1986; Dawson et al., 1992).

The incidence of metastases to the male breast is similar to that in the female (1% of all breast tumors) with the striking difference that the majority occur in patients with disseminated carcinoma of the prostate undergoing estrogen therapy (Berge, 1971).

The primary mode of therapy in men has remained surgical since the nineteenth century. The concept of cure in cases of breast cancer is less than 100 years old dating from Halsted's introduction of radical mastectomy (1894). Williams (1889) had outcome data on only 18 of 100 cases. In the five patients who underwent breast amputation, three survived from 117 to 273 months. In the remaining 13, survival ranged from 11 to 85 months from the time the disease was first noticed. By the time of Wainwright's (1927) paper, the importance of accurate follow-up had been appreciated, and of 418 cases studied, data was available in 111. The average survival in patients was 5.2 years without axillary involvement and fell to 2.8 years when this was present. Ulceration was also a poor prognostic sign. Using the criteria set out by Greenough (1925) for grading breast cancer in women, Wainwright was able to show a correlation between tumor grade and survival among male patients. He noted that recurrences could occur after very long latent periods, just as in women. He also felt that the prognosis in men was not as good as in women. The prognosis of breast cancer in men versus that of women has been a continuing controversy. Without going into a detailed review of this subject, survival, when tumor stage and age at onset are controlled for, appears to be the same between the sexes (Adami et al., 1985; van Geel et al., 1985).

HORMONAL INTERACTIONS

The seminal observation on the hormonal responsiveness of breast cancer was made by Beatson (1896). It was not, however, until Dodds and colleagues reported the synthesis of stilboestrol (1938), that interest in the endocrine manipulation of breast cancer developed. Generally speaking, the ablative techniques tried in men followed those employed in women. Farrow and Adair (1942) reported the effects of surgical castration on a 72-year-old man with breast cancer and noted regression of a large ulcerated breast tumor and clinical and objective improvement in osseous metastases. The patient survived three years and died of hypostatic pneumonia following a cerebral hemorrhage. Subsequently, Treves reported (1949) a series of 13 cases with similar good results. These observations led to orchiectomy as the treatment of choice for palliation of male breast cancer. The effects of adrenalectomy in seven cases of breast cancer were reported by Huggins and Bergenstal (1952). One patient was a man; by the third postoperative week, he showed a significant decrease in both pulmonary and intracranial metastases that was maintained for four months. Luft and Olivecrona (1955) reported their results with hypophysectomy in

30 cases of breast cancer, two of whom were men. Both had local and skeletal and pulmonary metastases and both were considerably improved.

The next obvious step was the use of the antiestrogen tamoxifen. The first reports appeared in 1978 (Abele et al., 1978; Jeffreys and Efthimiou, 1978; Morgan and Hong, 1978). In a series of 31 cases of male breast cancer including those mentioned above, Patterson and colleagues (1980) found a response rate of 48%, which was better than that seen in women, although this may be a reflection of the older age incidence in men. Tamoxifen remains an important agent in palliating male breast cancer.

Following the introduction of orchiectomy and stilboestrol therapy for prostatic carcinoma, a number of patients developed breast cancer that was attributed to the therapy (Benson, 1957). The majority of these occurred within a comparatively short time after the onset of therapy in patients with widely disseminated prostate cancer. Immunocytochemical studies (Kumar et al., 1986; Aziz et al., 1986) have confirmed that these so-called breast cancers were in fact intramammary metastases from the prostate. Nevertheless, there remain occasional reports of breast cancer in men following long-term administration of estrogens (Schlappack et al., 1986). Evidence that high doses of unopposed estrogen over a long period of time can induce breast cancer is provided by reports of breast cancer in three men who underwent transsexual change (Symmers, 1968; Pritchard et al., 1988). These individuals not only underwent bilateral orchiectomy but received very high doses of estrogen for long periods.

There is thus a considerable body of data linking the male and female sex hormones in a reciprocal way with breast cancer in men. Steroid hormone imbalance may also be the final common pathway associated with such disparate etiologies as a history of orchitis (in adults) and bilharzia infection. The latter has been associated with gynecomastia and the relatively high male-to-female ratio of cases of breast cancer in Egypt (El-Gazayerli and Adel-Aziz, 1963). Patients with Klinefelter's syndrome have low testosterone and elevated estradiol levels (Wang, et al., 1975), but there is also evidence linking breast cancer more directly with the extra X chromosome (v infra).

In a more general way, estrogen has been linked with male breast cancer via gynecomastia. The association between gynecomastia and breast cancer is, however, unclear. In one series, only 17 of 625 (less than 3%) of men with breast cancer gave a history of prior gynecomastia (Crichlow, 1972). Scheike and Visfeldt (1973) found a history of pre-existing gynecomastia in only 10 of 265 patients with male breast cancer. Histological features of concomitant gynecomastia were present in 10 of 79 cases where this could be evaluated. Although not proven, they felt that gynecomastia might be premalignant in certain circumstances. It should, however, be borne in mind that gynecomastia is a rather common finding in elderly men.

The presence of estrogen receptors in female breast cancer and its value as a predictor of a therapeutic response to endocrine ablation is well known. In fact, the percentage of estrogen receptor–positive breast cancers in men is even higher than in women, and has been reported in 85% of cases (Everson et al., 1980).

GENETIC FACTORS

The earliest report of familial breast cancer in men appears to be that of Williams (1889). In 29 of his 100 cases where inquiry was made, there was a history of cancer in seven families. In three families, the patient's sister had breast cancer and in an additional family the father had breast cancer. Kozak and colleagues (1986) were able to collect 12 additional families in which two related males had breast cancer and another family has been added since (Demeter, et al., 1990). In six of the families, the patients were brothers, and in four, they were father and son. In two families, three males were affected. In six of the families, a first-degree female relative also had breast cancer. These data suggest that men as well as women can inherit a predisposition to breast cancer.

The first suggestion of a link between a specific genetic abnormality and male breast cancer was made by Bauer and Erickson (1955). They reported an 80-year-old man with Klinefelter's syndrome who had long-standing gynecomastia and subsequently developed carcinoma of the breast and later carcinoma of the prostate. At autopsy, the patient also had an adrenal cortical adenoma.

Jackson and colleagues (1965) noted the occurrence of breast cancer in a man with extremely small testes. Screening of 21 men with breast cancer revealed that three had chromatin-positive buccal smears. Subsequent karyotyping revealed that two patients were 47XXY and the third was mosaic 46XY/47XXY/48XXXY. While only a small percentage of male breast cancer patients have Klinefelter's syndrome, Jackson's observation was the first to link breast cancer with a specific genetic defect, namely, the presence of a supernumerary X chromosome. More than 27 cases of Klinefelter's syndrome with breast cancer have been reported; while the majority have the XXY genotype, some, including one of Jackson's cases, were mosaics with a more complicated genotype. There are two major controversies in relation to breast cancer and Klinefelter's syndrome. The first concerns the incidence of breast cancer among these individuals, and the second concerns the mechanism of carcinogenesis. Lynch and colleagues (1974) have suggested the extra X chromosome present in men with Klinefelter's syndrome is responsible for elevating the risk of breast cancer to a level similar to that found in women. In a study of 466 X chromatin–positive Klinefelter's patients, Price and colleagues (1985) observed that deaths from breast cancer were comparable to those expected if female mortality rates were applied, although there were only 2 breast cancer deaths in the series. If, however, the problem is approached from the perspective of what percentage of men with breast cancer are chromatin-positive, then a different result has been obtained. From a series of 242 male breast cancer patients from Denmark screened for sex chromatin, Scheike and colleagues (1973) found nine with Klinefelter's syndrome. From this, they calculated that the chances of developing breast cancer in patients with the syndrome were one-fifth of those in women and 20 times those for normal genotypic males. Evans and Crichlow (1987) applied a similar calculation to American data and concluded that 3% of Klinefelter's patients might be expected to develop breast cancer. The precise role of the genotypic abnormality in the induction of breast cancer is unknown.

Not surprisingly, breast cancer has been described in true hermaphrodites (i.e., individuals with both ovaries and testes) with an XX karyotype (Decker et al., 1982). Preliminary data by flow cytometry indicate that the majority of breast cancers in men are diploid (Wolman et al., 1995). Only a few cases have been analyzed cytogenetically; the majority appeared pseudodiploid. In one case analyzed after Giemsa-banding (Mitchell, 1990) abnormalities were found on chromosomes #8 and #13 as well as alterations in the long arm of chromosome #1. These changes are similar to those described in diploid breast cancers in females.

Recently, genetic linkage studies have lead to the recognition of two breast cancer genes, *BRCA1* and *BRCA2*, located on chromosomes #17 and #13, respectively (Hall et al., 1990; Wooster et al., 1994). These have been subsequently cloned (Miki et al., 1994; Wooster et al., 1995). While both have been associated with heredity, breast cancer alterations in the *BRCA2* gene have been associated particularly with familial breast cancer in men. However, the incidence in the reported series varies rather widely (Thorlacius et al., 1996; Phelan et al., 1996; Couch et al., 1996), perhaps reflecting the diversity, or lack thereof, of the individual gene pool.

CONCLUSIONS

Male breast cancer remains an interesting rarity which, unlike female breast cancer, is *not* increasing in incidence in the United States (Parker et al., 1997). The disease in the two sexes is remarkable for its similarities rather than its differences. These include not only the clinical and morphologic appearances but tumor markers as well. While estrogen stimulation, either endogenous or exogenous, may play a role in some cases, it is clear that this is not necessary for breast cancer to develop in men. The association with Kleinfelter's syndrome suggests that one or more factors associated with the X chromosome may be involved. The discovery of specific breast cancer genes, particularly *BRCA2* with its strong association with male breast cancer, makes more likely the hope that the study of breast cancer in men will contribute to our understanding of the much more prevalent cancer in women.

ACKNOWLEDGMENTS

I am particularly grateful to the late Mrs. Owen H. Wangensteen of the Department of the History of Medicine, University of Minnesota and Mrs. Donna Doelling of the University of South Florida Health Sciences Library for help with the early references. I am indebted to Dr. Eric Wolman for reading the manuscript and making many helpful suggestions. I would like to thank Drs. S. Nicosia, A. Mansour, and J. Szakacs for help with translations and Mrs. Jutta Payant and Mr. David Montgomery for secretarial and library help. Figure 1 is reproduced from "Pathology of Human Neoplasms" with kind permission of the author, Henry A. Azar, M.D. and Lippincott-Raven Publishers, Philadelphia, PA. Figure 2 is reproduced courtesy of the National Library of Medicine, Bethesda, MD.

REFERENCES

Abele, R., Alberto, P., and Mégevand, R. (1978). Carcinoma of the male breast and oestrogen metabolism. British Medical Journal 1, 1697-1698.

Adami, H.-O., Holmberg, L., Malker, B., and Ries, L. (1985). Long-term survival in 406 males with breast cancer. British Journal of Cancer 52, 99-103.

Arcaeus, F. (1588). A most excellent and compendious method of curing woundes in the head, and in other partes of the body, with other precepts of the same arte, ... translated by John Read, London, Bk 2, pp33v-34r.

Aziz, D.C., Milne C., Solomon, S., and Jothy, S.P. (1986). Contribution of immunohistochemistry to the diagnosis of breast cancer in men. Canadian Journal of Surgery 29, 135-136.

Bartholin, T. (1671-1672). Acta Medica et Philosophica. Hafniensa. Hafnia 1, 58.

Bauer, D.DeF. and Erickson, R.L. (1955). Male breast cancer: Klinefelter syndrome with prostatic, adrenal and mammary tumors. Northwest Medicine 54, 472-476.

Beatson, G.T. (1896). On the treatment of inoperable cases of carcinoma of the mamma: Suggestions for a new method of treatment, with illustrative cases. Lancet 2, 104-107, 162-165.

Benson, W.R. (1957). Carcinoma of the prostate with metastases to breasts and testes: Critical review of the literature and report of a case. Cancer 10, 1235-1245.

Berge, T. (1971). Metastases to the male breast. Acta Pathologica et Microbiologica Scandinavica. Section A 79, 491-496.

Bidloo, G. (1715). Opera Omnia Anatomico-Chirurgica. Lugduni Batavorum, S. Luchtmans, pp 157-160.

Blodgett, A.N. (1897). Cancer of the breast in a child. Boston Medical and Surgical Journal 136, 611.

Breasted, J.H. (1930). The Edwin Smith Surgical Papyrus. University of Chicago Press. Chicago, pp 403-406.

Chase, H.C. (1964). Treatment of gynecomastia and cancer of the male breast. In: Treatment of Cancer and Allied Diseases. (Pack, G.T. and Ariel I.M., Eds.), 2nd ed. pp 231-232, Harper & Row, New York.

Chenet, J. (1876). Étude sur le cancer du sein chez l' homme. Paris, 44 pp.

Couch, F.J., Farid, L.M., DeShano, M.L., Tavtigian, S.V., Calzone, K., Campeau, L., Peng, Y., Bogden, B., Chen, Q., Neuhausen, S., Shattuck-Eidens, D., Godwin, A.K., Daly, M., Radford, D.M., Sedlacek, S., Rommens, J., Simard, J., Garber, J., Merajver, S., and Weber, B.L. (1996). BRCA2 germline mutations in male breast cancer cases and breast cancer families. Nature Genetics 13, 123-125.

Courtade, A. (1885). De l'épithélioma du sein chez l'homme. L'Union Medicale (Paris) 40, 1070-1073.

Crichlow, R.W. (1972). Carcinoma of the male breast. Surgery, Gynecology, and Obstetrics 134, 1011-1019.

Dawson, P.J., Paine, T.M., and Wolman, S.R. (1992). Immunocytochemical characterization of male breast cancer. Modern Pathology 5, 621-625.

Decker, J.P., Lerner, H.J., and Schwartz, I. (1982). Breast carcinoma in a 46, XX true hermaphrodite. Cancer 49, 1481-1484.

Demeter, J.G., Waterman, N.G., and Verdi, G.D. (1990). Familial male breast carcinoma. Cancer 65, 2342-2343.

Dodds, E.C., Goldberg, L., Lawson, W., and Robinson, R. (1938). Oestrogenic activity of certain synthetic compounds. Nature 141, 247-248.

El-Gazayerli, M.M. and Abdel-Aziz, A.S. (1963). On bilharziasis and male breast cancer in Egypt: A preliminary report and review of the literature. British Journal of Cancer 17, 566-571.

Erdheim, S. (1930). Über Gynäkomastie. Deutsche Zeitschrift fuer Chirurgie 20, 181-225.

Evans, D.B. and Crichlow, R.W. (1987). Carcinoma of the male breast and Klinefelter's syndrome: Is there an association? CA: A Cancer Journal for Clinicians 37, 246-251.

Everson, R.B., Lippman, M.E., Thompson, E.B., McGuire, W.L., Wittliff, J.L., De Sombre, E.R., Jensen, E.V., Singhakowinta, A., Brooks, S.C. Jr., and Neifeld, J.P. (1980). Clinical correlations of steroid receptors and male breast cancer. Cancer Research 40, 991-997.

Farrow, J.H. and Adair, F.E. (1942). Effect of orchidectomy on skeletal metastases from cancer of the male breast. Science 95, 654.

Forrest, R.W. (1880). Case of cancer of the mamma in the male, preceded by so-called eczema of the mammary areola, "Paget's disease of the nipple." Glasgow Medical Journal 14, 457-460.

Graham, H. (1939). The Story of Surgery. pp 118-120, Doubleday Doran, New York.

Greenough, R.B. (1925). Varying degrees of malignancy in cancer of the breast. Journal of Cancer Research 9, 453-463.

Hall, J.M., Ming, K.L., Newman, B., Morrow, .J.E, Anderson, L.A., Huey, B., and King, M-C. (1990). Linkage of early-onset familial breast cancer to chromosome 17q21. Science 250, 1684-1689.

Halstead, W.S. (1894). The results of operations for the cure of cancer of the breast performed at the Johns Hopkins Hospital from June, 1889, to January, 1894. Annals of Surgery 20, 497-555.

Heister, L. (1750). General System of Surgery. 4th Ed. (Translated from Latin), p. 229, London.

Hildanus, G.F. (1606). Observationum and Curationum Chirurgicarum Centuriae, p. 274, Basil.

Horteloup, P. (1872). Des Tumeurs du Sein Chez l'Homme. p. 102, Asselin, Paris.

Huggins, C., and Bergenstal, D.M. (1952). Inhibition of human mammary and prostatic cancers by adrenalectomy. Cancer Research 12, 134-141.

Imbert, L. (1891). Du cancer du sein chez l'homme. Gazette Hebdomadaire des Sciences Médicales de Montpelier 13, 541-544.

Jackson, A.W., Muldal, S., Ockey, C.H., and O'Connor, P.J. (1965). Carcinoma of male breast in association with the Klinefelter syndrome. British Medical Journal 1, 223-225.

Jefferys, D.B. and Efthimiou, J. (1978). Carcinoma of the male breast and oestrogen metabolism. British Medical Journal 1, 1697.

Karsner, H.T. (1946). Gynecomastia. American Journal of Pathology 22, 235-275.

Kozak, F.K., Hall, J.G., and Baird, P.A. (1986). Familial breast cancer in males: A case report and review of the literature. Cancer 58, 2736-2739.

Kumar, P.P., Good, R.R., and Newland, J.R. (1986). Immunodiagnosis by prostatic acid phosphatase to differentiate primary male breast cancer from metastatic prostate cancer. Journal National Medical Association 78, 782-786.

Luft, R. and Olivecrona, H. (1955). Hypophysectomy in man: Experiences in metastatic cancer of the breast. Cancer 8, 261-270.

Lundy, J., Mishriki, Y., Viola, M.V., Chao, S., Kasa, B., Oravez, S., and Schuss, A. (1986). A comparison of tumor-related antigens in male and female breast cancer. Breast Cancer Research and Treatment 7, 91-96.

Lunn, J.R. (1896-7). A case of cancer of the breast in a man aged ninety-one. Transactions of the Pathological Society of London 48, 247-248.

Lynch, H.T., Kaplan, A.R., and Lynch, J.F. (1974). Klinefelter syndrome and cancer: A family study. Journal American Medical Association 229, 809-811.

Melville, J.G. (1933). Gynecomastia. Archives of Surgery 26, 1054-1083.

Miki, Y., Swensen, Y., Shattuck-Eidens, D., Futreal, P.A., Harshman, K., Tavtigian, S., Liu, Q., Cochran, C., Bennett, L.M., Ding, W., Bell, R., Rosenthal, J., Hussey, C., Tran, T., McClure, M., Frye, C., Hattier, T., Phelps, R., Haugen-Strano, A., Katcher, H., Yakumo, K., Gholami, Z., Shaffer, D., Stone, S., Bayer, S., Wray, C., Bogden, R., Dayananth, P., Ward, J., Tonin, P., Narod, S., Bristow, P.K., Norris, F.H., Helvering, L., Morrison, P., Rosteck, P., Lai, M., Barrett, J.C.,

Lewis, C., Neuhausen, S., Cannon-Albright, L., Goldgar, D., Wiseman, R., Kamb, A., and Skolnick, M.H. (1994). A strong candidate for the breast and ovarian cancer susceptibility gene *BRCA1*. Science 266, 66-71.

Mitchell, E.L.D. (1994). A cytogenetic study of male breast cancer. Cancer Genetics and Cytogenetics 47, 107-112.

Morgagni, G.B. (1769). The seats and causes of diseases investigated by anatomy. (Translated by Benjamin Alexander), Vol. 3, Bk. 4, p. 64, London.

Morgan, D.A.L. and Hong, A. (1978). Carcinoma of the male breast and oestrogen metabolism. British Medical Journal 2, 206.

Norris, H.J. and Taylor, H.B. (1969). Carcinoma of the male breast. Cancer 23, 1428-1435.

Paré, A. (1634). The Collected works of Ambrose Paré: (Translated out of the Latin by Thomas Johnson T. From the First English Edition.), p. 280, London.

Parker, S.L., Tong, T., Bolden, S., and Wingo, P.A. (1997). CA: A Cancer Journal for Clinicians: Cancer Statistics 47, 5-27.

Patterson, J.S., Battersby, L.A., and Bach, B.K. (1980). Use of tamoxifen in advanced male breast cancer. Cancer Treatment Reports 64, 801-804.

Phelan, C.M., Lancaster, J.M., Tonin, P., Gumbs, C., Cochran, C., Carter, R., Ghadirian, P., Peret, C., Moslehi, R., Dion, F., Faucher, M-C., Dole, K., Karimi, S., Foulkes, W., Lounis, H., Warner, E., Goss, P., Anderson, D., Larsson, C., Narod, S.A., and Futreal, P.A. (1996). Mutation analysis of the *BRCA2* gene in 49 site-specific breast cancer families. Nature Genetics 13, 120-122.

Poirier, P. (1883). Contribution à l'étude des tumeurs du sein chez l'homme (tubercules, sarcomes, épithéliomes, carcinomes): Étude clinique du cancer. pp. 110. Paris.

Power, D'A. (1910). Introduction to: Treatises of fistula in ano, hemorrhoids, and clysters by John Arderne. pp. vii-viii. Oxford University Press, London.

Price, W.H., Clayton, J.F., Wilson, J., Collyer, S., and De Mey, R. (1985). Causes of death in X chromatin-positive males (Klinefelter's syndrome). Journal of Epidemiology and Community Health 39, 330-336.

Pritchard, T.J., Pankowsky, D.A., Crowe, J.P., and Abdul-Karim, F.W. (1988). Breast cancer in a male-to-female transsexual: A case report. Journal American Medical Association 259, 2278-2280.

Rigoni-Stern, D. (1842). Fatti statistici relativi alle mallattie cancerose. Fourth Congress of Italian Scientists. 23 Sept. 1842. Giornalle per Servire ai Progressi della Patologia e della Theraputica. Ser. 2:2, 507-517.

Sanchez, A.G., Villanueva, A.G., and Redondo, C. (1986). Lobular carcinoma of the breast in a patient with Klinefelter's syndrome: A case with bilateral, synchronous, histologically different breast tumors. Cancer 57, 1181-1183.

Sandison, A.T. and Wells, C. (1967). Diseases of the Reproductive System. In: Diseases in Antiquity. (Brothwell, D. and Sandison, A.T., Eds.) p. 512, C.C. Thomas, Springfield, Illinois.

Sandison, A.T. (1962). An autopsy study of adult human breast. National Cancer Institute Monograph. No. 8. p. 77. U.S. Department of Health, Education, and Welfare, Bethesda, MD.

Scheike, O. and Visfeldt, J. (1973). Male breast cancer: 4. Gynecomastia in patients with breast cancer. Acta Pathologica et Microbiologica Scandinavica, Section A, 81, 359-365.

Scheike, O., Visfeldt, J., Petersen, B. (1973). Male breast cancer: 3. Breast carcinoma in association with the Klinefelter syndrome. Acta Pathologica et Microbiologica Scandinavica. Section A, 81, 352-358.

Schlappack, O.K., Braun, O., and Maier, U. (1986). Report of two cases of male breast cancer after prolonged estrogen treatment for prostatic carcinoma. Cancer Detection and Prevention 9, 319-322.

Schuchardt, B. (1885) Weitere Mittheilungen zur Casuistik und Statistik der Neubildungen in der männlichen Brust. Archiv fuer Klinische Chirurgie 32, 277-322.

Schwartz, I.S. and Wilens, S.L. (1963). The formation of acinar tissue in gynecomastia. American Journal of Pathology 43, 797-807.

Symmers, W.St.C. (1968). Carcinoma of breast in transexual individuals after surgical and hormonal interference with the primary and secondary sex characteristics. British Medical Journal 2, 83-85.

Thorlacius, S., Olafsdottir, G., Tryggvadottir, L., Neuhausen, S., Jonasson, J.G., Tavtigian, S.V., Tulinius, H., Ogmundsdottir, H.M., and Eyfjörd, J.E. (1996). A single *BRCA2* mutation in male and female breast cancer families from Iceland with varied cancer phenotypes. Nature Genetics 13, 117-119

Treves, N. (1949). Castration as a therapeutic measure in cancer of the male breast. Cancer 2, 191-222.

van Geel, A.N., van Slooten, E.A., Mavrunac, M., and Hart, A.A.M. (1985). A retrospective study of male breast cancer in Holland. British Journal of Surgery 72, 724-727.

Velpeau, A. (1854). Traité des maladies du sein et de la règion mammaire. Deuxième Partie. pp. 713-715. V Masson, Paris.

Visfeldt, J. and Scheike, O. (1973). Male breast cancer: I. Histologic typing and grading of 187 Danish cases. Cancer 32, 985-990.

Wainwright, J.M. (1927). Carcinoma of the male breast: Clinical and pathologic study. Archives of Surgery 14, 836-859.

Wang, C., Baker, H.W.G., Burgen, H.G., de Kretser D.M., and Hudson, B. (1975). Hormonal studies in Klinefelter's syndrome. Clinical Endocrinology 4, 399-411.

Williams, W.R. (1889). Cancer of the male breast. Based on the records of one hundred cases; with remarks. Lancet 2, 261-263; 310-312.

Wolman, S.R., Sanford, J., Ratner, S., and Dawson, P.J. (1995). Breast cancer in males: DNA content and sex chromosome constitution. Modern Pathology 8, 239-243.

Wolff, J.M. (1742). Commercium Litterarium 44, 351-352.

Wooster, R., Bignell, G., Lancaster, J., Swift, S., Seal, S., Manglon, J., Collins, N., Gregory, S., Gumbs, C., Micklem, G., Barfoot, R., Hamoudi, R., Patel, S., Rice, C., Biggs, P., Hashim, Y., Smith, A., Connor, F., Arason, A., Gudmundsson, J., Ficenec, D., Kelsell, D., Ford, D., Tonin, P., Bishop, D.T., Spurr, N.K., Ponder, B.A.J., Eeles R., Peto, J., Devilee, P., Cornellisse, C., Lynch, H., Narod, S., Lenoh, G., Eglisson, V., Barkadottir, R.B., Easton, D.F., Bentley, D.R., Futreal, P.A., Ashworth, A., Stratton, M.R. (1995). Identification of the breast cancer susceptibility gene *BRCA2*. Nature 378, 789-792.

Wooster, R., Neuhausen, S.L., Mangion, J., Quirk, Y., Ford, D., Collins, N., Nguyen, K., Seal, S., Tran, T., Averill, D., Fields, P., Marshall, G., Narod, S., Lenoir, G.M., Lynch, H., Feunteun, J., Devilee, P., Cornelisse, C.J., Menko, F.H., Daly, P.A., Ormiston, W., McManus, R., Pye, C., Lewis, C.M., Cannon-Albright, L.A., Peto, J., Ponder, B.A.J., Skolnick, M.H., Easton, D.F., Goldgar, D.E., and Stratton, M.R. (1994). Localization of a breast cancer susceptibility gene, *BRCA2*, to Chromosome 13q12-13. Science 265, 2088-2090.

This Page Intentionally Left Blank

INDEX

www.ingramcontent.com/pod-product-compliance
Ingram Content Group UK Ltd.
Pitfield, Milton Keynes, MK11 3LW, UK
UKHW020801290926
15470UKWH00003B/894

9 780762 303885